TRAFFIC FLOW ·

FUNDAMENTALS

ADOLF D. MAY
University of California, Berkeley

PRENTICE HALL, Upper Saddle River, NJ 07458

Library of Congress Cataloging-in-Publication Data

May, Adolf D. (Adolf Darlington)
    Traffic flow fundamentals / Adolf D. May.
        p.   cm.
    Includes bibliographical references.
    ISBN   0–13–926072–2
    1. Traffic flow—Mathematical models.   I. Title.
HE336.T7M39   1990
388.3'1—dc20
                                            89–39687
                                                CIP

Editorial/production supervision: Virginia L. McCarthy
Cover design: Edsal Enterprises   .
Manufacturing buyer: Denise Duggan

© 1990 by Prentice-Hall, Inc.
A Pearson Education Company
Upper Saddle River, NJ 07458

Printed in the United States of America
10 9 8 7 6 5 4 3

ISBN 0-13-926072-2

Prentice-Hall International (UK) Limited,London
Prentice-Hall of Australia Pty. Limited, Sydney
Prentice-Hall Canada Inc., Toronto
Prentice-Hall Hispanoamericana, S.A., Mexico
Prentice-Hall of India Private Limited, New Delhi
Prentice-Hall of Japan, Inc., Tokyo
Pearson Education Asia Pte. Ltd., Singapore
Editora Prentice-Hall do Brasil, Ltda., Rio de Janeiro

Contents

7   MACROSCOPIC DENSITY CHARACTERISTICS   192

PART TWO: ANALYTICAL TECHNIQUES

8   DEMAND-SUPPLY ANALYSIS   227

9   CAPACITY ANALYSIS   247

10   TRAFFIC STREAM MODELS   283

11   SHOCK WAVE ANALYSIS   321

12   QUEUEING ANALYSIS   338

13   COMPUTER SIMULATION MODELS   376

APPENDIXES

A   HEADWAY TABULATIONS AND PARAMETER VALUES OF FOUR MEASURED TIME HEADWAY DISTRIBUTIONS   415

B   DERIVATION OF POISSON COUNT DISTRIBUTION   420

Preface

This book is designed for a one-semester first-year graduate course in traffic flow fundamentals, with approximately one week being spent on each chapter. Senior-level undergraduate courses and courses with fewer credit hours can use the book with certain modifications as noted. The book is also intended to be used as a ready reference by professionals and researchers because of its in-depth comprehensive coverage of research results and applications. People from other countries and those working with other modes of transportation may find the book valuable because of its coverage of basic principles and theories which cut across political and modal boundaries.

In the early 1970s the author developed a new first-year graduate course at the University of California devoted to traffic flow fundamentals. Over the years an extensive set of course notes was developed and continuously updated. In 1985 and 1987 draft chapters were tested in a classroom situation. Student comments were most helpful in improving the chapters and encouraging completion of the book. The author is indebted to the students, from whom he has learned so much and who have contributed in so many ways to this book.

The author also wishes to acknowledge the contributions of four of his colleagues who reviewed the final draft and offered many thoughtful suggestions: Professor Jan L. Botha, University of Alaska; Professor Paul P. Jovanis, University of California at Davis; Professor Robert Layton, Oregon State University; and Professor K. C. Sinha, Purdue University. A word of appreciation is also extended to Mr. Doug Humphrey and the staff of Prentice Hall for their continuous support and encouragement.

In closing, I would like to dedicate this book to my family. To my mother, who sacrificed and encouraged my education with love and affection. To my father, who introduced me to transportation engineering and always served as an example of a man I would aspire to become. To my wife, who has been my partner in life, sharing the sorrows and enjoying the happinesses. They made this book possible.

1

Introduction

Knowledge of fundamental traffic flow characteristics and associated analytical techniques is an essential requirement in the planning, design, and operation of transportation systems. Traffic flow characteristics include time headway, flow, time–space projectory, speed, distance headway, and density. Traffic flow analytical techniques include supply–demand modeling, capacity and level-of-service analysis, traffic stream modeling, shock wave analysis, queueing analysis, and simulation modeling. The purpose of this book is to provide an in-depth treatment of these fundamentals of traffic flow, and one chapter is devoted to each of these traffic flow characteristics and associated analytical techniques.

Transportation system planners, designers, and operators should have a basic knowledge of traffic flow fundamentals. Planners assess traffic and environmental impacts of proposed system modifications, and this can be accomplished only through a supply–demand framework that requires understanding of flow characteristics and their interrelations. Designers determine link sizes and configure systems, and must carefully evaluate the trade-off between traffic flow levels and levels of service. Operators identify locations and causes of existing system defects and generate operational improvement plans and predict their effects. For these reasons and others, knowledge of traffic flow fundamentals is required for all graduate transportation students at the University of California.

Although this book focuses primarily on the highway transportation system, traffic flow characteristics and associated analytical techniques are fundamental to all transportation systems. All transportation systems have particles moving along links that interact with each other, interact with the physical facility, and require analysis and control. In air transportation, for example, flow characteristics are studied and analysis made of air-side and land-side subsystems. On the air side, flow control schemes have been developed for the airways, simulation models developed for use in design of airports, and queueing models employed for allocating aircraft to gate positions. On the land side, analysis has been made of baggage handling, person flow at check-in counters and gate positions, and mode transfer at the terminal interface. In water transportation, simulation models have been developed to study line-haul ocean operations and ferry systems, and queueing analysis has been applied to lock operations and port facilities. In fixed-guideway systems, time–space diagrams are in continuous use for railroad line operations, simulation models for urban rail systems and freight yards, and queueing analysis for transit stations. One should not overlook special transportation systems which also require traffic flow fundamentals, such as for pedestrian-ways, escalators, elevators, and conveyor belts. For these reasons, the coverage in this book has multimodal implications, and specialists in other transportation systems should readily be able to adopt and transfer material to their area of interest. Some examples of modal traffic operations are shown in Table 1.1.

TABLE 1.1   Examples of Modal Traffic Operations

Transport Mode	Microscopic	Macroscopic
Air	Airways, airfields, gate positions (aircraft)	Terminals, check-in, gate positions, (persons)
Water	Port facilities, line-haul operations, locks (ships)	Terminals (freight)
Rail	Urban systems, railroads (trains)	Terminals (persons)
Highway	Intersections, trucks on grades (vehicles)	Route operations, parking (vehicles)
Other	Pedestrians, elevators (persons)	Pedestrians, escalators, conveyor belts (persons or freight)

1.1 AN ANALYTICAL APPROACH

The analytical process consists of predicting an output as a function of specified inputs. This analytical process can vary from a simple equation to a complex simulation model. The important issues for analysts are their knowledge of the system being considered, including its flow characteristics, and of analytical techniques and their appropriate selection for the problem at hand.

A flowchart is shown in Figure 1.1 to illustrate this analytical process and to emphasize the importance of knowing fundamental flow characteristics and basic analytical techniques. Although the process can be shown in many different ways, the most common approach is to predict performance as a function of a set of prespecified inputs. These inputs consist of traffic demand, transport supply, and traffic control. The predicted performance may include performance measures from a system and/or user perspective and may include environmental as well as traffic consequences.

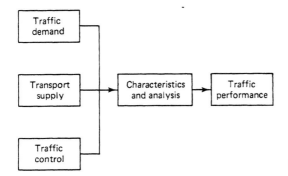

Figure 1.1   The Analytical Process

An understanding of flow characteristics is essential to convert prespecified demand, supply, and control information into mutually compatible quantitative expressions that can serve as direct input to the selected analytical technique. The demand must be quantified in terms of demand flow rates or time headway distributions for the location and time period selected for analysis. The demand may include mixed-vehicle flow conditions in which the effect of vehicle characteristics must also be evaluated. The transport supply features must be converted to capacity flow rates or minimum time headway distributions, and relationships between traffic loads and resulting operations established.

The resulting operations include predictions of speed projectories, operating speeds, distance headway distributions, and/or density levels. The control elements must be inserted into the supply–demand process and used to modify predicted performance. Without an understanding of flow characteristics, these tasks cannot be performed.

A knowledge of traffic analytical techniques and the ability to select the appropriate technique are required to predict the performance from developed input expressions. As mentioned earlier, the analytical process can vary from a simple equation to a complex simulation model. Traffic stream models can often be used for uninterrupted flow situations where demands do not exceed capacities. For interrupted oversaturated flow situations, more complex techniques, such as shock wave analysis, queueing analysis, and simulation modeling, are employed. Microscopic analysis may be selected for moderate-sized systems where the number of transport units passing through the system is relatively small and there is the need to study the behavior of individual units in the system. Macroscopic analysis may be selected for higher-density, larger-scale systems in which a study of the behavior of groups of units is sufficient. Without a knowledge of traffic analytical techniques and the ability to select the most appropriate microscopic or macroscopic technique for the problem at hand, these tasks cannot be performed.

1.2 TEXTBOOK ORGANIZATION

The format of the various chapters of the textbook is quite similar and the main body consists of introduction, empirical evidence, principles and existing theories, evaluation process, and applications. Each chapter is accompanied by selected problems and selected references. Appendices, including a table of metric conversion factors and a list of symbols and definitions, and an index are located at the end of the book.

The first part of the book, Chapters 2 through 7, deals with traffic characteristics. The second part of the book, Chapters 8 through 13, is concerned with analytical techniques. Chapter highlights are presented briefly in the next two sections of this chapter.

1.3 PART ONE: TRAFFIC CHARACTERISTICS

The fundamental characteristics of traffic flow are flow, speed, and density. These characteristics can be observed and studied at the microscopic and macroscopic levels. Table 1.2 provides a framework for these characteristics as presented in Chapters 2 through 7.

TABLE 1.2   Framework for Fundamental Characteristics of Traffic Flow

Traffic Characteristic	Microscopic (Individual Units)	Macroscopic (Groups of Units)
Flow	Time headways (Chapter 2)	Flow rates (Chapter 3)
Speed	Individual speeds (Chapter 4)	Average speeds (Chapter 5)
Density	Distance headways (Chapter 6)	Density rates (Chapter 7)

Chapter 2, devoted to microscopic flow characteristics, is concerned with individual time headways between vehicles, with particular emphasis on mean values and distribution forms. The chapter presents empirical measurements, describes pertinent mathematical distributions, suggests evaluation procedures, and provides selected applications.

Chapter 3 is concerned with macroscopic flow characteristics, which are expressed as flow rates, and attention is given to temporal, spatial, and modal patterns. Empirical evidence is provided, analytical techniques are described, and applications are presented.

Chapter 4 describes microscopic speed characteristics of individual vehicles passing a point or short segment during a specified time period. Particular attention is given to vehicular speed trajectories over time and space as well as to statistical analysis of individual speed measurements.

Chapter 5 is directed to macroscopic speed characteristics, which are concerned with the speed of groups of vehicles passing a point or short segment during a specified period. Particular attention is given to temporal, spatial, and modal variations as well as to statistical analysis of group speed measurements. Travel time and delay techniques are also included.

Chapter 6 is devoted to microscopic density characteristics, which are concerned with individual distance headways between vehicles, with particular emphasis on minimum and average values. The chapter includes an extensive coverage of car-following theories and automatic data collection systems.

Chapter 7 is concerned with macroscopic density characteristics, which are expressed as the number of vehicles occupying a section of roadway. Density measurement techniques are described and particular attention is given to density contour maps. Analytical techniques using density contour maps are described.

1.4 PART TWO: ANALYTICAL TECHNIQUES

Analytical techniques associated with individual traffic characteristics are discussed in earlier traffic characteristics chapters. The remaining portion of the book is devoted to analytical techniques involving the total traffic flow situation. Table 1.3 provides a framework of these analytical techniques as presented in Chapters 8 through 13.

Chapter 8 provides a transition from the treatment of individual traffic characteristics to a demand–supply framework, which permits the integration of analytical techniques presented later. The demand–supply framework is introduced followed by a rather comprehensive example of traffic control and traveler response interactions.

Chapter 9 is devoted to capacity and level-of-service analysis. The supply, demand, and control features are analyzed to determine the capacity and the trade-off between the quantity of traffic and the resulting level of service to the users. Attention is given to multilane facilities, including ramps and weaving sections as well as to signalized intersections.

Chapter 10 is concerned with traffic stream models which provide for the fundamental relationships of macroscopic traffic stream characteristics for uninterrupted flow situations. The relationships are for free-flow and congested-flow conditions. Extensive sets of field measurements are described and single- and multiregime models presented and evaluated.

Chapter 11 presents shock wave analysis, which is one of the techniques available for the analysis of oversaturated traffic systems. Traffic stream models serve as a

TABLE 1.3 Framework for Traffic Analysis Techniques[a]

Analytical Technique	Microscopic (Individual Units)	Macroscopic (Groups of Units)
Supply analysis	Minimum time headways (Chapter 2)	Capacity analysis (Chapter 9)
Flow relationships	Time–space diagrams (Chapter 10) .	Flow–speed–density relationships (Chapter 10)
Shock wave analysis	Time–space diagrams (Chapter 11)	Shock waves (Chapter 11)
Queueing analysis	Discrete analysis (Chapter 12)	Continuous analysis (Chapter 12)
Simulation modeling	Discrete models (Chapter 13)	Continuous models (Chapter 13)

[a]See also Chapter 8.

beginning point for shock wave analysis. The fundamentals of shock wave theory are presented. Examples of the application of shock wave theory to signalized intersections, uninterrupted highways, and pedestrian-ways are illustrated.

Chapter 12 is devoted to queueing analysis, one of the techniques available for the analysis of oversaturated traffic systems. A classification of queueing analysis techniques is formulated which covers the gamut from deterministic situations to completely stochastic situations. Queueing theory is presented followed by a variety of applications.

Chapter 13 describes computer simulation modeling techniques and their application to traffic systems. Both microscopic and macroscopic simulation models are covered. The chapter includes a step-by-step procedure for developing a simulation model, an example of actually developing a simulation model, and a description of some models currently available.

1.5 FUTURE CHALLENGES

Like many other transportation and infrastructure systems, highway transportation passes through periods of expansion and consolidation. In this century the 1920s and 1930s marked the rapid expansion of the highway system in both road mileage and route quality. The construction of the interstate highway system, including the urban freeway system, were major accomplishments of the 1950s and 1960s which provided for high-quality intercity and intracity travel never before obtained. These accomplishments improved the quality of travel so much that phenomenon growth in traffic demand occurred, and soon demands approached or exceeded capacities at many locations, particularly in the urban and suburban areas.

The 1970s ushered in a new era of consolidation and with it transportation system management. As demand continued to approach or exceed capacities, attention turned to controlling the demand while making only spot improvements on the capacity side.

During the 1980s, transportation system management has continued but has been enhanced and broadened by the anticipation of innovative vehicle, computer, and electronic technology. Two closely related challenges are envisioned for the remainder of this century and the beginning of the twenty-first century: a more systematic and comprehensive transportation system management activity and a gradual implementation of the most promising new technologies. Both will be directed toward making maximum use of the existing transportation system. Traffic flow fundamentals will play an important role in meeting these two challenges.

Planners, designers, and operators of the transportation system all have a role to play in developing a more systematic and comprehensive transportation system management activity. The skills of planners will be needed to develop and apply improved techniques for evaluating the impacts of land use changes and in developing more precise behavior models of the effects of system changes on spatial, temporal, modal, and total traveler responses. The ingenuity of the designer will be required to identify critical links in the system where capacity increases are urgently needed and to develop design plans that meet the needs but with serious constraints on available right-of-way and environmental impacts. The tenacity of the operator will be essential to identify operational defects and to develop plans to balance system operation on a real-time basis.

Innovated vehicle, computer, and electronic technology are on the threshold of developments that have the potential of making maximum use of the existing transportation system. Technologies with the greatest potential must be identified and a gradual implementation plan developed. New vehicle technologies include in-vehicle longitudinal and lateral information warning systems, radar brakes, and perhaps ultimately, fully automatic controlled guidance systems. Computers can play an even greater role in the future in both off-line and on-line operations. Off-line computer packages are becoming faster, more flexible, and user friendly. The use of on-line computer systems provides the opportunity of engaging improved control theory algorithms, such as artificial intelligence, expert systems, fuzzy sets, and the like, to make maximum use of the highway system under normal and unusual traffic conditions. New electronic technologies interact strongly with vehicle and computer technologies. New detectors, communication links, and control processors are being researched intensively. New roadside and in-vehicle information systems are being developed that may lead toward navigation systems and route selection under dynamic traffic conditions.

The highway system of today carries a more significant number of vehicle-miles of travel than ever before—greater than that for which it was designed. Demands continue to grow at faster rates than improvements are being made. The movement of persons and goods has gradually deteriorated. Transportation systems management and new technologies offer the greatest challenge and hope for improving the quality of movement. The ability to understand and apply traffic flow fundamentals is an essential ingredient in working toward improving the transportation system. It is hoped that this book will stimulate and involve the reader in the challenges that lie ahead.

1.6 SELECTED PROBLEMS

1. Define the six traffic flow characteristics identified in this chapter.
2. Describe the six traffic flow analytical techniques identified in this chapter.
3. Expand the examples in the text on how planners, designers, and operators use traffic flow fundamentals.
4. Develop an expanded Table 1.1.
5. Describe the elements contained in Figure 1.1. Consider extensions such as environmental conditions, feedback loops, and other elements that you feel should be included.
6. Why do interrupted oversaturated flow situations require more complex analytical techniques?
7. What are some of the factors to be considered in selecting between microscopic and macroscopic analyses?
8. Study the organization of the book. Are there other organizational schemes that you would consider?
9. Select one chapter of the book that interests you especially and write a half-page summary describing its contents.
10. During the original work on the book a separate chapter on time–space diagrams was considered, to illustrate another analytical technique. It was finally decided to include this material in other chapters. What other chapters contain materials on time–space diagrams, and do you feel that the coverage is adequate?
11. Define transportation system management and provide a historical perspective.
12. Write a new Section 1.5 and discuss what you think the future challenges will be.
13. The year is 2000. What new technologies would you expect to see on your new automobile?
14. What impact do you feel the following new vehicle technologies will have on capacity?
    (a) Longitudinal and lateral information warning systems
    (b) Radar brakes
    (c) Fully automatic controlled guidance systems
15. Conduct a brief literature search of one of the following control theory algorithms.
    (a) Artificial intelligence
    (b) Expert systems
    (c) Fuzzy sets
16. Conduct a brief literature search of in-vehicle information systems.
17. Do you feel that in-vehicle information systems will prove to be cost-effective? Why ?
18. In relation to the materials covered in this chapter, what professional activities do you see for yourself in the first quarter of the twenty-first century?

1.7 SELECTED REFERENCES

1. W. D. Ashton, *The Theory of Road Traffic Flow*, John Wiley & Sons, Inc., New York, 1966, 178 pages.

2. Paul C. Box and J. C. Oppenlander, *Manual of Traffic Engineering Studies*, 4th Edition, Institute of Transportation Engineers, Washington, D.C., 1976, 233 pages.

3. Per Bruun, *Port Engineering*, 3rd Edition, Gulf Publishing Company, Book Division, Houston, Tex., 1981, 787 pages.

4. Don Drew, *Traffic Flow Theory and Control*, McGraw-Hill Book Company, New York, 1968, 467 pages.

5. John J. Fruin, *Pedestrian Planning and Design*, New York Metropolitan Association of Urban Designers and Environmental Planners, Inc., New York, 1971, 206 pages.

6. Denos Gazis, *Traffic Science*, John Wiley & Sons, Inc., New York, 1974, 293 pages.

7. Daniel L. Gerlough and Matthew J. Huber, *Traffic Flow Theory—A Monograph*, Transportation Research Board, Special Report 165, TRB, Washington, D.C., 1975, 222 pages.

8. Daniel L. Gerlough and Matthew J. Huber, *Statistics with Applications to Highway Traffic Analyses*, Eno Foundation, Saugatuck, Conn., 1978, 179 pages.

9. Frank A. Haight, *Mathematical Theories of Traffic Flow*, Academic Press, Inc., New York, 1963, 242 pages.

10. William W. Hay, *Railroad Engineering*, 2nd Edition, John Wiley & Sons, Inc., New York, 1982, 758 pages.

11. Wolfgang S. Homburger and James H. Kell, *Fundamentals of Traffic Engineering*, 12th Edition, University of California Press, Berkeley, Calif., 1988.

12. Robert Horonjeff and Francis X. McKelvey, *Planning and Design of Airports*, 3rd Edition, McGraw-Hill Book Company, New York, 1983, 616 pages.

13. J. J. Leeming, *Statistical Methods for Engineers*, Blackie & Son Ltd., Glasgow, 1963, 146 pages.

14. Wilhelm Leutzbach, *Introduction to the Theory of Traffic Flow*, Springer-Verlag, Berlin, 1988, 204 pages.

15. M. H. Lighthill and G. B. Whitham, On Kinematic Waves: A Theory of Traffic Flow on Long Crowded Roads, *Proceedings of the Royal Society, Series A*, Vol. 229, 1957, pages 317–345.

16. Louis J. Pignataro, *Traffic Engineering—Theory and Practice*, Prentice-Hall, Inc., Englewood Cliffs, N.J., 1973, 502 pages.

17. Vukan R. Vuchic, *Urban Public Transportation: Systems and Technology*, Prentice-Hall, Inc., Englewood Cliffs, N.J., 1981, 673 pages.

18. F. V. Webster and B. M. Cobbe, *Traffic Signals*, Her Majesty's Stationery Office, London, 1966, 111 pages.

19. Martin Wohl and Brian V. Martin, *Traffic System Analysis for Engineers and Planners*, McGraw-Hill Book Company, New York, 1967, 558 pages.

20. American Association of State Highway and Transportation Officials, *A Policy on Geometric Design of Highways and Streets*, AASHTO, Washington, D.C., 1984, 1087 pages.

21. Organization for Economic Cooperation and Development, *Route Guidance and In-Car Communications Systems*, OECD, Paris, February 1988, 104 pages.

22. Transportation Research Board, *Highway Capacity Manual*, Special Report 209, TRB, Washington, D.C., 1985, 474 pages.

23. Institute of Transportation Engineerings, *Manual on Uniform Traffic Control Devices*, Washington, D.C., 1989, 562 pages.

24. Great Britain Road Research Laboratory, *Research on Road Traffic*, Her Majesty's Stationery Office, London, 1965, 505 pages.

25. Organization for Economic Cooperation and Development, *Dynamic Traffic Management in Urban and Suburban Road Systems*, OECD, Paris, April 1987, 104 pages.

26. Organization for Economic Cooperation and Development, *Traffic Capacity of Major Routes*, OECD, Paris, July 1983, 120 pages.

27. Transportation Research Board, *Traffic System Management*, Special Report 172, TRB, Washington, D.C., 1977, 163 pages.

28. Institute of Transportation Engineers, *Transportation and Traffic Engineering Handbook, 2nd Edition*, Prentice-Hall, Inc., Englewood Cliffs, N.J., 1982, 883 pages.

2

Microscopic Flow Characteristics*

The time headway between vehicles is an important flow characteristic that affects the safety, level of service, driver behavior, and capacity of a transportation system. A minimum time headway must always be present to provide safety in the event that the lead vehicle suddenly decelerates. The percentage of time that the following vehicle must follow the vehicle ahead is one indication of level or quality of service. The distribution of time headways determines the requirement and the opportunity for passing, merging, and crossing. The capacity of the system is governed primarily by the minimum time headway and the time headway distribution under capacity-flow conditions. For these reasons it is important that the designer and operational manager of transportation systems have a theoretical knowledge of time headways and their distributions and the ability to apply this knowledge to real-life traffic problems. The purpose of this chapter is to provide a theoretical base with practical applications of this important microscopic flow characteristic.

Although primary attention will be given to highway traffic systems, time headways and time headway distributions are also important in other modes of transportation, such as in pedestrian, rail, water, and air transportation. The theoretical principles and practical applications contained in this chapter can, in large measure, be utilized in these other modes, particularly where the human beings control the movement. Transportation modes in which the individual vehicles are controlled or scheduled automatically provide special cases for study.

This chapter begins by defining and describing time headways and time headway distributions. Next, a classification scheme is presented which provides a systematic

*If limited time is available, the more advanced materials included in Section 2.5 can be scanned or omitted.

structure for all of the most relevant theoretical distributions. Then the mathematical formulations and characteristics of each such theoretical distribution are given. At this point real-world time headway data are introduced and techniques are presented for evaluating and selecting most appropriate mathematical distributions to represent the real-world time headway distributions. In the final section examples are provided for applying the principles covered in this chapter to a variety of highway traffic problems.

2.1 TIME HEADWAYS AND TIME HEADWAY DISTRIBUTIONS

A microscopic view of traffic flow is shown in Figure 2.1 as several individual vehicles traverse a length of roadway in single file for a certain period of time. The arrival time of each vehicle at the observation point is noted as t_1, t_2, t_3, and t_4. The elapsed time between the arrival of pairs of vehicles is defined as the time headway, and thus the time headways can be shown as

$$(h)_{1-2} = t_2 - t_1, \quad (h)_{2-3} = t_3 - t_2, \quad \text{etc.} \tag{2.1}$$

Observe that the time headway (h) actually consists of two time intervals: the *occupancy time* for the physical vehicle to pass the observation point and the *time gap* between the rear of the lead vehicle and the front of the following vehicle. In theory, an individual time headway does not necessarily have to be the elapsed time from the passage of the leading edges of two consecutive vehicles but only the elapsed time between the passage of identical points on two consecutive vehicles. However, in practice the leading edges are used whether the measurements are taken automatically by detectors or manually by observers. Occupancy time is discussed in Chapter 6 where

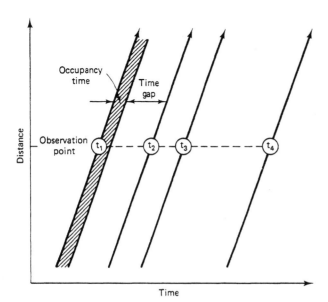

Figure 2.1   A Microscopic View of Traffic Flow

microscopic density characteristics are covered. The time gap parameter is used infrequently in highway traffic analysis.

An observer could continue to record individual time headways at a specific location for periods of time representing different flow situations. Then the individual time headways for each flow situation could be sorted into time headway intervals and plotted as shown in Figure 2.2.* The four time headway distributions are for four traffic flow levels: 10–14, 15–19, 20–24, and 25–29 vehicles per minute. The vertical scale is minute flow rate and the four measured time headway distributions are positioned on this scale at their average minute flow rate values. The horizontal scale is time

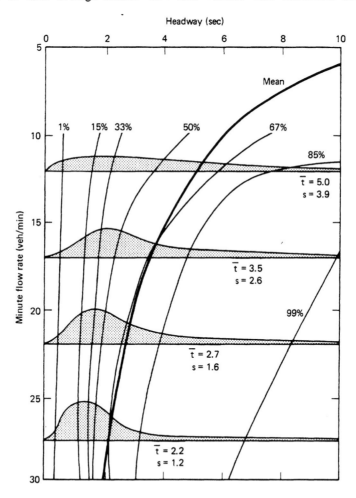

Figure 2.2   Measured Time Headway Distributions (From Reference 5)

*The time headway distributions should normally be shown as bar graphs, but for illustration purposes only are shown in Figure 2.2 as continuous distributions.

headway in seconds, and each of the four distributions is plotted on this scale based on
0.5-second time headway intervals. The height of the shaded areas represents the proportion of observed headways in each 0.5-second headway interval. Finally, there are
two sets of contour lines superimposed on top of the distributions, which represent the
mean time headway and the cumulative percentage of headways (1, 15, 33, 50, 67, 85,
and 99%). Headway tabulations and parameter values for the four measured time headway distributions are summarized in Appendix A.

This extensive data set of some 14,570 individually measured time headways
presented in Figure 2.2 and Appendix A provides a basis for the reader to obtain
insights and identify patterns of measured time headway distributions [5]. Some of the
pertinent observations are listed below.

- Individual time headways are rarely less than 0.5 second (on the order of 1 to 2
  percent).

- Individual time headways are rarely over 10 seconds unless the minute flow rate
  is below 15 vehicles per minute.

- The time headway mode is always less than the median, which is always less than
  the mean. However, they tend to converge as the minute flow rate increases
  toward capacity.

- The mean time headway tracks the 67 cumulative percentile curve for the entire
  minute flow rate range.

- The ratio of the standard deviation to the mean time headway approaches 1 under
  low flow conditions but decreases continuously as the minute flow rate increases.

The subject of time headways and related mathematical distributions has been studied at least since the 1930s and the reader is encouraged to review some of the early
references to get a historical perspective [10, 15, 16, 35, 40, 42, 44, 45]. Many data
sets of individually measured time headways are available in the literature and the
reader is encouraged to study them to validate the observations identified above [5, 6,
31, 32, 34, 36, 38, 39, 58, 62, 68].

2.2 CLASSIFICATION OF TIME HEADWAY DISTRIBUTIONS

The shape of the time headway distribution varied considerably as the traffic flow rate
increased. This was observed in Figure 2.2 and was due to the increasing interactions
between vehicles in the traffic stream. For example, under very low flow conditions,
there is very little interaction between the vehicles and the time headways appear to be
somewhat random. As the traffic flow level increases, there are increasing interactions
between vehicles. In simple terms, some headways appear to be random, while other
headways are of vehicles, that are following one another. As the traffic flow level
approaches capacity, almost all vehicles are interacting and are in a car-following process. In this process all time headways are approximately constant. A classification

scheme is proposed for time headway distributions consisting of a random distribution state for low flow levels, an intermediate distribution state for moderate flow levels, and a constant distribution state for high flow levels. The following three paragraphs describe this three-level classification scheme.

Under very low flow conditions, all the vehicles may be thought of as traveling independent of one another. In other words, any point in time is as likely to have a vehicle arriving as any other point in time. The only modification in real life would be due to a minimum time headway always being required from a safety point of view. Except for the minimum headway specification, these headways could be considered as *random* time headways and the time headway distribution as a random time headway distribution. This situation will be classified as the random headway state and can be considered as one of the two boundary conditions.

The other boundary condition is that when the traffic flow level is near capacity. This situation is discussed next, and the intermediate case is discussed in the following paragraph. Under heavy-flow conditions, almost all vehicles are interacting, and if an observer stood at a point on the roadway, the time headways would be almost constant. The only difference from real life would be that while the driver would attempt to maintain a constant headway, driver error would cause some variation regarding this constant headway. Except for this variation, these headways could be considered as constant time headways, and the time headway distribution as a constant time headway distribution. This situation will be classified as the constant headway state and can be considered as the other boundary condition.

The intermediate headway state lies between these two boundary conditions. That is, some vehicles are traveling independent of one another, while other vehicles are interacting. It will be shown in the next section that this is the most difficult to analyze, yet it is this situation that is most often encountered in the real world. Several different analytical approaches are employed for the intermediate headway state. One approach is to select a mathematical distribution that contains parameters that will modify the mathematical distribution in a continuous spectrum from a random distribution to a constant distribution. A variety of other approaches have been proposed and will be presented. This situation will be classified as the intermediate headway state and can be considered the state between the two boundary conditions.

2.3 RANDOM HEADWAY STATE

The negative exponential distribution is the mathematical distribution that represents the distribution of random intervals such as time headways. For time headways to be truly random, two conditions must be met. First, any point in time is as likely to have a vehicle arriving as is any other point in time. Second, the arrival of one vehicle at a point in time does not affect the arrival time of any other vehicle.

The negative exponential distribution can be derived from the Poisson count distribution. Note that the negative exponential distribution is an interval distribution; That is, it is the distribution of the number of individual time headways in various time headway intervals. This can be seen graphically in Figure 2.2. The Poisson count distribution, on the other hand, is a count distribution; That is, it is the distribution of the number of time periods which contain different flow levels.

The Poisson count distribution derived in Appendix B will be utilized in Chapter 3. The equation for the Poisson count distribution is

$$P(x) = \frac{m^x e^{-m}}{x!} \qquad (2.2)$$

where $P(x) =$ probability of exacting x vehicles arriving in a time interval t
   $m =$ average number of vehicles arriving in a time interval t
   $x =$ number of vehicles arriving in a time interval being investigated
   $e =$ a constant, Napierian base of logarithms ($e = 2.71828...$)
   $t =$ selected time interval

Consider the special case where $x = 0$. That is, consider the case when no vehicles arrive in a time interval. Then equation (2.2) can be rewritten

$$P(0) = e^{-m} \qquad (2.3)$$

What is the significance of $P(0)$ in terms of individual time headways? If no vehicles arrive in time interval (t), the individual time headway must be equal to or greater than t. Therefore,

$$P(0) = P(h \geq t) \qquad (2.4)$$

$$P(h \geq t) = e^{-m} \qquad (2.5)$$

Recall that m is defined as the average number of vehicles arriving in time interval t. If the hourly flow rate is specified as V and t is expressed in seconds, then

$$m = \left(\frac{V}{3600}\right)t \qquad (2.6)$$

and equation (2.5) becomes

$$P(h \geq t) = e^{-Vt/3600} \qquad (2.7)$$

The mean time headway in seconds, $\bar{t}$, can be determined from the hourly flow rate, V, by

$$\bar{t} = \frac{3600}{V} \qquad (2.8)$$

Equation (2.7) can be modified by equation (2.8), and an alternative formulation can result:

$$P(h \geq t) = e^{-t/\bar{t}} \qquad (2.9)$$

In this form, given the mean time headway, $\bar{t}$, the analyst can select various time intervals t, such as $0, 0.5, 1.0, 2.0,...,\infty$, and $P(h \geq 0, 0.5, 1.0, 2.0, ... , \infty)$ can be calculated. Note that when $t = 0$, $P(h \geq t) = 1.0$ which is obviously correct. As the value of t increases, $P(h \geq t)$ decreases until finally when $t \rightarrow \infty$, $P(h \geq t) \rightarrow 0$.

Although analysts may be interested in probability distributions, such as $P(h \geq t)$, it is often desirable to calculate the probability for specific time headway intervals, such as the probability of a time headway between t and $t + \Delta t$ seconds. This can be accomplished by calculating $P(h \geq t)$ and $P(h \geq t + \Delta t)$ and subtracting in the following manner:

$$P(t \leq h < t + \Delta t) = P(h \geq t) - P(h \geq t + \Delta t) \qquad (2.10)$$

For illustration purposes and to obtain insights into the strengths and weaknesses of this random-headway state, the negative exponential distribution derived will be applied to the measured time headway distributions shown in Figure 2.2. Detailed calculations of the theoretical time headway distribution for only the lowest traffic flow level will be presented, but results for all traffic flow levels will be summarized a little later in a figure like Figure 2.2 and also included in Appendix A.

The mean time headway for the lowest traffic flow level was shown in Figure 2.2 to be 5 seconds per vehicle, and the time headway distribution was presented in 0.5-second intervals. Using equation (2.9) and substituting 5 for the mean time headway, $\bar{t}$, the equation becomes

$$P(h \geq t) = e^{-0.2t} \qquad (2.11)$$

Substituting t values of $0, 0.5, 1.0, \ldots, 9.5$ seconds into equation (2.11), the $P(h \geq t)$ values are calculated and are shown in Table 2.1. Exponential function tables are included in Appendix C to aid the analyst in solving equations (2.9) and (2.11). The next step is to calculate the probabilities for each time headway interval by utilizing equation (2.10), and these results are also shown in Table 2.1. The final step is to convert the calculated probabilities of each time headway interval into the frequency of headways for each time headway group. This is accomplished using the equation

$$F(t \leq h < t + \Delta t) = N[P(t \leq h \leq t + \Delta t)] \qquad (2.12)$$

where $N = $ total number of observed headways (in the lowest volume case, 1320)

$F(t \leq h \leq t + \Delta t) = $ predicted number of time headways in the time headway group $(t \leq h < t + \Delta t)$

The results are also shown in Table 2.1.

This procedure can be followed for all traffic flow levels as cited in Figure 2.2 and the theoretical results plotted in a format similar to Figure 2.2. The theoretical results are shown in Figure 2.3 superimposed on the measured time headway distributions of Figure 2.2. The standard deviation for each distribution is also indicated. Careful study of Figure 2.3 provides insights as to the strengths and weaknesses of representing measured time headway distributions with the random (negative exponential) distribution. Some of the pertinent observations are listed below.

- The random (negative exponential) distribution has the inherent characteristic that the smallest headways are most likely to occur and the probabilities consistently decrease as time headway increases.

TABLE 2.1   Random Time Headway Distribution Calculations[a]

t	$0.2t$	$P(h \geq t)$	$P(t \leq h < t + \Delta t)$	$F(t \leq h < t + \Delta t)$
0.0	0.00	1.000		
			0.095	125
0.5	0.10	0.905		
			0.086	114
1.0	0.20	0.819		
			0.078	103
1.5	0.30	0.741		
			0.071	94
2.0	0.40	0.670		
			0.063	83
2.5	0.50	0.607		
			0.058	77
3.0	0.60	0.549		
			0.052	69
3.5	0.70	0.497		
			0.048	63
4.0	0.80	0.449		
			0.042	55
4.5	0.90	0.407		
			0.039	51
5.0	1.00	0.368		
			0.035	46
5.5	1.10	0.333		
			0.032	42
6.0	1.20	0.301		
			0.028	37
6.5	1.30	0.273		
			0.026	34
7.0	1.40	0.247		
			0.024	32
7.5	1.50	0.223		
			0.021	28
8.0	1.60	0.202		
			0.019	25
8.5	1.70	0.183		
			0.018	24
9.0	1.80	0.165		
			0.015	20
9.5	1.90	0.150		
			0.150	198

[a]Based on measured headway distribution data presented in Figure 2.2.

- The two distributions are quite different, particularly under higher-flow conditions.
- The comparison between distributions is best under the lowest flow level.
- Even for the lowest flow level, the two distributions are quite different for time headway groups of less than 1 second. The theoretical probabilities are higher for

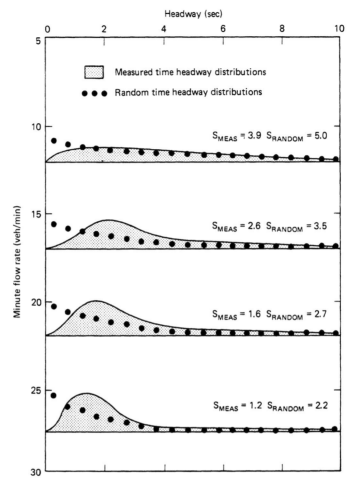

Figure 2.3   Random Time Headway Distributions (based on measured headway distribution data presented in Figure 2.2)

time headway intervals of less than 1 second and lower for time headway intervals between 1.5 and 4.5 seconds.

- The standard deviation for the measured distribution is always less than the standard deviation of the corresponding random distribution but appears to be converging at the lower flow levels.

2.4 CONSTANT HEADWAY STATE

The normal distribution is a mathematical distribution that can be used when either the time headways are all constant or when drivers attempt to drive at a constant time headway but driver errors cause the time headways to vary about the intended constant time headway. A unique normal distribution is defined by specifying the mean time head-way and the standard deviation of the time headway distribution. In the constant-time-headway case, the mean time headway is simply

$$\bar{t} = \frac{3600}{V} \qquad (2.13)$$

while the standard deviation is zero. For the second case, the mean time headway is also calculated from equation (2.13), but the standard deviation is greater than zero. To solve this case, the analyst must specify the value of the standard deviation. Although some research has been directed to this issue, only general guidelines can be given. For example, if the capacity of the stream of traffic is on the order of 1800 vehicles per hour, the mean time headway would be 2 seconds per vehicle. Since time headways cannot be negative (in fact, normally cannot be less than 0.5 seconds), the lowest theoretical headway should be approximately

$$\alpha = \bar{t} - 2s \qquad (2.14)$$

where $\alpha =$ minimum expected time headway

$\bar{t} =$ mean time headway (seconds per vehicle)

$s =$ standard deviation of the time headway distribution

$2 =$ constant, specifying that 2 standard deviations below the mean time head-way should be very near the lowest theoretical headway

Rearranging equation (2.14) and solving for s, the standard deviation yields

$$s = \frac{\bar{t} - \alpha}{2} \qquad (2.15)$$

Putting in numerical values for $\bar{t}$ and α of 2 seconds and 0.5 seconds respectively, the standard deviation is estimated to be 0.75 second, and hence the corresponding variance is 0.56.

For illustration purposes and to obtain insights into the strengths and weaknesses of this constant-headway state, the normal distribution, including driver error, will be applied to the measured time headway distributions shown in Figure 2.2. Detailed cal-culations of the theoretical time headway distribution for only the highest traffic flow levels will be presented, but results for all traffic flow levels will be summarized a little later in a figure like Figure 2.2 and also included in Appendix A.

The mean time headway for the highest traffic flow was shown in Figure 2.2 to be 2.2 seconds per vehicle, and the time headway distribution is presented in 0.5-second intervals. Using equation (2.15) and substituting numerical values for mean time head-way and minimum time headway of 2.2 and 0.5, respectively, the standard deviation is found to be 0.85. Using the procedures described in Appendix D for normal distribu-tion calculations, and the numerical values of 2.2 for mean time headway and 0.85 for standard deviation, the resulting calculations are summarized in Table 2.2 for a normal distribution.

This procedure can be followed for all traffic flow levels as specified in Figure 2.2, and the theoretical results plotted in a format similar to Figure 2.3. The theore-tical results are shown in Figure 2.4 superimposed on the measured time headway

TABLE 2.2   Normal Time Headway Distribution Calculations[a]

t	z	$\dfrac{z}{s}$	Probability of Headway between t and $\bar{t}$	$P(t \leq h < t + \Delta t)$	$F(t \leq h < t + \Delta t)$
0.0	2.2	2.59	0.4952		
				0.0179	62
0.5	1.7	2.00	0.4773		
				0.0566	198
1.0	1.2	1.41	0.4207		
				0.1268	443
1.5	0.7	0.82	0.2939		
				0.1991	695
2.0	2.2	0.24	0.0948		
				0.2316	809
2.5	0.3	0.35	0.1368		
				0.1896	662
3.0	0.8	0.94	0.3264		
				0.1106	386
3.5	1.3	1.53	0.4370		
				0.0460	161
4.0	1.8	2.12	0.4830		
				0.0136	47
4.5	2.3	2.71	0.4966		
				0.0029	10
5.0	2.8	3.29	0.4495		
				0.0004	1
5.5	3.3	3.88	0.4999		
				0.0001	0
6.0	3.8	4.47	0.5000		
				0.0000	0
6.5	4.3	5.06	0.5000		
				0.0000	0
7.0	4.8	5.65	0.5000		
				0.0000	0
7.5	5.3	6.24	0.5000		
				0.0000	0
8.0	5.8	6.82	0.5000		
				0.0000	0
8.5	6.3	7.41	0.5000		
				0.0000	0
9.0	6.8	8.00	0.5000		
				0.0000	0
9.5	7.3	8.59	0.5000		
				0.0000	0

[a]Based on measured headway distribution data presented in Figure 2.2.

distributions of Figure 2.2. Careful study of Figure 2.4 provides insights of the strengths and weaknesses of representing measured time headway distributions with the normal distribution. Some of the more pertinent observations are listed below.

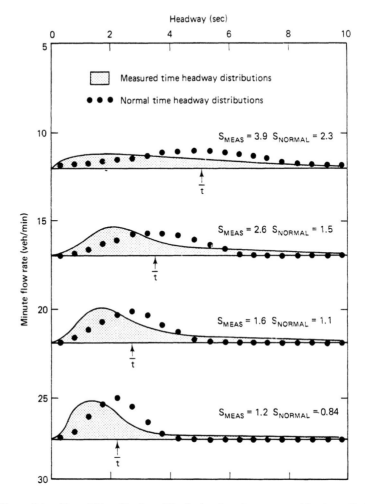

Figure 2.4   Normal Time Headway Distribution (based on measured headway distribution data presented in Figure 2.2)

- The normal distribution has the inherent characteristic of being symmetrical about the mean time headway and having a bell-shaped distribution.

- The two distributions are quite different, particularly under lower flow conditions.

- The comparison between distributions is best under the highest flow level.

- Even for the highest flow level, the normal distribution is shifted to the right of the measured distribution on the order of 0.5 to 1.0 seconds.

- The standard deviation of the measured distribution is always greater than the standard deviation of the normal distribution but appears to be converging at higher flow levels.

2.5 INTERMEDIATE HEADWAY STATE

The intermediate headway state lies between the two boundary conditions exemplified by the random- and constant-headway states, which were discussed in previous sections of this chapter. Much less is known about the intermediate state, yet this is the situation that is most often encountered in practice. The purpose of this section is to present the state of the art as to describing time headway distributions mathematically for the intermediate headway state, including both boundary conditions. Three approaches are described: a generalized mathematical model approach, a composite model approach, and a variety of other approaches.

2.5.1 Generalized Mathematical Model Approach

The Pearson type III distribution is an example of a generalized mathematical model approach that has been proposed. The probability density function of this distribution is given by the equation

$$f(t) = \frac{\lambda}{\Gamma(K)}[\lambda(t - \alpha)]^{K-1} e^{-\lambda(t-\alpha)} \qquad (2.16)$$

where $f(t)$ = probability density function

λ = parameter that is a function of the mean time headway and the two user-specified parameters, K and α

K = user-selected parameter between 0 and ∞ that affects the shape of the distribution

α = user-selected parameter greater than or equal to zero that affects the shift of the distribution (seconds)

t = time headway being investigated (seconds)

e = constant parameter, 2.71828

$\Gamma(K)$ = gamma function, equivalent to $(K-1)!$

A graphical example of a probability density function for a Pearson type III distribution is shown in Figure 2.5a. Note that the integration of $f(t)$ from 0 to ∞ (the area under the curve) represents all possible outcomes and therefore is equal to a probability of 1.00. The probability of a headway greater than some specified t value, is given in the following equation and denoted in Figure 2.5b:

$$P(h \geq t) = \int_{t}^{\infty} f(t)\, dt \qquad (2.17)$$

Therefore, the probability of a headway lying between t and $t + \Delta t$ can be formulated as

$$P(t \leq h < t + \Delta t) = \int_{t}^{\infty} f(t)\, dt - \int_{t+\Delta t}^{\infty} f(t)\, dt \qquad (2.18)$$

and can be denoted in Figure 2.5c as the shaded area underneath the probability density curve. If an approximate solution is acceptable, and particularly if the time headway group intervals Δt are small, the formulation can be simplified to the following:

$$P(t \leq h < t + \Delta t) = \left[\frac{f(t) + f(t + \Delta t)}{2}\right]\Delta t \qquad (2.19)$$

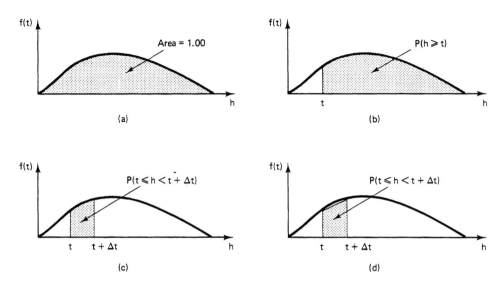

Figure 2.5   An Example of a Probability Function of a Pearson Type III Distribution

and the results shown in Figure 2.5d. Note that the assumption is that the probability density curve is a straight line over the Δt interval.

Returning to the original probability density function given in equation (2.16) and considering equation (2.19), the probability of time headways in any time headway group can be calculated using the following equation:

$$P(t \leq h < t + \Delta t) = \left\{ \frac{\lambda}{\Gamma(K)} [\lambda(t - \alpha)]^{K-1} e^{-\lambda(t-\alpha)} \right.$$

$$\left. + \frac{\lambda}{\Gamma(K)} \left[\lambda(t + \Delta t) - \alpha\right]^{K-1} e^{-\lambda[(t+\Delta t) - \alpha]} \right\} \frac{\Delta t}{2} \tag{2.20}$$

The Pearson type III distribution is actually a family of distribution models that can be telescoped down into a nested subset of simpler distribution models. Figure 2.6 is a graphical illustration of this nested subset of simpler distribution models. The total K, α plane of Figure 2.6 represents the complete Pearson type III distribution family, and each point in the plane defines a unique Pearson type III distribution. The first subset of distribution models, those in which $\alpha = 0$, is represented graphically in Figure 2.6 as the y-axis. The models in this subset of distribution models are called gamma distribution models, and the probability density function shown in equation (2.16) can be simplified as shown by the equation

$$f(t) = \frac{\lambda}{\Gamma(K)} (\lambda t)^{K-1} e^{-\lambda t} \tag{2.21}$$

Essentially, a Pearson type III distribution model becomes the simpler gamma distribution model when $\alpha = 0$ and K takes on any positive value.

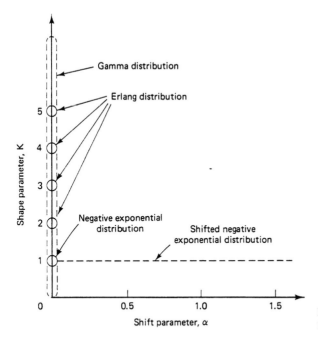

Figure 2.6   The Pearson Type III Distribution Model Family

The next subset of distribution models are those in which $\alpha = 0$ but K is a positive integer value. This is represented graphically in Figure 2.6 as integer K values along the y-axis. The models in this subset of distribution models are called Erlang distribution models, and the probability density function shown in equation (2.21) can be simplified as follows:

$$f(t) = \frac{\lambda}{K-1!}(\lambda t)^{K-1}e^{-\lambda t} \tag{2.22}$$

Essentially, a gamma distribution model becomes the simpler Erlang distribution model when K takes on any positive integer value.

The next distribution subset is when $\alpha = 0$ and $K = 1$. This is represented graphically in Figure 2.6 as a single point on the y-axis. This single distribution model is called the negative exponential (or random) distribution model, and the probability density function given in equation (2.22) can be simplified as shown by the following equation:

$$f(t) = \lambda e^{-\lambda t} \tag{2.23}$$

This probability density function can be transformed into the cumulative probability form of the negative exponential distribution as shown in equation (2.17):

$$P(h \geq t) = \int_{t}^{\infty} f(t)\,dt$$

Substitution yields

$$P(h \geq t) = \int_{t}^{\infty} (\lambda e^{-\lambda t})\,dt \tag{2.24}$$

Then

$$P(h \geq t) = e^{-\lambda t} = e^{-t/\bar{t}} \tag{2.25}$$

Since λ is equivalent to the reciprocal of the mean time headway when $K = 1$ and $\alpha = 0$, $\lambda = 1/\bar{t}$ and equation (2.25) is equivalent to equation (2.9). In summary, the negative exponential distribution is a special case of the Erlang distribution, which is a subset of special cases of the gamma distribution, which is a subset of special cases of the Pearson type III distribution.

There is one other subset of distribution models that can be derived from the Pearson type III distribution family. This subset of models occurs when $K = 1$ and $\alpha > 0$ and is represented graphically in Figure 2.6 as a horizontal line with $K = 1$. The models in this subset of distribution models are called shifted negative exponential models, and the probability density function given in equation (2.16) can be simplified as follows:

$$f(t) = \lambda e^{-\lambda(t-\alpha)} \tag{2.26}$$

This probability density function can be transformed into the cumulative probability form starting with equation (2.17):

$$P(h \geq t) = \int_t^\infty f(t)\,dt$$

Substitution gives

$$P(h \geq t) = \int_t^\infty [\lambda e^{-\lambda(t-\alpha)}]\,dt$$

Then

$$P(h \geq t) = e^{-\lambda(t-\alpha)} = e^{-(t-\alpha)/(\bar{t}-\alpha)} \tag{2.27}$$

Since λ is equivalent to the reciprocal of the mean time headway minus the shift parameter when $K = 1$ and $\alpha \geq 0$, $\lambda = 1/(\bar{t} - \alpha)$ and equation (2.27) is formulated in a fashion similar to equation (2.9).

Table 2.3 is a summary of the most important equations required in analyzing Pearson type III, Gamma, Erlang, negative exponential, and shifted negative exponential distribution models. These equations are used in the following illustration.

Eight steps will now be followed in applying the Pearson type III distribution to measured time headway distributions. The first step is to calculate the mean time headway $(\bar{t})$ and the standard deviation (s) of the measured time headway distribution. For the examples given in Figure 2.2, the mean time headways (and standard deviations) for the four traffic flow levels are 5.0 (3.9), 3.5 (2.6), 2.7 (1.6), and 2.2 (1.2), respectively.

The second step is to select an appropriate value of α that effects the shift of the distribution. The α value can be greater than or equal to zero. If α is selected to be zero, there will be no shift, which implies that time headways may approach zero. If α is selected to be 0.5, time headways of less than 0.5 seconds are assumed to have a probability of zero. Normal values for α range between 0.0 and 1.0 second, and later for example purposes, α will be assumed to be 0.5.

TABLE 2.3 Pearson Type III Distribution Model Family of Equations

Distribution Family	Estimating $\hat{K}$	Calculating λ	Probability Density Function, $f(t)$	Probability Distribution, $P(h \geq t)$
Pearson type III (K, α)	$\dfrac{\bar{t} - \alpha}{s}$	$\dfrac{K}{\bar{t} - \alpha}$	$\dfrac{\lambda}{\Gamma(K)}[\lambda(t-\alpha)]^{K-1} e^{-\lambda(t-\alpha)}$	$\displaystyle\int_t^\infty f(t)\,dt$
Gamma $(K, \alpha = 0)$	$\dfrac{\bar{t}}{s}$	$\dfrac{K}{\bar{t}}$	$\dfrac{\lambda}{\Gamma(K)}[\lambda t]^{K-1} e^{-\lambda t}$	$\displaystyle\int_t^\infty f(t)\,dt$
Erlang $K = 1, 2, 3, ..., \alpha = 0$	$\dfrac{\bar{t}}{s}$	$\dfrac{K}{\bar{t}}$	$\dfrac{\lambda}{(K-1)!}[\lambda t]^{K-1} e^{-\lambda t}$	$e^{-\lambda t} \displaystyle\sum_{h=0}^{n=K-1} \dfrac{(\lambda t)^n}{n!}$
Negative exponential $(K = 1, \alpha = 0)$	$\dfrac{\bar{t}}{s}$	$\dfrac{1}{\bar{t}}$	$\lambda e^{-\lambda t}$ or $\lambda e^{\frac{-t}{\bar{t}}}$	$e^{-\lambda t}$ or $e^{-t/\bar{t}}$
Shifted negative exponential $(K = 1, \alpha > 0)$	$\dfrac{\bar{t} - \alpha}{s}$	$\dfrac{1}{\bar{t} - \alpha}$	$\lambda e^{-\lambda(t-\alpha)}$ or $\lambda e^{-(t-\alpha)/(\bar{t}-\alpha)}$	$\lambda e^{-\lambda(t-\alpha)}$ or $e^{-(t-\alpha)/(\bar{t}-\alpha)}$

The third step is to select an appropriate value of K that affects the shape of the distribution. The K value can vary from 0 to ∞. If K is selected to be 1, the resulting distribution takes the form of a negative exponential (random) distribution. As the K value selected approaches infinity, the resulting distribution approaches a constant headway distribution. An approximate K value can be determined using the equation

$$\hat{K} = \frac{\bar{t} - \alpha}{s} \tag{2.28}$$

The fourth step is to calculate the resulting λ value, which is a function of the mean time headway and the user-selected parameters, K and α, shown in the equation

$$\lambda = \frac{K}{\bar{t} - \alpha} \tag{2.29}$$

Note that λ is equal to the flow rate (reciprocal of the mean time headway) only when $K = 1$ and $\alpha = 0$. Also note that if $\hat{K}$ in equation (2.28) is substituted for K in equation (2.29), λ becomes the reciprocal of the standard deviation.

The fifth step is to calculate the gamma function [$\Gamma(K)$]. The $\Gamma(K)$ is equal to $(K-1)!$. If K is a position integer such as 1, 2, 3, etc. the gamma function is simply 0!, 1!, 2!, etc. However, if K is a noninteger value such as 4.785, the gamma function is 3.785!, which is not so easily calculated. A gamma function table and an example problem are given in Appendix E to aid the analyst [17].

The sixth step is to solve for $f(t)$ for various desired values of t (i.e., $t = 0.0, 0.5,$ 1.0, 1.5, etc.) using equation (2.16). The seventh step is to calculate time headway group probabilities using equation (2.19). The eighth and final step is to calculate time headway group frequencies using equation (2.12).

For illustration purposes and to understand the strengths and weaknesses of this intermediate headway state, the Pearson type III distribution will be applied to the measured time headway distributions shown in Figure 2.2. Detailed calculations of the theoretical time headway distribution for only the lowest traffic flow level will be presented, but results for all traffic flow levels will be summarized a little later in a figure like Figure 2.2 and included in Appendix A.

The previously described eight-step process was applied to the measure time headway distribution of 10 to 14 vehicles per minute, represented graphically in Figure 2.2 and given in tabular form in Appendix A. The results are summarized in Table 2.4. The same eight-step process was applied to the other three measured time headway distributions presented graphically in Figure 2.2. These theoretical results are shown in Figure 2.7 superimposed on the measured time headway distributions of Figure 2.2. The theoretical results are also tabulated in Appendix A. Careful study of Figure 2.7 provides insights as to the strengths and weaknesses of representing measured time headway distributions with the Pearson type III distribution. Some of the more pertinent observations are listed below.

- The Pearson type III distribution has the inherent characteristic that the first nonzero time headway group will have a probability approximately of one-half the highest probability, and that the next time headway group will have the highest probability. Thereafter, the probabilities decrease steadily as the time headway increases.

- A comparison of the Pearson type III and measured time headway distributions for the four flow levels indicates that qualitatively the two are about the same.

- The probabilities of the theoretical distributions are almost always less than the corresponding measured distributions when time headways exceed 4 seconds.

- The probabilities of the theoretical distributions are almost always less than the corresponding measured distributions for headway groups of less than 1 second.

- The probabilities of the theoretical and measured distributions are most inconsistent when time headways between 1.0 and 4.0 seconds are encountered.

- It should be kept in mind that only a single Pearson type III distribution was investigated for each traffic flow level, and the K and α values are only approximate estimates of the "best" Pearson type III parameters. Sensitivity analysis of a matrix of K and α parameter values is required to find the "best" Pearson Type III distribution.

2.5.2 Composite Model Approach

Observations and insights from the analysis contained in Sections 2.3 and 2.4 indicated that the random headway state was best suited for very low flow conditions, while the nearly-constant headway state was best suited for very high flow conditions.

TABLE 2.4   Pearson Type III Time Headway Distribution Calculations[a]

t	$\dfrac{\lambda}{\Gamma(K)}$	$t-\alpha$	$\lambda(t-\alpha)$	$\lambda(t-\alpha)^{K-1}$	$e^{-\lambda/t-\alpha}$	$f(t)$	Probability, $P(t \le h < t+\Delta t)$	$F(t \le h < t+\Delta t)$
0.0	0.274	—	—	—	—	—		
							0.000	0
0.5	0.274	0.0	0.000	0.000	1.000	0.000		
							0.044	58
1.0	0.274	0.5	0.128	0.735	0.880	0.177		
							0.088	116
1.5	0.274	1.0	0.256	0.815	0.774	0.173		
							0.084	111
2.0	0.274	1.5	0.384	0.866	0.681	0.162		
							0.078	103
2.0	0.274	2.0	0.512	0.904	0.599	0.148		
							0.071	93
3.0	0.274	2.5	0.640	0.935	0.527	0.135		
							0.064	84
3.5	0.274	3.0	0.768	0.961	0.464	0.122		
							0.058	77
4.0	0.274	3.5	0.896	0.984	0.408	0.110		
							0.052	69
4.5	0.274	4.0	1.024	1.004	0.359	0.099		
							0.047	62
5.0	0.274	4.5	1.152	1.021	0.316	0.088		
							0.042	55
5.5	0.274	5.0	1.280	1.038	0.278	0.079		
							0.038	50
6.0	0.274	5.5	1.408	1.053	0.245	0.071		
							0.034	45
6.5	0.274	6.0	1.536	1.066	0.215	0.063		
							0.030	40
7.0	0.274	6.5	1.664	1.079	0.189	0.056		
							0.026	34
7.5	0.274	7.0	1.792	1.091	0.167	0.050		
							0.024	32
8.0	0.274	7.5	1.920	1.103	0.147	0.044		
							0.021	28
8.5	0.274	8.0	2.048	1.114	0.129	0.039		
							0.018	24
9.0	0.274	8.5	2.176	1.124	0.113	0.035		
							0.016	21
9.5	0.274	9.0	2.304	1.133	0.100	0.031		
							0.165	218

[a] $s = 3.9$, $\Delta t = 0.5$, $N = 1320$, $\alpha = 0.5$, $K = 1.15$, $\lambda = 0.256$, $\Gamma(K) = 0.933$, $\bar{t} = 5.0$.

Figure 2.8 is a graphical illustration of the relationship between the standard deviation and the mean time headway for the three distributions: random, normal, and measured. Standard deviation–mean time headway linear equations are given for the three

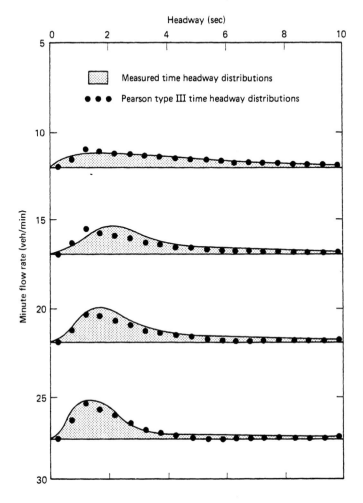

Figure 2.7   Pearson Type III Time Headway Distributions (based on measured head-way distribution data presented in Figure 2.2)

distributions. Comparing the normal and measured distributions, the two lines are seen to converge as the mean time headway decreases and to intersect when the time mean headway is 1.43 seconds (hourly flow rate slightly over 2500 vehicles per hour). Comparing the random and measured distributions, the two lines are almost parallel, with the line representing the measured distribution shifted about 1 second to the right. This is an interesting observation considering the shifted random (negative exponential) distribution. In Table 2.3 the following equation was given relating the mean time headway and the standard deviation:

$$\hat{K} = \frac{\bar{t} - \alpha}{s} \qquad (2.30)$$

The K value for the shifted negative exponential distribution is equal to 1; therefore, solving for s:

$$s = \overline{t} - \alpha \qquad (2.31)$$

The reader can observe that as α increases from 0 to 1.0, a line can be superimposed on Figure 2.8 for the shifted negative exponential distribution, and it will lie parallel to and between the lines representing the random and measured distributions.

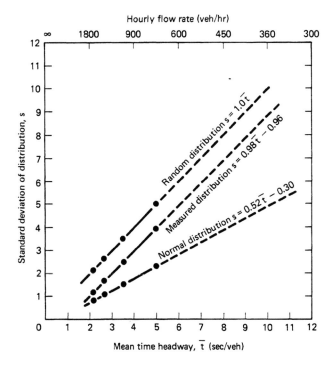

Figure 2.8   Relationship Between Standard Deviation and Mean Time Headway

The composite model approach utilizes the combination of a normal headway distribution for those vehicles that are in the car-following or platoon mode and a shifted negative exponential distribution for those vehicles that are not interacting. Four independent parameters must be specified for such a composite distribution: mean and standard deviation of the normal distribution, the proportion of vehicles in platoons, and the minimum time headway for vehicles not in platoons. Three other parameters must be known but are dependent on the initial four independent parameters:

- The proportion of vehicles not in platoons is equal to 1.00 minus the proportion of vehicles in platoons.
- The mean time headway of the vehicles not in platoon can be computed from the equation

$$\overline{t} = \overline{t}_p \, P_p + \overline{t}_{Np} \, P_{Np} \qquad (2.32)$$

where P_p = proportion of vehicles in platoon
$\quad\quad\; P_{Np}$ = proportion of vehicles not in platoon

• The standard deviation of the time headways of vehicles not in platoons can be calculated using equation (2.31).

For illustrating a composite distribution and based on qualitative results of the preceding sections, assume that the mean and standard deviation of the time headways of those vehicles in platoons are 1.5 and 0.5 second per vehicle, respectively. Assuming that all vehicles traveling at headways of less than 1.5 seconds are in platoons and represent 50 percent of all such vehicles (recall that the proposed mean is 1.5 seconds and the normal distribution is symmetrical about the mean), the proportion of vehicles in platoons can be determined from the measured distribution (see Appendix A) and plotted against the mean time headway as shown in Figure 2.9. Assuming that 100 percent of the vehicles are in platoons when the time mean headway is 1.5 seconds, and using a fairly simple equation,

$$P_p = \frac{1.50}{\bar{t}}, \tag{2.33}$$

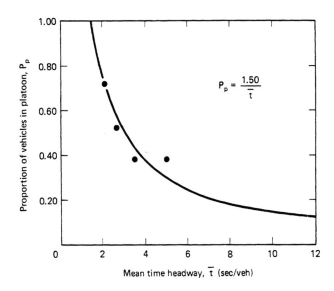

Figure 2.9   The Proportion of Vehicles in Platoon as a Function of Mean Time Headway

the resulting equation can be plotted on Figure 2.9. For mean time headways of 5.0, 3.5, 2.7 and 2.2 seconds per vehicle, the percent of vehicles in platoons can be estimated to be 30, 43, 56, and 68%, respectively. The last independent parameter is the minimum time headway of vehicles not in platoons. Inspection of the differences between the probabilities of the measured and normal time headway distributions indicates that 2.0 seconds per vehicle might be an appropriate minimum time headway.

The three dependent parameters can now be determined. The proportion of vehicles not in platoons would be 70, 57, 44 and 32%, respectively. The mean time headway of vehicles not in platoons can be computed using equation (2.32) and the results are 6.50, 5.00, 4.23, and 3.69 seconds per vehicle, respectively. The standard deviation of vehicles not in platoon can be computed using equation (2.31) and the results are 4.50, 3.00, 2.23, and 1.69 seconds per vehicle, respectively.

To understand the strengths and weaknesses of this approach, the composite distribution model was applied to the four measured time headway distributions shown in Figure 2.2. Detailed calculations of the theoretical time headway for only the lowest traffic flow level is shown in Table 2.5, but results for all four traffic flow levels are presented graphically in Figure 2.10 and in tabular form in Appendix A. The normal distribution probabilities were calculated following the procedures described in Appendix D, and the shifted negative exponential distribution probabilities were calculated using equations given in Table 2.3. Careful study of Figure 2.10 provides insights as to the strengths and weaknesses of representing measured time headway distributions with the composite model distribution. Some of the more pertinent observations are listed below.

TABLE 2.5 Composite Time Headway Distribution Calculations[a]

t	$t - \alpha$	$\bar{t} - \alpha$	$\dfrac{t - \alpha}{\bar{t} - \alpha}$	$e^{-(t - \alpha)/(\bar{t} - \alpha)}$	Vehicles Not in Platoons 100%	70%	Platoon Vehicles 100%	30%	Composite Distribution (P)	(F)
0.0	—	—	—	—	0.000	0.000	0.023	0.007	0.007	9
0.5	—	—	—	—	0.000	0.000	0.136	0.041	0.041	54
1.0	—	—	—	—	0.000	0.000	0.341	0.102	0.102	135
1.5	—	—	—	—	0.000	0.000	0.341	0.102	0.102	135
2.0	0.0	4.5	0.00	1.000	0.104	0.073	0.136	0.041	0.114	150
2.5	0.5	4.5	0.11	0.896	0.094	0.065	0.022	0.007	0.072	95
3.0	1.0	4.5	0.22	0.802	0.084	0.059	0.001	0.000	0.059	78
3.5	1.5	4.5	0.33	0.719	0.075	0.052	0.000	0.000	0.052	69
4.0	2.0	4.5	0.44	0.644	0.073	0.051	0.000	0.000	0.051	67
4.5	2.5	4.5	0.56	0.571	0.059	0.042	0.000	0.000	0.042	55
5.0	3.0	4.5	0.67	0.512	0.054	0.037	0.000	0.000	0.037	49
5.5	3.5	4.5	0.78	0.458	0.047	0.033	0.000	0.000	0.033	44
6.0	4.0	4.5	0.89	0.411	0.043	0.030	0.000	0.000	0.030	40
6.5	4.5	4.5	1.00	0.368	0.038	0.027	0.000	0.000	0.027	36
7.0	5.0	4.5	1.11	0.330	0.035	0.024	0.000	0.000	0.024	32
7.5	5.5	4.5	1.22	0.295	0.031	0.022	0.000	0.000	0.022	29
8.0	6.0	4.5	1.33	0.264	0.027	0.019	0.000	0.000	0.019	25
8.5	6.5	4.5	1.44	0.237	0.027	0.019	0.000	0.000	0.019	25
9.0	7.0	4.5	1.56	0.210	0.022	0.015	0.000	0.000	0.015	20
9.5	7.5	4.5	1.67	0.188	0.188	0.132	0.000	0.000	0.132	173

[a] $\bar{t}_{Np} = 6.5$ sec, $\alpha = 2.0$ sec, $s_{Np} = 4.5$ sec, $P_{Np} = 0.70$, $E_p = 1.5$ sec, $s_p = 0.5$ sec, $P_p = 0.30$.

- The two distributions appear to have the same general shape.
- The two distributions are most different under low-flow conditions, but become more similar as the flow level increases.

- The probabilities of the theoretical distributions are almost always greater than the corresponding measured distributions when time headways exceed 4 seconds.
- The probabilities of the theoretical distributions are almost always less than the corresponding measured distributions when time headways between 2.5 and 4 seconds are encountered.
- Larger differences occur when the time headways lie between 1 and 2.5 seconds (particularly for lower traffic flow levels).
- It should be kept in mind that only a single composite model distribution was investigated for each traffic flow level. Sensitivity analysis of a matrix of the four independent parameters is required to find the "best" composite model distribution.

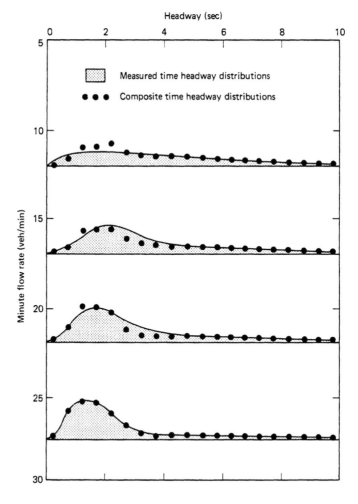

Figure 2.10 Composite Time Headway Distributions (based on measured headway distribution data presented in Figure 2.2)

2.5.3 Other Approaches

A number of other approaches have been proposed for describing time headway distributions mathematically for the intermediate headway state. Space does not permit detailed coverage of these other approaches, but they will be identified, described briefly, and referenced.

Schuhl was one of the first investigators to propose a composite model approach [3]. In 1955 he proposed that the traffic flow consisted of two classes of vehicles: constrained vehicles and free-moving vehicles. He proposed further that the time headway distributions for constrained vehicles could be represented by a shifted negative exponential distribution while the time headway distributions for free-flowing vehicles could be represented by a negative exponential distribution. The combining of these two mathematical distributions were used to represent the total traffic stream. The mathematical formulation is as follows:

$$P(h \geq t) = Pe^{-(t - \alpha)/(\bar{t} - \alpha)} + (1 - P)e^{-(t/\bar{t})} \tag{2.34}$$

where P is the proportion of total traffic stream made up of constrained vehicles (all other symbols identical to those defined in earlier equations).

In 1962, Kell modified this formulation by representing free-flow vehicles with a shifted negative exponential distribution rather than by a negative exponential distribution, and incorporated this modified formulation in a microscopic simulation model [41]. In 1968, Grecco and Sword applied the Schuhl composite model to a set of time headway measurements [21]. Particular attention was given to finding numerical values for three equation parameters: α, $\bar{t}$ for constrained vehicles, and $\bar{t}$ for free-flowing vehicles. Wohl and Martin [1], Gerlough et al. [3], and Gerlough and Huber [4] summarized Schuhl's work and the more recent modifications.

The log-normal distribution for describing time headway distributions was proposed by Daou [18, 19] and Greenberg [14]. The works of Hald [43] and Aitchison and Brown [28] discuss the log-normal distribution particularly from a statistical point of view. Further applications of the log-normal distribution for describing time headway distributions were undertaken by May [5], Gerlough and Huber [4], and Tolle [20]. One quick qualitative method for evaluating the appropriativeness of using the log-normal distribution is by plotting the cumulative measured headway distribution on special graph paper. A perfect cumulative log-normal distribution curve will appear as a straight line if the y-axis is a log scale and the x-axis is a normal probability scale. The measured time headway distributions contained in Figure 2.2 for the four traffic flow levels are plotted on such a graph in Figure 2.11.

Other approaches for describing time headway distributions include the binomial model proposed by Allan [30], the hyperlang model proposed by Dawson [13, 22], the semirandom model proposed by Buckley [6, 24], and several others [9, 25, 26, 29, 64, 65]. Allan extended the binomial count model to a continuous interval model. Dawson investigated the use of a shifted Erlang model which is a special case of the Pearson type III distribution. Buckley proposed a so-called semirandom model, which is a composite model somewhat similar to the composite model described in Section 2.5.2.

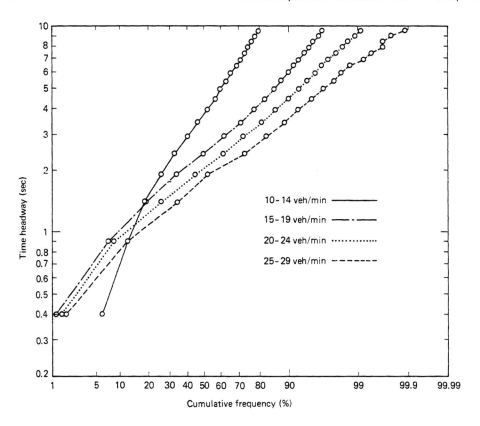

Figure 2.11   Measured Time Headway Distributions Plotted on a Log-Normal Probability Graph (From Reference 5)

2.6 EVALUATING AND SELECTING MATHEMATICAL DISTRIBUTIONS

A number of mathematical distributions have been proposed in Section 2.5 to describe measured time headway distributions. Some appear to represent measured time headway distributions rather well, whereas, others appear not to be appropriate. However, this evaluation has been qualitative, and quantitative evaluation techniques are needed. Two statistical techniques, chi-square and Kolmogorov–Smirnov (with certain restrictions), can be applied to evaluate how well a measured distribution can be represented by a mathematical distribution. The chi-square test is described in the following paragraphs, and the reader is referred to other references for the Kolmogorov–Smirnov Test [3, 4, 12, 47].

The chi-square test is a technique that can be used to assess statistically the likelihood that a measured distribution has the attributes of a mathematical distribution. The chi-square test can also be used to assess statistically how closely the measured distribution is similar to another measured distribution. The results of such tests are statistical in nature. That is, probabilities of similarity (or the lack thereof) are determined and statements can be made as to the outcome. However, there is always a risk of error, and the conclusions must be stated very carefully. For example, assume the hypothesis that there is no difference between a measured time headway distribution and a specific mathematical distribution. If the outcome were to accept the hypothesis, the concluding statement would be: "There is no evidence of a statistical difference between the two distributions and the measured distribution could be identical to the mathematical distribution." If, on the other hand, the outcome were to reject the hypothesis, the concluding statement would be; "There is evidence of a statistical difference between the two distributions, and it is unlikely that the measured distribution is identical to the mathematical distribution." Either concluding statement can "in truth" be correct or false.

Consider the possible outcomes by studying Table 2.6. Outcomes I and IV are favorable outcomes. That is, the predicted situation is the true situation. Outcomes II and III are unfavorable outcomes. That is, the predicted situation is *not* the true situation. Of course, the analyst rarely knows the true situation and hence must rely on the statistical nature of the process. Fortunately, the chi-square test permits the analyst to specify the allowable error for the unfavorable outcome II. That is the situation although the analyst rejects the hypothesis when in truth the two distributions are identical. The analyst's specified allowable error (usually referred to as significance level) is expressed as the probability of this outcome occurring. Significance levels of 0.01, 0.05, or 0.10 are frequently specified. For example, a selected significance level of 0.05 means that only 1 time in 20 will the hypothesis be rejected when in fact the two distributions are identical. The error associated with the unfavorable outcome (III) is usually not addressed. A number of statistical references are available, and the reader should consult them for additional background and descriptions of the chi-square test [3, 4, 23, 43, 47].

TABLE 2.6   Possible Outcomes of Chi-Square Tests

		Truth Situation	
		Two Distributions Identical	Two Distributions Different
Predicted Situation	No evidence of difference (hypothesis accepted)	I	III
	Evidence of difference (hypothesis rejected)	II	IV

A chi-square value is calculated for each comparison of two distributions using the equation

$$\chi^2_{CALC} = \sum_{i=1}^{I} \frac{(f_o - f_t)^2}{f_t} \tag{2.35}$$

where χ^2_{CALC} = calculated chi-square value

  f_o = observed number or frequency of observations in time headway interval i

  f_t = theoretical (or other observed) number or frequency of expected observations in time headway interval i

  i = any time headway interval

  I = number of time headway intervals

Inspection of equation (2.35) reveals that if the observed and theoretical frequencies are exactly the same, the calculated chi-square value is zero. This would indicate a perfect fit between the two distributions. If there are large differences between the observed and theoretical frequencies, the calculated chi-square value would be very large and the two distributions quite different. Consider the diagram shown in Figure 2.12. If the calculated chi-square value is very small, there is a good fit between the two distributions and the hypothesis is likely to be accepted. On the other hand, if the calculated chi-square value is very large, the fit between the two distributions is poor and the hypothesis is likely to be rejected. What is needed is a pointer on the chi-square scale to differentiate between these two outcomes, and this is the chi-square table value. If the chi-square calculated value is smaller than the chi-square table value, the hypothesis is accepted, whereas if the calculated value is greater, the hypothesis is rejected.

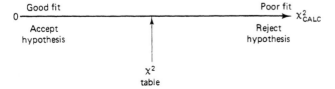

Figure 2.12   Comparing Chi-Square Calculated and Table Values

Consider the chi-square table value further by an example. Assume that there is an urn filled with numbered balls, with each numbered ball representing an individual time headway and the total collection of numbered balls representing a perfect negative exponential distribution. Select a sample of numbered balls from the urn and treat this sample as a sample of observed time headways. Calculate the chi-square value using equation (2.35) based on the selected sample frequencies and the known negative exponential distribution frequencies. Repeat this experiment a large number of times. A distribution of calculated chi-square values would result, and the distribution would follow a chi-square distribution, such as that shown in Figure 2.13. For example, 95 percent of the calculated chi-square values would be less than 14.0 and 5 percent would be greater. In other words, the chances of obtaining a chi-square calculated value greater than 14.0 when the observed distribution is in fact taken from a negative

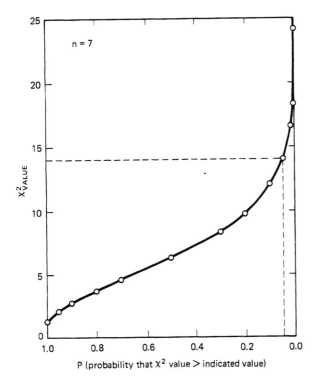

Figure 2.13   Example of Chi-Square Distribution (From the Data in Appendix F)

exponential distribution is 1 in 20, or a probability of 0.05. This is the so-called significance level, which is user specified and represents the probability of a false conclusion. That is, the hypothesis is rejected when in fact the distributions are identical.

Actually, there is a family of chi-square distributions depending on the number of degrees of freedom in the particular chi-square experiment. The number of degrees of freedom is calculated by the equation

$$n = (I - 1) - p \tag{2.36}$$

where n = number of degrees of freedom

I = number of time headway intervals being compared

1 = constant

p = number of parameters estimated in defining the mathematical distribution

As the number of compared time headway intervals increases, the number of degrees of freedom increases and the chi-square distribution values increase. A constant 1 is subtracted from the number of time headway groups since the total frequencies of the two distributions are set equal, and therefore the theoretical frequency of the last group is not independent of the $(I - 1)$ frequencies. The number of parameters that are required to define the mathematical distribution, in fitting it to the measured distribution reduces the number of degrees of freedom. The following distributions have the following numbers of parameters:

Distribution	p Value
Measured	0
Negative exponential	1
Shifted negative exponential	2
Normal	2
Pearson type III	2
Composite model	4

Two final points should be kept in mind about testing distributions with the chi-square test. First, the comparison is made between frequencies, not between probabilities [note equation (2.35)]. Second, each time headway group must have a frequency of some minimum number usually defined as 5 [note the denominator in equation (2.35)].

For illustration purposes, the measured time headway distribution for the lowest traffic flow level shown in Figure 2.2 and Appendix A will be compared with a negative exponential distribution (shown in Table 2.1). When theoretical frequencies of less than 5 are encountered, time headway intervals must be combined. This usually occurs on the "tails" of the theoretical distribution, and the initial (or final) time headway intervals are combined to represent a time headway interval less than (or greater than) a selected time headway.

The chi-square test calculations are shown in Table 2.7. The observed and theoretical frequencies are obtained by multiplying the probabilities shown in Appendix A, by the total headway frequency of 1320. Then equation (2.35) is followed and $(f_o - f_t)$, $(f_o - f_t)^2$, and $(f_o - f_t)^2/f_t$ are calculated for each time headway interval. The chi-square equation can also be formulated as shown in equation (2.37), which results in a slightly simpler calculation procedure.

$$\chi^2_{CALC} = \left(\sum_{i=1}^{I} \frac{f_o^2}{f_t} \right) - N \tag{2.37}$$

However, the calculation procedure exhibited in Table 2.7 permits the analyst to inspect the chi-square contributions from each time headway interval and to identify those intervals that make large chi-square contributions. Note, for example, the large chi-square contributions made by the 0.0 to 0.5 and > 9.5-second time headway intervals.

The individual chi-square contributions are summed, and the calculated chi-square value is found to be 67.5. The number of degrees of freedom is determined to be 18 based on 20 time headway intervals and one parameter required for the negative exponential distribution. Assuming a 0.05 significance level, the table chi-square value is determined to be 28.9 in Appendix F. Since the calculated chi-square value is larger than the table chi-square value, the hypothesis is rejected and the concluding statement would be: "There is evidence of a statistical difference between the two distributions, and it is unlikely that the measured distribution is identical to the negative exponential distribution."

TABLE 2.7 Chi-Square Test Calculations[a]

Time Headway Group	f_o	f_t	$f_o - f_t$	$(f_o - f_t)^2$	$\dfrac{(f_o - f_t)^2}{f_t}$
0.0–0.5	80	125	−45	2025	16.2
0.5–1.0	87	114	−27	729	6.4
1.0–1.5	90	103	−13	169	1.6
1.5–2.0	102	94	8	64	0.7
2.0–2.5	87	83	4	16	0.2
2.5–3.0	102	77	25	625	8.1
3.0–3.5	83	69	14	196	2.8
3.0–4.0	80	63	17	289	4.6
4.0–4.5	65	55	10	100	1.8
4.5–5.0	36	51	−15	225	4.4
5.0–5.5	40	46	−6	36	0.8
5.5–6.0	41	42	−1	1	0.0
6.0–6.5	33	37	−4	16	0.4
6.5–7.0	32	34	−2	4	0.1
7.0–7.5	26	32	−6	36	1.1
7.5–8.0	20	28	−8	64	2.3
8.0–8.5	22	25	−3	9	0.4
8.5–9.0	24	24	0	0	0.0
9.0–9.5	17	20	−3	9	0.4
> 9.5	253	198	55	3025	15.3
	1320	1320	0		$\chi^2_{CALC} = 67.5$

[a] $n = (I - 1) - p = (20 - 1) - 1 = 18$, significance level = 0.05, $\chi^2_{TABLE} = 28.9$, $\chi^2_{CALC} > \chi^2_{TABLE}$, 67.5 > 28.9; therefore, reject hypothesis.

2.7 SELECTED APPLICATIONS

Knowledge of time headway characteristics and time headway distributions is essential in analyzing many highway traffic problems. In this section several examples will be given of such applications for uninterrupted and interrupted traffic situations. References are cited that provide the reader opportunities to study these applications in greater depth.

Examples of uninterrupted traffic situations include driver behavior studies of minimum time headways, freeway and nonfreeway simulation models, ramp merging and entry control, and level-of-service measures of effectiveness. Examples of interrupted traffic situations include gap acceptance studies, intersection simulation models, saturation flow studies, and traffic signal control.

2.7.1 Uninterrupted Traffic Applications

Minimum time headways are critical from a safety point of view, on the one hand, and from a capacity viewpoint, on the other [7]. Larger time headways provide greater margins of safety, while smaller time headways result in greater capacities. The minimum time headway should always exceed the driver's reaction time, assuming that the braking deceleration rate of the following vehicle is as great as that of the lead vehicle. The reaction time varies considerably between drivers and for the same driver depending on the complexity of the decision, information content, and the expectation of the driver. The American Association of State Highway and Transportation Officials (AASHTO) gives considerable attention to this issue in their design policies and provides estimates of reaction time for a variety of situations [48]. These results are shown in Figure 2.14 and indicate that for very simple decision situations when the driver is expecting to take some action, the average reaction times are on the order of $\frac{2}{3}$ seconds. Those concerned with facility capacity have attempted to develop information and control systems that alert the driver to a vehicle-closing situation and to provide such information in the simplest possible way. On most facilities it is difficult to maintain a mean time headway of much less than 2 seconds (1800 vehicles per hour per lane). Attention to this area of

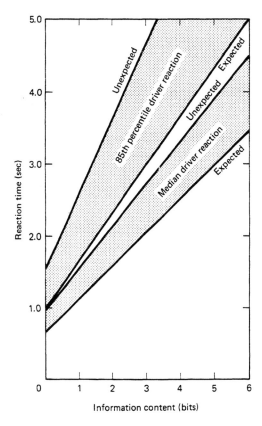

Figure 2.14   Median and 85th Percentile Driver Reaction Time to Expected and Unexpected Information (From Reference 48)

study will continue to grow as concern for safety and capacity increases in the years ahead [57, 59, 60, 66, 67].

A number of microscopic simulation models have been developed for freeways and two-lane highways. A key component of these models is the generation of vehicle arrival times as input into the simulation process. Much of the research work on time headway distributions and fitting such distributions to known mathematical distributions is due to researchers developing such simulation models. Gerlough [8] developed one of the first freeway simulation models, and a more recent freeway model is the INTRAS model [49]. At least five microscopic simulation models have been developed for two-lane highways and the Midwest and North Carolina models are typical examples [50, 63]. Several of these models require a warm-up section in the simulation model between the point where vehicles are generated and where the facility in question is being simulated because the mathematical distributions employed are inadequate.

A critical element on a freeway system is in the merging area, where new traffic is entering a freeway lane that is shared by traffic already on the freeway. Two situations have received study: the uncontrolled situation and the controlled situation. In the uncontrolled situation, drivers entering at an on-ramp select their own gaps in the adjacent freeway lane. The availability of gaps in the adjacent freeway lane is dependent on the time headway distribution while the acceptance of gaps depends on their size, merging design, and driver characteristics. May [5] and others have undertaken gap availability studies while Drew [2] and others have performed gap acceptance studies. Wohl and Martin [1] have developed a ramp merging microscopic simulation model which incorporates gap availability and gap acceptance submodels for evaluating the effect of merge design on capacity and level of service. Traffic entering at several thousand on-ramps in the United States is controlled by metering devices that regulate the allowable input rate to the freeway in terms of minimum time headway between entering vehicles. Some of these entry control systems are fixed-time systems based on historical data, while others are real-time traffic-responsive systems. Some traffic-responsive entry control systems have been rather sophisticated in that ramp vehicles are released and/or guided down the ramp based on real-time measurement of acceptable available time headways in the adjacent freeway lane. The Texas Transportation Institute [2] and Raytheon [51] have developed and tested such systems.

The new *Highway Capacity Manual* has incorporated a new measure of effectiveness for determining level of service on two-lane rural highways [52]. This new level-of-service measure is called percent time delayed and "is defined as the average percent of the total travel time that all motorists are delayed in platoons while traveling a given section of highway. For field measurement purposes, percent time delayed in a section is approximately the same as the percentage of all vehicles traveling in platoons at headways of less than 5 seconds." In other words, the percent time that a vehicle is in platoon is indicative of the level of service a driver is experiencing. This is related directly to time headway distributions and the concept of the intermediate headway state developed in Section 2.5. Under low-traffic-flow conditions, few vehicles are in platoons and the users have a high level of service. As traffic flow increases, more vehicles are in platoons and the users have a lower level of service. For example, the percent vehicles in platoon (defined as the percent of vehicles with time headways < 5 seconds) is plotted against hourly flow rate (vehicles per hour) in Figure 2.15. The solid curve represents the probabilities of headways equal to or less than 5 seconds based on the negative exponential distribution. The dashed curve represents the probabilities of headways equal to or less than 5 seconds based on the measured time headway distribution shown in Figure 2.2. The five levels of service (A, B, C, D, and E) proposed in the *Highway Capacity Manual* for two-lane rural highways are shown along the *y*-axis [52]. This new level of service measure based on time headways and their distribution will undoubtedly receive further study in the future and will be extended to other types of facilities.

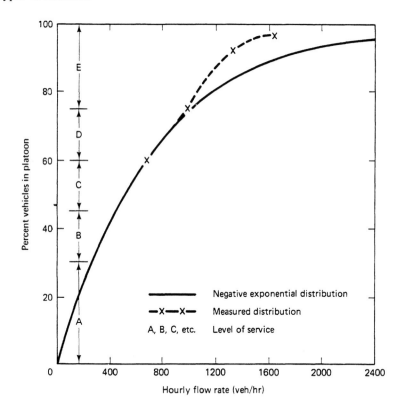

Figure 2.15   Percent Vehicles in Platoon as a Function of Hourly Flow Rate

2.7.2 Interrupted Traffic Applications

Headway (sometimes referred to as "gap") acceptance studies have been undertaken at signalized and sign-controlled intersections [11, 61]. The purpose of these studies has been to assess performance and capacity at intersections. Similar to the ramp merging process, the availability of headways (gaps) and their acceptance by drivers are the two main analytical components for such intersection studies.

For example, Bissell undertook a study of gap acceptance at an interesection controlled by a stop sign [27]. Through the use of detectors Bissell obtained the actual headways on the through street automatically, and an observer recorded when these various-sized headways were accepted or rejected by drivers from the stop-sign approach. From these measurements the percent acceptance was calculated for each available headway interval. The median acceptance headway was found to be 5.5, 5.8, and 6.6 seconds for right turns, straight-through, and left-turn traffic, respectively. Some results of his investigation are shown graphically in Figure 2.16.

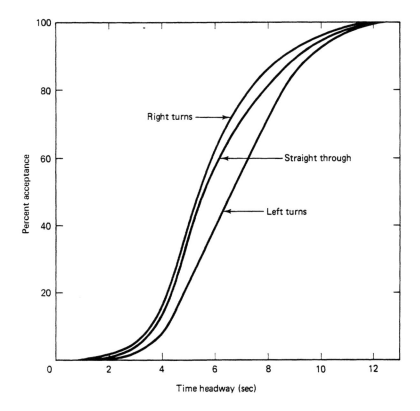

Figure 2.16   Gap Acceptance from a Stop Sign (From Reference 27)

The 1985 *Highway Capacity Manual* contains a chapter for computing capacities and levels of service for unsignalized intersections [56]. A key step in this procedure is the estimation of the critical gap, which is defined as the median time headway between two successive vehicles in the major street traffic stream that is accepted by drivers who must cross and/or merge with the major street flow. Suggested values for the median time headway for various situations are given in Table 2.8.

A number of microscopic simulation models have been developed for signalized and sign-controlled intersections. Like freeway and rural highway models, a key component of these models is the generation of vehicle arrival times as input into the

TABLE 2.8 Critical Gap Criteria for Unsignalized Intersection[a]

BASIC CRITICAL GAP FOR PASSENGER CARS (sec)

Vehicle Maneuver	Average Running Speed, Major Road[b]			
	30 miles/hr		55 miles/hr	
and	Number of Lanes on Major Road			
Type of Control	2	4	2	4
RT from minor road				
Stop	5.5	.5.5	6.5	6.5
Yield	5.0	5.0	5.0	5.0
LT from major road	5.0	5.5	5.5	6.0
Cross major road				
Stop	6.0	6.5	7.5	8.0
Yield	5.5	6.0	6.5	7.0
LT from minor road				
Stop	6.5	7.0	8.0	8.5
Yield	6.0	6.5	7.0	7.5

ADJUSTMENTS AND MODIFICATIONS TO CRITICAL GAP (sec)

Condition	Adjustment
RT from minor street: curb radius > 50 ft or turn angle < 60[c]	−0.5
RT from minor street: acceleration lane provided	−1.0
All movements: population ≥ 250,000	−0.5
Restricted sight distance[a]	Up to +1.0

[a]Maximum total decrease in critical gap = 1.0 sec; maximum critical gap = 8.5 sec.
[b]For values of average running speed between 30 and 55 miles/hr, interpolate.
[c]This adjustment is made for the specific movement impacted by restricted sight distance.
Source: Reference 56.

simulation process. Simulation modelers have devoted considerable effort to mathematical descriptions of time headway distributions. Kell [41] and others [33] developed some of the early models; more recent intersection simulation models include the TEXAS [53] and NETSIM [54] models.

The discharge headways of vehicles crossing a stop line at a signalized intersection during the green phase when a queue is present is a measure of intersection capacity [37]. Consider an observer standing at a stop line for a single-lane approach. The instant the signal turns green a stopwatch is started and as each vehicle crosses the stop line, the time is recorded. The time headways are computed and can be plotted as shown in Figure 2.17. The first time interval recorded is from the time the signal turns green until the passage of the first vehicle and thereafter the time headway between

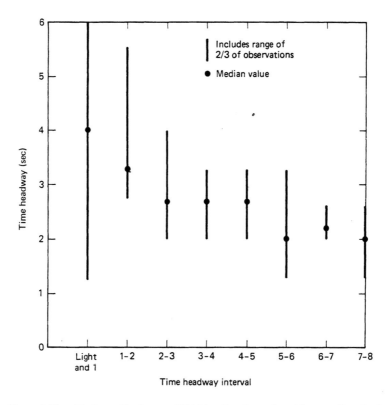

Figure 2.17   Discharge Headways of Vehicles Crossing a Stop Line at a Signalized Intersection (From Reference 16)

consecutive vehicles. The vertical bars indicate the range of two-thirds of the observations and the median is located on each bar. As can be seen from the figure, the time headways continue to decrease until a time headway on the order of 2 to 2.2 seconds is reached. If the queue of vehicles is discharged before the end of the green phase, the time headways will significantly increase, indicating that the intersection is no longer being taxed to its capacity.

The 1985 *Highway Capacity Manual* [55] proposes that discharge headways be recorded in order to estimate saturation flows and capacities at signalized intersections. "The period defined as saturation flow begins when the rear axle of the fourth vehicle in queue crosses the stopline or reference point and ends when the rear axle of the last queued vehicle at the beginning of the green crosses the same point." For example, if the fourth vehicle was observed at 10.2 seconds, and the time for the fifteenth and last vehicle in the queue was 36.5 seconds, the average saturation headway per vehicle would be

$$\frac{36.5 - 10.2}{(15 - 4) - 1} = \frac{26.3}{10} = 2.63 \text{ seconds/vehicle} \tag{2.38}$$

and the saturated flow rate would be estimated to be 1370 vehicles per hour.

Some isolated signalized intersections are traffic actuated. The passage of a vehicle or the lack thereof at a specific point and critical time can either extend or terminate the green phase on an approach. Such control algorithms can be very complicated, but consider a very simple situation to illustrate how time headways are utilized. A detector is placed some distance in advance of the stop line and detects the passage of vehicles approaching the signalized intersection. The travel time for the average vehicle to travel from the detector to the stop line is t seconds. It is considered potentially unsafe if the amber phase begins when a vehicle is between the detector and the stop line. These time headways are measured automatically at the detector, and the green phase is extended as long as each individual time headway is less than t seconds. When an individual time headway exceeds t seconds, the amber phase can begin and the next approaching vehicle will be upstream of the detector when the signal changes and can safely stop.

2.8 SUMMARY

This chapter has attempted to provide a theoretical base with practical applications for time headway analyses and time headway distributions. Field measurements of time headways and a critique of resulting time headway distributions are presented. A classification of time headway distributions is proposed consisting of random, intermediate, and constant headway states. Relevant mathematical distributions are presented in each classification and include the following theoretical distributions: negative exponential, Pearson type III, gamma, Erlang, shifted negative exponential, composite, normal, and several others. The chi-square test for comparing distributions is described and applied for analyzing measured and theoretical time headway distributions. Selected applications are discussed to illustrate the importance of this microscopic flow characteristics in analyzing traffic flow problems. The chapter concludes with a variety of sample problems and a list of cited references.

2.9 SELECTED PROBLEMS

1. Time headways and their distributions are important microscopic flow characteristics in other modes of transportation. Through library study and/or field experiments consider one of the following modal questions.
   (a) Pedestrians crossing a lightly traveled uncontrolled street
   (b) Time headway distributions for an urban rapid rail system
   (c) Estimating arrival times of river barges at a lock
   (d) Minimum time headways of aircrafts at airports
2. Measure individual time headways at a freeway off-ramp or on a local street away from intersection control under light-flow conditions. Record approximately 200 to 300 time headways to the nearest second and plot in a fashion similar to Figure 2.2 for one volume level. Identify patterns and compare with results shown in Section 2.1.

3. Apply the negative exponential distribution to a flow level of 300 vehicles per hour and prepare a table similar to Table 2.1 and plot on a figure similar to Figure 2.3.

4. Apply the normal distribution to a traffic flow level of 2000 vehicles per hour and prepare a table similar to Table 2.2 and plot on a figure similar to Figure 2.4.

5. Plot relationships between standard deviation and mean time headway for the following Pearson type III distributions: $K = 1$ with $\alpha = 0.0$, 0.5, and 1.0; $K = 2$ with $\alpha = 0.0$, 0.5, and 1.0; and $K = 3$ with $\alpha = 0.0$, 0.5, 1.0. Superimpose on Figure 2.8. Which Pearson type III distributions compare most favorably with measured distribution relationships?

6. Using the measured time headway distributions for the traffic flow level of 15–19 vehicles per minute, apply one of the following mathematical distributions and prepare a table similar to Table 2.4 and plot on a figure similar to Figure 2.7.

    (a) Shifted negative exponential distribution ($\alpha = 0.5$ second)
    (b) Erlang distribution ($K = 2$)
    (c) Gamma distribution ($K = \hat{K}$)

    Compare results with the measured and Pearson type III distributions shown in Figure 2.7.

7. Attempt to improve the composite time headway distribution for a traffic flow level of 25 to 29 vehicles per minute as shown in Figure 2.10 by modifying one or more of the four independent parameters.

8. Apply one of the following proposed distributions to one of the measured time headway distributions shown in Figure 2.2.

    (a) Schuhl model
    (b) Log-normal model
    (c) Binomial model
    (d) Hyperlang model
    (e) Semirandom model

9. Review all proposed distribution models and their applications to the four measured time headway distributions shown in Figure 2.2. Select the distribution model that you feel best represents one of the measured time headway distributions and perform a chi-square test.

10. The Kolmogorov–Smirnov test has been mentioned as a method for quantitative evaluation of representing measured distributions by mathematical distributions. Study the Kolmogorov–Smirnov test procedures and identify strengths and weaknesses of this method compared to the chi-square test.

11. Undertake a library study of minimum time headways on freeway-type facilities. Consider the following issues in your study:

    (a) Effect of viewing several vehicles ahead, not just the one immediately ahead
    (b) Effect of available parallel lanes
    (c) New innovations for improving information transfer to drivers
    (d) Possible effects of automatic vehicle control

12. Undertake a library study of microscopic simulation models and identify the mathematical distributions employed for describing vehicle arrivals. Include the authors' justification for their use.

13. Undertake a library investigation of gap acceptance studies at freeway ramps and intersections. Identify the mathematical distributions employed for describing vehicle arrivals. Include the authors' justification for their use.

14. The new *Highway Capacity Manual* incorporates percent time delayed as the measure of level of service on two-lane rural highways. Consider the following issues in your study:

(a) Evidence that percent time delay is closely related to percent vehicles with headway < 5 seconds.

(b) Appropriateness of using this measure of level of service for other types of facilities, such as for freeways

(c) Effect of using this measure of level of service on selecting an appropriate mathematical distribution

15. Measure individual time headways of vehicles being discharged in a single through lane at the stop line of a heavily traveled signalized intersection. Attempt to record time headways to the nearest tenth of a second and observe at least 20 green phases. Plot a figure similar to Figure 2.17 and estimate the saturation flow rate using equation (2.38).

16. Select an isolated traffic-actuated signalized intersection with detectors several hundred feet upstream of the stop line. Estimate the travel time (t) from the detector to the stop line for vehicles that are not required to stop. Then observe approaching traffic when the signal is green. As long as time headways are less than t seconds, the green phase should be continued (unless the green phase has reached a preset maximum value). When a time headway greater than t seconds occurs, the green phase should be terminated (provided that there is a call on the side street). Record the results and evaluate signal operation.

2.10 SELECTED REFERENCES

1. Martin Wohl and Brian V. Martin, *Traffic System Analysis for Engineers and Planners,* McGraw-Hill Book Company, New York, 1967, pages 352–362.

2. Donald Drew, *Traffic Flow Theory and Control,* McGraw-Hill Book Company, New York, 1968, pages 153–159, 166–167, and 215–219.

3. Daniel L. Gerlough, Frank C. Barnes, and André Schuhl, *Poisson and Other Distributions in Traffic,* Eno Foundation, Saugatuck, Conn., 1971, pages 35–49 and 64–70.

4. Daniel G. Gerlough and Matthew J. Huber, *Traffic Flow Theory — A Monograph,* Transportation Research Board, Special Report 165, TRB, Washington, D.C., 1975, pages 21–31.

5. Adolf D. May, Jr., *Gap Availability Studies,* Highway Research Board Record 72, HRB, Washington, D.C., 1965, pages 105–136.

6. D. J. Buckley, Road Traffic Headway Distributions, *Australian Road Research Board Proceedings,* Vol. 1, Part 1, 1962, pages 153–187.

7. Louis J. Pignataro, *Traffic Engineering — Theory and Practice,* Prentice-Hall Inc., Englewood Cliffs, N.J., 1973, pages 178–179.

8. D. L. Gerlough, Traffic Inputs for Simulation on a Digital Computer, *Highway Research Board Proceedings,* Vol. 38, 1959, pages 480–492.

9. F. A. Haight, B. F. Whisler, and W. Mosher, New Statistical Method for Describing Highway Distribution of Cars, *Highway Research Board Proceedings,* 1961, pages 557–564.

10. William F. Adams, Road Traffic Considered as a Random Series, *Journal of Institution of Civil Engineers (London),* Vol. 4, November 1936, pages 121–130.

11. W. R. Blunden, C. M. Clissold and R. B. Fisher, Distribution of Acceptance Gaps for Crossing and Turning Maneuvers, *Australian Road Research Board Proceedings,* Vol. 1, Part 1, 1962, pages 188–205.

12. F. J. Massey, Jr., The Kolmogorov–Smirnov Test for Goodness of Fit, *Journal of American Statistical Association,* Vol. 46, 1951, pages 68–78.

13. R. F. Dawson and L. A. Chimini, *The Hyperlang Probability Distribution — A Generalized Traffic Headway Model*, Highway Research Board Record 230, HRB, Washington, D.C., 1968, pages 1–114.

14. I. Greenberg, The Log-Normal Distribution of Headways, *Australian Road Research*, Vol. 2, No. 7, March 1966, pages 14–18.

15. John P. Kinzer, *Application of the Theory of Probability to Problems of Highway Traffic*, Thesis, Polytechnic Institute of Brooklyn, 1933. (Abstracted in *Institute of Traffic Engineers Proceedings*, Vol. 5, 1934, pages 118–124).

16. Bruce D. Greenshields, Donald Schapiro, and Elroy L. Erickson, *Traffic Performance at Urban Street Intersections*, Technical Report No. 1, Yale Bureau of Highway Traffic, New Haven, Conn., 1947.

17. K. Pearson, *Tables of the Incomplete Gamma Function*, Cambridge University Press, Cambridge, 1965.

18. A. Daou, On Flow within Platoons, *Australian Road Research*, Vol. 2, No. 7, 1966, pages 4–13.

19. A. Daou, The Distribution of Headways in a Platoon, *Operations Research*, Vol. 12, No. 2, 1964, pages 360–361.

20. J. E. Tolle, The Lognormal Headway Distribution Model, *Traffic Engineering and Control*, Vol. 13, No. 1, 1971, pages 22–24.

21. W. L. Grecco and E. C. Sword, Prediction of Parameters for Schuhl's Headway Distribution, *Traffic Engineering*, Vol. 38, No. 5, 1968, pages 36–38.

22. R. F. Dawson, The Hyperlang Probability Distribution — A Generalized Traffic Headway Model, *International Symposium on Theory of Traffic Flow Proceedings*, Karlsruhe, West Germany, 1968, pages 30–36.

23. F. A. Haight, *Mathematical Theories of Traffic Flow*, Academic Press, Inc., New York, 1963.

24. D. J. Buckley, A Semi-Poisson Model of Traffic Flow, *Transportation Science*, Vol. 2, No. 2, 1968, pages 107–133.

25. R.J. Serfling, Non-Poisson Models for Traffic Flow, *Transportation Research*, Vol. 3, No. 3, 1969, pages 299–306.

26. B. T. Pak-Poy, The Use and Limitation of the Poisson Distribution in Road Traffic, *Australian Road Research Board Proceedings*, Vol. 2, 1964, pages 223–247 (with discussion by R. T. Underwood).

27. H. H. Bissell, *Traffic Gap Acceptance from a Stop Sign*, University of California Research Report, University of California, Berkeley, Calif., May 1960.

28. J. Aitchison and J. A. C. Brown, *The Lognormal Distribution*, Cambridge University Press, Cambridge, 1963

29. A. M. Miller, An Empirical Model for Multilane Road Traffic, *Transportation Science*, Vol. 4, No. 2, 1970, pages 164–186.

30. R. R. Allan, Extension of the Binomial Model of Traffic to the Continuous Case, *Australian Road Research Board*, Volume 3, 1966, pages 276–316.

31. L. Breiman, A. V. Gafarian, R. Lichtenstein, and V. K. Murthy, An Experimental Analysis of Single-Lane Headways in Freely Flowing Traffic, *International Symposium on Theory of Traffic Flow Proceedings*, Karlsruhe, West Germany, 1968, pages 22–29.

32. W. Leutzbach, Distribution of Time Gaps between Successive Vehicles, *International Road Safety Traffic Review*, Vol. 3, No. 3, 1957, pages 31–36.

33. R. M. Lewis, A Proposed Headway Distribution for Traffic Simulation Studies, *Traffic Engineering*, Vol. 33, No. 5, 1963, pages 16–19, and 48.

34. A. D. May, Jr. and F. A. Wagner, Jr., Headway Characteristics and Interrelationships of the Fundamental Characteristics of Traffic Flow, *Highway Research Board Proceedings*, Vol. 39, 1960, pages 524–547.

35. K Moskowitz, Waiting for a Gap in a Traffic Stream, *Highway Research Board Proceedings*, Vol. 33, 1954, pages 385–394.

36. T. Suzuki, Some Results in Road Traffic Distributions, *Operations Research Society (Japan)*, Vol. 4, No. 1, 1966, pages 16–25.

37. B. Widermuth, Average Vehicle Headways at Signalized Instructions, *Traffic Engineering*, Vol. 33, No. 2, 1963, page 14–16.

38. P. Athol, Headway Groupings, *Highway Research Board, Record 72*, HRB, Washington, D.C., 1965, pages 137–155.

39. Adolf D. May, Traffic Characteristics and Phenomena on High Density Controlled Access Facilities, *Traffic Engineering*, Vol. 31, No. 6, March 1961, pages 15–18.

40. Lloyd F. Rader, Review of Applications of the Theory of Probability to Problems of Highway Traffic, *Institute of Traffic Engineers Proceedings*, Vol. 5, 1934, pages 118–123.

41. J.H. Kell, Analyzing Vehicular Delay at Intersections through Simulation, *Highway Research Board Bulletin 356*, HRB, Washington, D.C., 1962, pages 28–39.

42. G. F. Newell, Statistical Analyses of the Flow of Highway Traffic through a Signalized Intersection, *Applied Mathematics Quarterly*, Vol. 13, No. 4, 1956, pages 353–364.

43. A. Hald, *Statistical Theory with Engineering Applications*, John Wiley & Sons, Inc., New York, 1952.

44. Institute of Transportation Engineers, *Transportation and Traffic Engineering Handbook*, 2nd Edition, Prentice-Hall, Inc., Englewood Cliffs, N. J., 1982, pages 443–448 and 534–538.

45. W. Leutzbach, *Ein Beitrag zur Zeitlückenverteilung gestörter Strassenwerkehrsströme* (On the Distribution of Time Intervals in Impeded Road Traffic Flows), Thesis, Aachen Technical University, August, 1956.

46. K. Krell, *Beitrag zur Theorie der Zeitlücken und deren Bedeutung für die Leistungsfähigkeit von Strassen und Knoten* (On the Theory of Time Intervals and their Influence on the Capacity of Roads and Road Junctions), Thesis, Darmstadt Technical University, November 1956.

47. Daniel L. Gerlough and Matthew J. Huber, *Statistics with Applications to Highway Traffic Analyses*, Eno Foundation, Saugatuck, Conn., 1978.

48. American Association of State Highway and Transportation Officials, *A Policy on Geometric Design of Highways and Streets*, AASHTO, Washington, D.C., 1984

49. D. A. Wicks and E. B. Lieberman, *Development and Testing of INTRAS, A Microscopic Freeway Simulation Model*, Final Report, October 1980, 229 pages.

50. A. D. St. John and D. R. Kobett, *Grade Effects on Traffic Flow Stability and Capacity*, NCHRP Report 185, Transportation Research Board, Washington, D.C., 1978.

51. *Merging Control System*, Final Report, Raytheon Company, Wayland, Mass., 1970.

52. Transportation Research Board Special Report, *Highway Capacity Manual*, Special Report 209, TRB, Washington, D.C., 1985, Chapter 8.

53. Clyde E. Lee and Vivek S. Savur, *Analysis of Intersection Capacity and Level of Service by Simulation*, Transportation Research Board, Record 699, TRB, Washington, D.C., 1979, pages 34–41.

54. Federal Highway Administration, *Traffic Network Analysis with NETSIM—A User Guide*, FHWA, Washington, D.C., 1980, 216 pages.

55. Transportation Research Board, *Highway Capacity Manual*, Special Report 209, TRB Washington, D.C., 1985, Chapter 9.

56. Transportation Research Board, *Highway Capacity Manual*, Special Report 209 TRB, Washington, D.C., 1985, Chapter 10.

57. R. E. Fenton, R. J. Mayhan, G. M. Takaski, and J. Glinim, *Fundamental Studies in Automatic Vehicle Control*, Ohio State University, Columbus, Ohio, 1981, 296 pages.

58. Raymond G. C. Fuller, Nicholas J. McDonald, Patrick A. Holahan, and Edward P. Bolger, Technique for Continuous Recording of Vehicle Headway, *Perceptual and Motor Skills*, Vol. 47, No. 2, 1978, pages 515–521.

59. R.D. Helliar-Symods, *Automatic Close-Following Warning sign at ASCOT*, Transport and Road Research Laboratory Report LR 1095, TRRL, Crowthorne, England, 1983, 16 pages.

60. R. C. Postans and W. T. Wilson, Close-Following on the Motorway, *Ergonomics*, Vol. 26, No. 4, April 1983, pages 317–327.

61. M. R. C. McDowell, J. Wennell, P. A. Storr, and J. Dirgentas, *Gap Acceptance and Traffic Conflict Simulation as a Measure of Risk*, Transport and Road Research Laboratory Report 116, TRRL, Crowthorne, England, 1983, 19 pages.

62. P. Wasielewski, The Effect of Car Size on Headways in Freely Flowing Freeway Traffic, *Transportation Science*, Vol. 15, No. 4, November 1981, pages 364–378.

63. S. Khasnabis and C. L. Heinbach, *Headway-Distribution Models for Two-Lane Rural Highways*, Transportation Research Board Record 772, TRB Washington, D.C., 1980, pages 44–51.

64. G. C. Ovaworie, J. Darzentas, and M. R. C. McDowell, Free-Movers, Followers, and Others: A Reconsideration of Headway Distribution, *Traffic Engineering and Control*, Vol. 21, No. 819, August 1980, pages 425–428.

65. P. Wasielewski, Car-Following Headways on Freeways Interpreted by the Semi-Poisson Headway Distribution Model, *Transportation Science*, Vol. 13, No. 1, February 1979, pages 36–55.

66. J. Glimm and R. E. Fenton, Accident-Severity Analysis for a Uniform Spacing Headway Policy, *29th IEEE Vehicular Technology Conference Proceedings*, 1979, pages 271–278.

67. R. E. Fenton, Headway Safety Policy for Automated Highway Operations, *IEEE Transactions on Vehicular Technology*, Vol. VT2, No. 1, February 1979, pages 22–28.

68. G. N. Steuart and B. T. Shin, Effect of Small Cars on the Capacity of Signalized Urban Intersections, *Transportation Science*, Vol. 12, No. 3, August 1978, pages 250–263.

3

Macroscopic Flow Characteristics

Traffic flow represents the traffic load on the transportation system and the interaction between these "loadings," and the facility capacity determines the operational performance of the system. Hence it is extremely important to know the flow rates, their temporal, spatial, and modal variations, and the composition of the traffic stream. The measurement of traffic flow has many uses in the planning, design, and operations of highway facilities. In planning, flow measurements are used in the classification of streets, identification of traffic trends, in origin–destination studies, in revenue prediction, and in many other ways. The design of new facilities and the redesign of existing facilities require traffic flow information. Operational analysis requires flow measurements for control warrants, accident analysis, and in the investigation of operational improvements such as priority lanes, reversible lanes, and flow restrictions.

Flow rate (or volume) is the important macroscopic flow characteristic and is defined as the number of vehicles passing a point in a given period of time usually expressed as an hourly flow rate. It is related to the microscopic flow characteristic, time headway, in the following manner:

$$q_{60} = \frac{3600}{\bar{h}} \tag{3.1}$$

where q_{60} = flow rate (in this case hourly flow rate)
   3600 = number of seconds per hour
   $\bar{h}$ = average time headway (seconds per vehicle)

In the past the term "volume" has been used, but gradually the term "flow" or "flow rate" has taken its place. Because of the variation in the flow rate over time, not only does the flow rate need to be defined but also the period of time over which the flow rate was measured. For example, assume that a 15-minute rate of flow of 1000 vehicles per hour was observed. This means that 250 vehicles were counted during a 15 minute interval and the flow rate is expressed as an hourly flow rate.

There are four important parameters of this macroscopic flow characteristic: existing traffic demand, service volume, capacity, and saturation flow rate. Existing traffic demand is a unique flow rate value that plays an important role when analyzing the oversaturated or congested situation and represents the flow rate at which vehicles would like to be serviced. The existing traffic demand is numerically equal to the measured flow rate if there is no oversaturation upstream or congestion at the site in question. If, however, oversaturation or congestion is encountered, the flow rate is indicative only of the flow rate level that can be handled and is *not* an indication of the existing traffic demand. Under these situations special traffic demand studies are required. Service volume is defined as the maximum hourly rate at which persons or vehicles can reasonable be expected to traverse a point or short section of the lane or roadway during a given time period (usually, 15 minutes) under prevailing roadway, traffic, and control conditions *while maintaining a designated level of service* [1]. Capacity is defined as the *maximum* hourly rate at which persons or vehicles can reasonably be expected to traverse a point or uniform section of a lane or roadway during a given period (usually, 15 minutes) under prevailing roadway, traffic, and control conditions [1]. Saturation flow rate is defined as the equivalent maximum hourly rate at which vehicles can traverse a lane or intersection approach under prevailing traffic and roadway conditions assuming that the green signal was available at all times and that no loss times are experienced [1].

This chapter contains seven sections plus selected problems and references. The first three sections deal with temporal, spatial, and modal flow patterns. In the next two sections mathematical count distributions and techniques for estimating hourly flow rates are presented. The last two sections are concerned with uninterrupted and interrupted traffic applications.

3.1 TEMPORAL FLOW PATTERNS

Traffic flow rates vary over time, and there are monthly, daily, hourly, and within-hour temporal flow variations. It is important for traffic analysts to understand these temporal flow patterns in order to estimate traffic flow rates for selected periods of time based on known flow rates from other periods of time. Although estimation of traffic flow rates is not an exact science and individual routes may have unique temporal flow patterns, flow rate estimates can be made in lieu of actual field counts.

Different types of highway facilities have different monthly flow variations. Monthly flow patterns differ primarily between urban and rural facilities, between recreational and nonrecreational facilities, and between facilities in mild and severe climatic conditions. Figure 3.1 graphically displays typical monthly flow patterns for different types of highway facilities. April–May and October–November normally have average traffic flow conditions, and many surveys are conducted during these months of the year. Urban freeways located in a mild climate have the most uniform

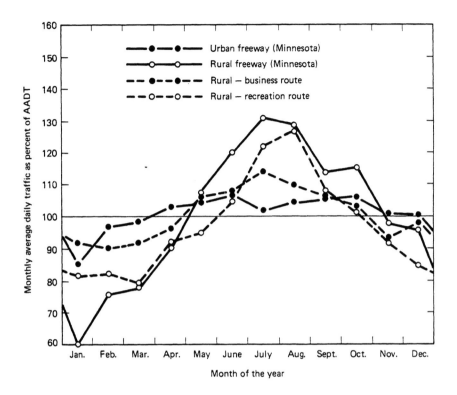

Figure 3.1   Monthly Flow Patterns for Various Types of Highway Facilities (From Reference 1)

traffic flows during the year while rural highways serving a highly seasonal recreation area have the greatest monthly variations.

Traffic flow rates also vary among days of the week. Daily flow patterns differ primarily between urban and rural facilities and between recreational and nonrecreational facilities. Figure 3.2 displays typical daily flow patterns for different types of highway facilities. Urban freeways and major arterials have heavier-than-average traffic on weekdays and lower-than-average traffic on weekends. Rural highways, particularly rural recreational routes, have heavier traffic on weekends and less on weekdays.

Traffic flow rates also vary by time of day. These hourly peaking patterns vary by day of week, direction of flow, urban–rural setting, and type of route. Figure 3.3 graphically presents the hourly flow patterns on the San Francisco–Oakland Bay Bridge by direction of flow. Note the single peaks when flow is separated by direction and the two peak flows when directional flow is combined. Flat peaks may indicate that demands exceed capacities and the peak flow period is being extended. Daily flow patterns on rural highways are usually quite different, as shown in Figure 3.4. Hourly flows are combined for both directions, and single (not two) peak flow periods can be observed during the late afternoon. Figure 3.4 also indicates that hourly flow patterns may differ by day of the week.

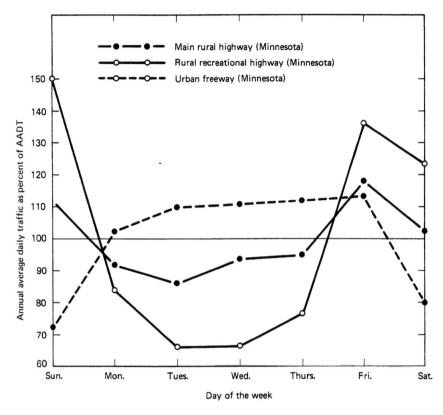

Figure 3.2   Daily Flow Patterns for Various Types of Highway Facilities (From Reference 1)

Traffic flow rates also vary within the hour period. Three distinct within-the-hour traffic flow patterns are encountered. These three traffic flow patterns can be described somewhat in parallel with the three headway states presented in Chapter 2, that is, random, constant, and intermediate traffic flow patterns. Random traffic flow patterns are observed when the traffic demand is small in comparison with the facility capacity, and there are no nearby demand inputs that are schedule dependent, such as locations near a factory during the beginning or end of work periods. This random pattern is most often encountered on low-volume rural highways. The Poisson count distribution can be used to describe the flow variations, such as the distribution of minute flow rates within the hour. The Poisson count distribution is be discussed later in this chapter.

Constant flow rates are observed when the traffic demand exceeds the facility capacity and the flow rates are indicative of capacity rather than demands. Nearly constant flow rates can be observed downstream of bottlenecks where the output is being metered or upstream of bottlenecks in the queueing process. The degree of uniformity of the flow rates is dependent primarily on the period of time during which demand exceeds capacity. For example, the minute flow rate during the peak hour will be very uniform if the oversaturated situation lasts for more than an hour.

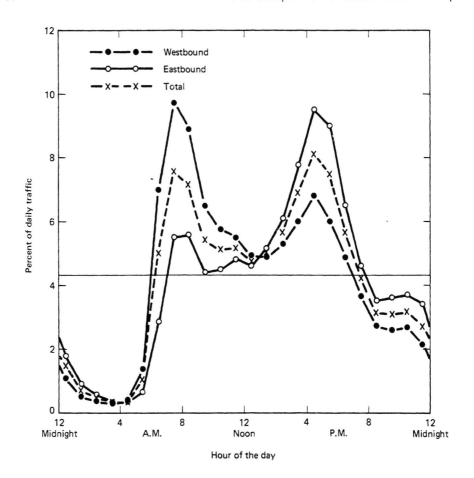

Figure 3.3   Hourly Flow Patterns on the San Francisco–Oakland Bay Bridge (From Reference 2)

The third traffic flow pattern could be considered the intermediate case where demands do not exceed capacities (at least not for the entire period) but flow is not random due to predominant trip scheduling. This is the most difficult to analyze, yet is the case encountered most frequently in a real-world situation. The *Highway Capacity Manual* proposes to handle this issue by introducing a peak-hour factor which is used as an indicator of the traffic flow fluctuations within the hour [1]. The following equation demonstrates how this concept is applied:

$$\text{PHF} = \frac{q_{60}}{q_{15}} \tag{3.2}$$

where q_{15} = peak 15-minute rate of flow expressed as an hourly flow

q_{60} = hourly flow rate

PHF = peak-hour factor

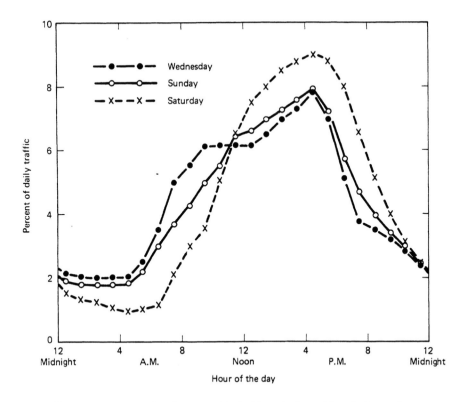

Figure 3.4   Hourly Flow Patterns on Typical Intercity Route (From Reference 1)

Thus, to estimate the peak 15-minute rate of flow, the hourly flow is measured (or estimated) and a peak-hour factor is selected. Theoretically, the PHF could vary from 0.25 (which means that all the traffic within the hour passes during one 15-minute period) to 1.00 (which means that each 15-minute period carries the same amount of traffic). In practice, however, the PHF normally is not less than about 0.80 assuming random flow and not more than 0.98 assuming nearly uniform flow.

3.2 SPATIAL FLOW PATTERNS

Traffic flow rates vary over space and consequently, linear, network, directional, and lane use variations are important concepts to understand and apply. Traffic flow along urban radial routes is analogous to water flow along a river. During the early portion of the morning peak period, motorists with origins farthest from the central business district (CBD) begin their trip along the urban radial routes, and as time passes, other motorists with origins closer to the CBD join them. The traffic flow can be thought of as waves, and the higher-intensity traffic flows move toward the CBD, resulting in a linear traffic demand pattern. Thus the inbound peak periods are experienced first on the outskirts and as time passes the inbound peak periods move toward the CBD. The afternoon peak period occurs in a similar fashion but in the reverse direction. Workers leave downtown during the early portion of the afternoon peak period and as they move outward along urban radial routes, additional traffic joins at later periods. Thus the outbound peak periods are experienced first near the CBD, and gradually these waves of intensive flows move outward to the suburbs. This wave action may be magnified if demands exceed capacities and unmet traffic demands are delayed to later periods. Again, the analogy of flow along rivers with a series of dams (bottlenecks) and inflows exceeding possible outflows is shown.

Most highway facilities, particularly in urban areas, are connected to dense networks of highway facilities. Since motorists have the freedom of choosing their route and are guided primarily by following the route of least resistance (usually minimum travel time), there can be strong interactions between parallel routes. One spatial pattern found is that as higher-quality routes begin to become oversaturated, motorists divert to parallel lower-quality routes. For example, motorists may change from freeways to frontage roads and major arterials, and motorists on major arterials may divert to collector streets and neighborhood streets. This spatial phenomenon affects the flow patterns on particular routes in the network and should be considered when oversaturation is encountered in part of the network.

Another spatial flow pattern is the directional flow pattern along a specific route. Figure 3.3 showed that the predominant traffic flow during the morning peak period was westbound, while in the afternoon peak period the predominant traffic flow was eastbound. Traffic data for this same route were used to calculate the directional split, and the results are depicted graphically in Figure 3.5. More traffic goes toward San Francisco from 4 A.M. until 1 P.M. and away from San Francisco from 2 P.M. until 4 A.M. The directional splits during the morning and afternoon peak periods are quite different (72 to 28 percent in the morning peak hour but only 59 to 41 percent in the afternoon peak hour).

The final spatial flow pattern to be considered is the distribution of traffic among lanes when several lanes are available for use. This is particularly important on freeways where multilanes are also available and the locations of on-ramps and off-ramps may cause overloading of individual lanes. Figure 3.6 presents typical lane volume distributions for a German autobahn and a U.S. freeway [3, 4]. Note that as the total directional flow increases, the percentage of traffic in the median lane increases while the percentage of traffic in the shoulder lane decreases. The median lane carries a higher percentage of traffic in Germany than in the United States because of German regulations prohibiting passing on the right, the significant difference between passenger car speed and truck speeds, and the greater distances between access points.

3.3 MODAL FLOW PATTERNS

Various modes of transportation use highway facilities such as truck transport, bus transport, rail transport, and passenger car transport. A newly developing means of passenger car transport is multioccupancy vanpooling and carpooling. It is important to

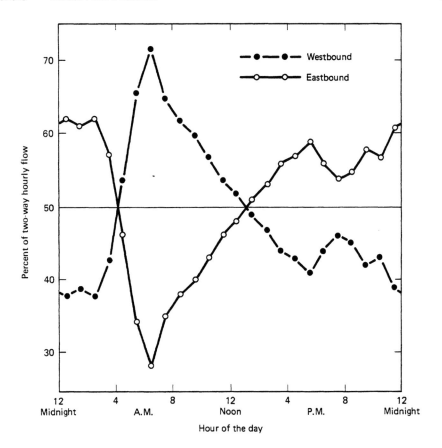

Figure 3.5   Directional Split on the San Francisco–Oakland Bay Bridge (From Reference 2)

study these modal flow patterns because of their demand-related interactions, their influence on facility capacity, and the possibility of improving the movement of people through preferential treatment solutions for specific modes.

The San Francisco–Oakland Bay Bridge is presented as an example of a multimodal transportation facility. The two-deck bridge carries passenger cars, buses, and trucks and provides special entry lanes and toll-free passage for priority vehicles. Priority vehicles consist of carpool vehicles (vehicles carrying three or more persons), vanpool vehicles, and buses. In addition, a rail transit system runs parallel to the bridge between San Francisco and Oakland.

Figure 3.7 graphically indicates the distribution of two-way hourly vehicular flows across the San Francisco–Oakland Bay Bridge by time of day and vehicle mode. Passenger cars (including carpool and vanpool vehicles) make up 80 to 90 percent of all vehicular trips, with a daily average of 87 percent. Trucks account for 10 to 20 percent of all vehicular trips, with a daily average of 12 percent. Buses and trains account for only 1 percent of all vehicular trips and rise to 3 percent during peak periods.

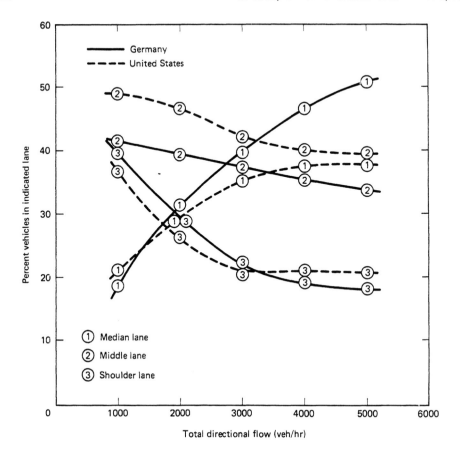

Figure 3.6   Typical Lane Volume Distributions (From References 3 and 4)

Figure 3.8 graphically presents the distribution of two-way hourly person-trips across the San Francisco–Oakland Bay Bridge by time of day and vehicle mode. Passenger cars (including carpool and vanpool vehicles) account for 66 percent of all daily person-trips but drop to 50 percent during peak hours. Trucks account for 5 to 10 percent of all person-trips with a daily share of 7 percent. Buses account for 11 percent of all daily person-trips, but this percentage varies on an hourly basis from near 1 percent to a high during peak hours of 25 percent. Trains carry 16 percent of all daily person-trips, but this percentage varies on an hourly basis from 0 percent to a high of 23 percent during the peak hour.

Figure 3.9 is a presentation of average vehicle occupancy across the San Francisco–Oakland Bay Bridge by time of day and vehicle mode. An unusual vertical scale is required because of the great differences in vehicle occupancies between vehicle modes. For example, during the day, train occupancies vary from 100+ persons per train during the middle of the day to over 400 persons per train during the morning peak period. Bus occupancies follow the same pattern, with 10± persons per bus during

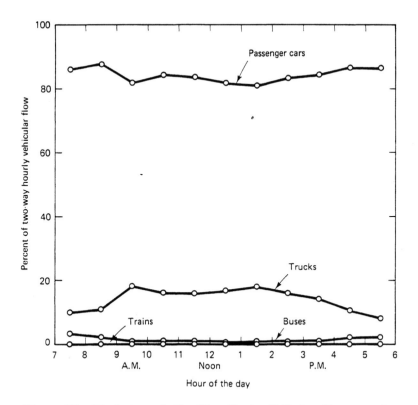

Figure 3.7 Distribution of Two-Way Hourly Vehicular Flows on San Francisco–Oakland Bay Bridge (From Reference 2)

the midday and 20 to 30 persons per bus during the peak hours. Passenger car occupancies vary from 1.3 persons per vehicle during the midday to 1.5 persons per vehicle during peak hours. Truck occupancies are assumed to be a constant 1.1 persons per vehicle throughout the day.

The presence of larger and lower-performance vehicles in the traffic stream, such as trucks, recreational vehicles, and buses, reduces the capacity of highway facilities. The 1985 *Highway Capacity Manual* [1] proposes passenger car equivalents for these larger and lower-performance vehicles under uninterrupted flow situations and a percent adjustment factor under interrupted flow situations. The reader may also wish to consult other references dealing with passenger car equivalencies [38–40].

The passenger car equivalents for freeways and two-lane rural highways proposed in the *Highway Capacity Manual* are given in Table 3.1. Three types of vehicles are considered operating in three types of terrain conditions. More detailed procedures for the analysis of the influence of larger and lower-performance vehicles under specific grade situations are presented in the 1985 *Highway Capacity Manual* [1].

The adjustment factors for heavy trucks operating at signalized intersections on grades are presented in Table 3.2. The unadjusted saturation flow is multiplied by the heavy vehicle adjustment factor to account for the effect of the proportion of heavy

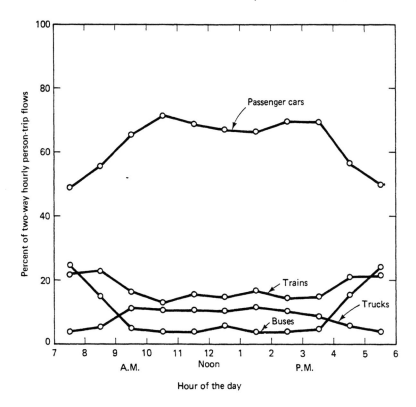

Figure 3.8 Distribution of Two-Way Hourly Person-Trip Flows on San Francisco–Oakland Bay Bridge (From Reference 2)

TABLE 3.1 Average Passenger Car Equivalents for Trucks, Buses, and Recreational Vehicles on Freeways and Two-Lane Rural Highways[a]

Vehicle Type	Type of Terrain		
	Level	Rolling	Mountainous
Trucks	1.7	4.0	8.0
	2.0	5.0	12.0
Buses	1.5	3.0	5.0
	1.6	2.9	6.5
Recreational vehicles	1.6	3.0	4.0
	1.6	3.3	5.2

[a]The first value given in each case is for freeways; the second is for two-lane rural highways under levels of service D and E.

Source: Reference 1.

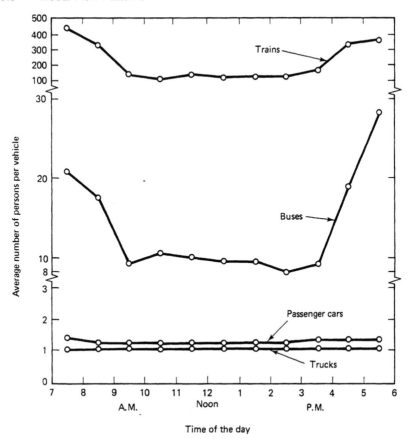

Figure 3.9 Average Hourly Vehicle Occupancy by Mode on the San Francisco–Oakland Bay Bridge (From Reference 2)

TABLE 3.2 Adjustment Factors for Heavy Vehicles and Grades at Signalized Intersections

Percent Heavy Vehicles	0	2	4	6	8	10	15	20	25	30
Heavy Vehicle Factor	1.00	0.99	0.98	0.97	0.96	0.95	0.93	0.93	0.89	0.87

	Downhill				Uphill		
Percent Grade	-6	-4	-2	0	+2	+4	+6
Grade Factor	1.03	1.02	1.01	1.00	0.99	0.98	0.97

Source: Reference 1.

vehicles in the traffic stream. The effects of grades are handled separately with the adjustment factors for grades, also presented in Table 3.2. The unadjusted saturation flow is multiplied by both of these adjustment factors to account for the combined effects of heavy vehicles and grade.

3.4 MATHEMATICAL COUNT DISTRIBUTIONS

Mathematical count distributions can be used to represent boundary conditions for field-measured count distributions. The Poisson count distribution has been applied for low-flow conditions, and the single-valued count distribution has been utilized for near-capacity conditions. The intermediate state between these two boundary states is very complex, and mathematical count distributions have not been employed successfully. In the following paragraphs the Poisson count distribution and the single-valued count distribution are presented, followed by a qualitative discussion of count distributions under intermediate-flow conditions.

The Poisson count distribution is the mathematical distribution that represents the count distribution of random events. For a count distribution to be truly random, two conditions must be met. First, any point in time is as likely to have a vehicle arriving as is any other point in time. Second, the arrival of one vehicle at a point in time does not affect the arrival time of any other vehicle. The Poisson count distribution is derived in Appendix B, and the equation is as follows:

$$P(x) = \frac{m^x e^{-m}}{x!} \tag{3.3}$$

where $P(x) =$ probability of exactly x vehicles arriving in a time interval t
$m =$ average number of vehicles arriving in a time interval t
$x =$ number of vehicles arriving in a time interval being investigated
$e =$ a constant, Napierian base of logarithms ($e = 2.71828$)
$t =$ selected time interval

Consider a 1-hour flow of 120 vehicles, during which time the analyst is interested in obtaining the distribution of 1-minute counts. The average minute count in this case is 2 vehicles per minute since $t = 1$ minute and the hourly flow is 120 vehicles per hour. Substituting 2 for m, equation (3.3) becomes:

$$P(x) = \frac{2^x e^{-2}}{x!} \tag{3.4}$$

It is now possible to enter values of 0, 1, 2, 3,... for x. Note that x must be zero or a positive integer value. For each x value, a $P(x)$ value is determined. Knowing the number of count periods per study period (60 one-minute periods per hour), the $P(x)$ value can be used to calculate the $F(x)$ value, which is the number of minutes expected to have a minute flow of exactly x vehicles. This procedure was followed, and the results are shown in Table 3.3 for x values from 0 to 10. The probabilities of x or more and x or less can easily be calculated from the probabilities of exactly x.

TABLE 3.3 Poisson Count Distribution Calculations

x	2^x	e^{-2}	$x!$	$P(x)$	$F(x)$
0	1	0.1353	1	0.1353	8.1
1	2	0.1353	1	0.2706	16.2
2	4	0.1353	2	0.2706	16.2
3	8	0.1353	6	0.1804	10.8
4	16	0.1353	24	0.0902	5.4
5	32	0.1353	120	0.0361	2.2
6	64	0.1353	720	0.0120	0.7
7	128	0.1353	5,040	0.0034	0.2
8	256	0.1353	40,320	0.0009	0.1
9	512	0.1353	362,880	0.0002	0.0
10	1,024	0.1353	3,628,800	0.0000	0.0

The calculations above are rather tedious, and two simple procedures can be employed. The first is mathematical and utilizes the recursive natural of equation (3.3). The other procedure utilizes a nomograph from which probabilities of x or fewer counts can be determined. Returning to equation (3.3), the equation can be rewritten with x values of 0, 1, 2, and 3 as follows:

$$P(0) = \frac{m^0 e^{-m}}{0!} = e^{-m} \tag{3.5}$$

Recall that m^0 is equal to 1, and 0! is equal to 1.

$$P(1) = \frac{m^1 e^{-m}}{1!} = \left(\frac{m}{1}\right) e^{-m} = \frac{m}{1} \left[P(x=0)\right] \tag{3.6}$$

$$P(2) = \frac{m^2 e^{-m}}{2!} = \left(\frac{m^2}{2}\right) e^{-m} = \frac{m}{2} \left[P(x=1)\right] \tag{3.7}$$

$$P(3) = \frac{m^3 e^{-m}}{3!} = \left(\frac{m^3}{6}\right) e^{-m} = \frac{m}{3} \left[P(x=2)\right] \tag{3.8}$$

Note from equations (3.5) through (3.8) that after calculating $P(0)$, which is simply e^{-m}, each following $P(x)$ can be calculated as follows:

$$P(x) = \frac{m}{x} \left[P(x-1)\right] \tag{3.9}$$

This procedure can be checked against the $P(x)$ values shown in Table 3.3. This recursive equation also permits a very easy method for predicting the mode of the Poisson count distribution. Note the term m/x in equation (3.9). As soon as x becomes larger than m, the $P(x)$ begins to decrease. If m is an integer value, there will be two modes: $P(x = m - 1)$ and $P(x = m)$. If m is a noninteger value, there will be a single mode: the last x value before $x > m$. In the example, $m = 2$ and the two modes in the Poisson count distribution occur at $P(x = 1)$ and $P(x = 2)$. If m had been slightly larger than 2, the single mode would have occurred at $P(x = 2)$.

The nomograph procedure utilizes the graphical illustration of Figure 3.10. The procedure is to calculate the average number of vehicles per interval (m), enter with this value on the horizontal axis, and read off the probabilities of x or fewer vehicles on the y-axis for various values of x. For example, with $m = 2$ (calculations shown in Table 3.3), the $P(x \le 0, 1, 2,$ etc.) would be 0.14, 0.41, 0.68, and so on.

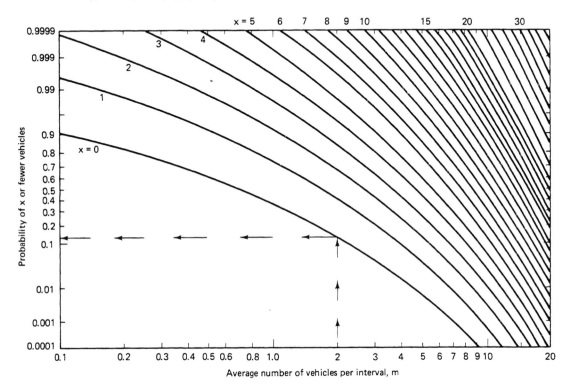

Figure 3.10   Cumulative Probabilities from the Poisson Distribution (From Reference 5)

Figure 3.11 illustrates the shape of the Poisson count distributions for various hourly flow values of 120, 240, and 360. The horizontal scale is shown as minute volumes, and the average minute volumes are noted on each of the three distributions. There are two vertical scales. The vertical scale on the left is hourly flow, and the baseline for the three distributions are plotted at the hourly flow values of 120, 240, and 360. The height of the individual distributions is the probability of exactly x vehicles per minute [$P(x)$] and values are shown on the right vertical scale. Note the relationship of modes and average minute flow rates, and the shifting to the right and the spreading of the distributions as hourly flow increases.

As flows get heavier and heavier and approach the capacity of the roadway, the measured count distribution approaches a single-valued count distribution. That is, for example, every minute during an hour contains the same number of vehicles. If a section of highway is operating near capacity and the capacity is 1800 vehicles per hour,

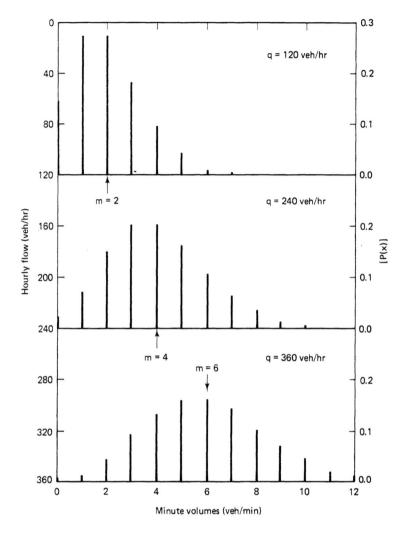

Figure 3.11   Poisson Count Distributions

each minute would have a count of approximately 30 vehicles. The variance of this distribution would approach zero.

Some insights as to count distributions between these two extremes of random and single-valued count distributions can be obtained by studying the relationship between variances and volume–capacity ratios. Figure 3.12 displays such a relationship. Two rays are constructed starting at the origin and extending to the right. One ray represents the relationship if a Poisson count distribution is assumed with the variance equal to the average mean flow ($s^2 = m$). The second ray represents the relationship if a single-valued count distribution is assumed with a variance equal to zero. Considering that Poisson count distributions are more likely to occur under low flow conditions

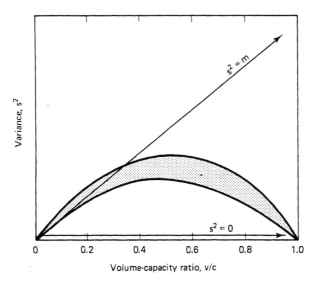

Figure 3.12 Conceptual Relationship between Variance of Count Distribution and Volume–Capacity Ratio

and uniform count distributions are more likely to occur under near-capacity conditions, the shaded area indicates the range of likely variances over the volume–capacity range. This is only an hypothesis but gives some qualitative insights as to intermediate count distributions. The reader is encouraged to study other references for coverage of other mathematical count distributions [5–7, 8, 12, 35]. Specific count distributions have been proposed by Buckeley [18], Haight et al. [11, 33, 34], Oliver [30, 31], Pak-Poy [20], Underwood [21], and Vaughan [32]. Trip scheduling results in very unusual measured count distributions and provide great complexity for analysis. In addition, the operational analyst is often not only interested in the count distribution, but the time sequence of the interval counts. The peak-hour factor approach discussed in Section 3.1 is currently used in practice to describe numerically the variation of flow within an hour period. When mathematical count distributions are considered for use, the chi-square test described in Section 2.6 can be used to evaluate how well the mathematical count distribution represents a measured count distribution.

3.5 ESTIMATING HOURLY FLOW RATES

Planners and designers of new and existing facility improvements are normally provided with anticipated traffic demand data on an annual average daily traffic (AADT) basis. The AADT cannot be used directly for design and planning analysis because of the great variations in monthly, daily, hourly, and directional traffic flow patterns discussed previously. Therefore, one of the initial tasks is to convert the AADT for the design year to a directional design hour volume. The following equation is normally used for this purpose:

$$DDHV = \left(\frac{DDHV}{DHV} \right) \left(\frac{DHV}{AADT} \right) AADT \qquad (3.10)$$

where DDHV = design hour volume in major direction
       DHV = design hour volume combining both directions
      AADT = annual average daily traffic combining both directions

The equation is simplified by substituting D for the first term and K for the second term, and equation (3.10) becomes

$$DDHV = (D)(K)AADT \tag{3.11}$$

where D = ratio of design hour volume in the major direction to the two-way design hour volume
      K = ratio of the two-way design hour volume to the two-way AADT

Given the two-way AADT, the accuracy of the directional design hour volume is dependent on the D and K values selected.

Fortunately, many state departments of transportation have established permanent count stations and have estimates for the D and K factors for different types of facilities in different geographic areas [16]. For example, in Minnesota, the hourly flow rates for each hour of the year for a given location are placed in rank order starting with the highest hourly flow rate and continuing to lower hourly flow rates and then plotted against the K factor for each hourly flow rate. The results are shown in Figure 3.13 for four types of routes. The thirtieth highest hour has been selected historically as the design hour volume because on rural highways the "knee" of the curve is generally located there, and it is somewhat of a balance between designing for the very highest hourly flow rate and significantly underdesigning [14, 41]. In urban areas the curves are very flat (see Figure 3.13), and there is little difference between the 30th and the 200th highest hour. Approximate ranges in K adjustment factors for different geographic areas and types of facilities are given in the top portion of Table 3.4.

Permanent count station data are also analyzed to provide estimates for the D adjustment factors. Approximate ranges in D adjustment factors for different geographic areas and types of facilities are given in the lower portion of Table 3.4. The values given in Table 3.4 are for example purposes only and such factors should be carefully developed before applying them to real-life problems.

The importance of these factors become more apparent by inspecting Figure 3.14 and considering the following examples. Assume that an intercity route is being upgraded to freeway standards and the AADT in the design year is estimated to be 40,000 vehicles per day. Inspecting Table 3.4, the ranges of K and D adjustment factors for an intercity route are 0.12 to 0.18 and 0.60 to 0.70, respectively. If two investigations are made, one using the lower K and D factors and the other using the higher K and D factors, the resulting products of K and D would be 0.072 and 0.126, respectively. Now returning to Figure 3.14 and entering the graph with an AADT of 40,000, the estimated DDHV would be 2880 and 5040 for KD combined factors of 0.072 and 0.126, respectively. If the design volume were limited to 1500 vehicles per hour per lane, the number of lanes required would vary from two to four, which is obviously a very significant difference. The example attempts to illustrate the importance of understanding the variations in traffic flow, and the need for detailed and careful study for selecting K and D adjustment factors.

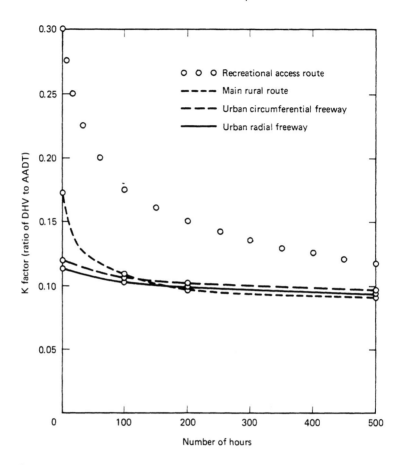

Figure 3.13   Ranked Hourly Volumes on Minnesota Highways (From Reference 1)

3.6 UNINTERRUPTED TRAFFIC APPLICATIONS

A freeway-type facility will be used as the example for uninterrupted traffic applications. The purpose of this section is to present the concept of capacity and service
volume, and briefly their relationship to level of service. The reader is referred to the
1985 *Highway Capacity Manual* for greater in-depth coverage and for analysis of other
types of highway facilities. The following paragraphs are highlights selected from the
1985 *Highway Capacity Manual*.

Capacity is defined as "the maximum hourly rate at which persons or vehicles can
reasonably be expected to traverse a point or uniform section of a lane or roadway during a given time period (usually, 15 minutes) under prevailing roadway, traffic, and control conditions." The capacity for freeways having design speeds of 60 or 70 miles per
hour and operating under ideal conditions is 2000 vehicles per hour per lane. "Ideal
conditions" imply 12-foot lanes and adequate lateral clearances; no trucks, buses, or
recreational vehicles in the traffic stream; and weekday or commuter traffic. When ideal

TABLE 3.4   *K* and *D* Adjustment Factors

K FACTORS

Area	Facility	*K* Factor
Urban	Heavily loaded	0.07–0.10
	Moderately loaded	0.11–0.14
Rural	Intercity route	0.12–0.18
	Recreational route	>0.15

D FACTORS

Area	Facility	*D* Factor
Urban	Circumferential route	0.50–0.55
	Radial route	0.55–0.60
Rural	Intercity route	0.60–0.70
	Recreational route	> 0.70

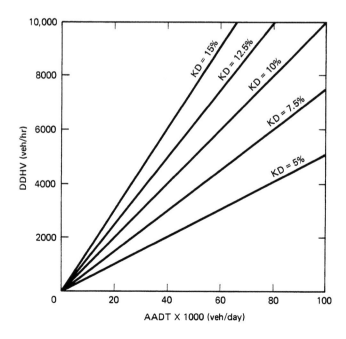

Figure 3.14   Influence of Flow Variations on Design

conditions do not exist, the capacity is reduced and the following equation is used to calculate the actual capacity:

$$c = c_j N f_w f_{HV} f_p \qquad (3.12)$$

where c = capacity (vehicles per hour)
   c_j = capacity under ideal conditions with design speed of j; c_j = 2000 vehicles per hour per lane with design speeds of 60 or 70 miles per hour and c_j = 1900 vehicles per hour per lane with design speeds of 50 miles per hour.
   N = number of lanes in one direction
   f_w = lane width and lateral clearance factor
   f_{HV} = heavy vehicle factor
   f_p = driver population factor

For example, if ideal conditions existed along a three-lane directional freeway having a design speed of 60 or 70 miles per hour, the capacity would be

$$c = 2000(3)(1.00)(1.00)(1.00) \qquad (3.13)$$

$$= 6000 \text{ vehicles/hour}$$

Normally, ideal conditions do not exist, and typical lane capacities of around 1800 vehicles per hour per lane are often encountered.

Capacity is a measure of maximum route productivity but does not directly address quality or level of service to the users. Consequently, the concept of service volume is introduced, which has a definition exactly like capacity except that a phrase is added at the end: "while maintaining a designated level of service." Six levels of service and corresponding service volumes are proposed and range from level of service A (service volume A) to level of service F (service volume F). Levels of service, service volume, and resulting operational performance for a freeway assuming ideal conditions and a 70-mile per hour design speed is shown in Table 3.5. For this example, if one wishes to operate this particular section of freeway at level of service C, the

TABLE 3.5 Levels of Service for Basic Freeway Sections[a]

Level of Service	Flow Condition	v/c Limit	Service Volume (veh/hr)	Speed (miles/hr)	Density (veh/mile/lane)
A	Free	0.35	700	≥ 60	≤ 12
B	Stable	0.54	1100	≥ 57	≤ 20
C	Stable	0.77	1550	≥ 54	≤ 30
D	High density	0.93	1850	≥ 46	≤ 40
E	Near capacity	1.00	2000	≥ 30	≤ 67
F	Breakdown	Unstable		< 30	> 67

[a]Assuming ideal conditions and 70-mile per hour design speed.

Source: Reference 1.

volume–capacity ratio should be limited to 0.77 (1550 vehicles per hour per lane), and speeds over 54 miles per hour and lane densities of less than 30 vehicles per mile per lane should result. Speed characteristics, density characteristics, and the relationship between these characteristics are discussed in later chapters.

3.7 INTERRUPTED TRAFFIC APPLICATIONS

The signalized intersection will be used as an example for interrupted traffic applications. The purpose of this section is to present the concept of saturation flow rate and platoon diffusion. British researchers have been leaders in developing these concepts, and the reader may wish to refer to selected British references on signal capacity and signal network models for greater coverage [9]. Researchers in several other countries have developed signalized intersection capacity techniques [15, 36, 37].

Saturation flow rate is defined as the equivalent maximum hourly rate at which vehicles can traverse a lane or intersection approach under prevailing traffic and roadway conditions assuming that the green signal is available at all times and no lost times are experienced. The saturation flow rate for a single lane under ideal conditions is 1800 vehicles per lane per hour of green according to the 1985 *Highway Capacity Manual* [1]. When ideal conditions do not exist, the saturation flow rate is reduced and the following equation is used to calculate the actual saturation flow rate:

$$s = 1800 N f_w f_{HV} f_g f_p f_{pb} f_{bb} f_a f_{RT} f_{LT} \tag{3.14}$$

where s = saturated flow rate
  N = number of lanes
  f_w = lane width factor
  f_{HV} = heavy vehicle factor
  f_g = approach grade factor
  f_p = parking adjustment factor
  f_{pb} = parking blockage factor
  f_{bb} = local bus blocking factor
  f_a = area type factor
  f_{RT} = right-turn factor
  f_{LT} = left-turn factor

For example, if ideal conditions existed at a two-lane approach to a signalized intersection, the saturation flow rate would be

$$s = 1800(2)(1.00)(1.00)(1.00)(1.00)(1.00)(1.00)(1.00)(1.00)(1.00) \tag{3.15}$$

$$= 3600 \text{ vehicles/approach/hour of green}$$

Normally, ideal conditions do not exist, and typical lane saturation flow rates on the order of 1500 to 1600 vehicles per hour of green per lane are encountered.

Figure 3.15 is presented to emphasize the importance of the saturation flow rate concept and to illustrate how capacity can be obtained from the saturation flow rate.

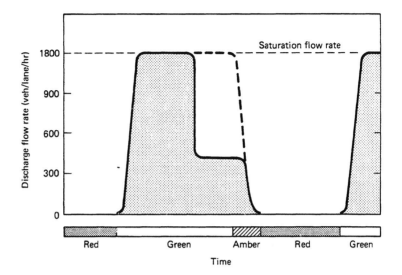

Figure 3.15   Hypothetical Discharge Flow Rate at an Approach to a Signalized Intersection

The horizontal scale is time, and the signal timing for this approach is shown on this
scale. The vertical scale is discharge flow rate, and the maximum value is saturation
flow rate. During the red period no vehicles are discharged, but soon after the signal
changes to green the vehicles begin to discharge across the stop line. After the first few
vehicles are discharged, the discharge rate remains fairly constant at its maximum rate,
which is the saturation flow rate. This maximum discharge is maintained until the
queue dissipates or until the signal changes to amber. If the queue is dissipated, the
discharge rate will decrease and maintain a discharge flow rate equal to the arrival flow
rate until the signal changes to amber. This is shown in Figure 3.15 as a solid line. If
the signal changes to amber before the queue is dissipated, the green phase is fully utilized and the discharge flow rate follows the dashed line in Figure 3.15. When the signal changes to amber, the discharge flow rate decreases and approaches a discharge flow
rate equal to zero near the beginning of the red period. During each signal cycle the
pattern repeats itself, depending on the arrival flow rate.

In the example given in Figure 3.15, the green period is not fully utilized. Note
that the shaded area represents the number of vehicles discharged since the horizontal
scale is time and the vertical scale is vehicles discharged per unit time. The area
between the solid and dashed lines represents the additional number of vehicles that
could be handled if an upstream queue were always present. Therefore, the shaded area
divided by the sum of the shaded area plus the area between the solid and dashed lines
represents the degree of utilization of this particular green period.

To calculate capacity, both the saturation flow rate and the fraction of the time
that saturation flow rates can occur must be known. For example if one-half of the time
the saturation flow can occur, the capacity is equal to one-half the saturation flow rate.
This relationship in terms of signal timing is shown as follows:

$$c = \left(\frac{g}{C}\right)s \qquad (3.16)$$

where c = capacity (vehicles per hour)
   g = effective green time in seconds (not necessarily the duration of the green period)
   C = signal cycle length (seconds)
   s = saturation flow rate

The greatest difficulty in calculating capacity from a known saturation flow rate is determining the effective green time. One technique employed is to convert the trapezoidal figure shown in Figure 3.15 (following the dashed line) to an equivalent rectangular figure, but keeping the height in both cases equal to the saturation flow rate. Thus the horizontal dimensions of the rectangular figure would be the effective green time. Effective green time can be related to the signal timing in the following manner:

$$g = G + a - \left(l_s + l_e\right) \qquad (3.17)$$

where g = effective green time (seconds)
   G = actual green time (seconds)
   a = amber time (seconds)
   l_s = starting lost time (seconds)
   l_e = ending lost time (seconds)

Starting and ending lost times vary considerably depending on the local situation, but often are approximately equal to the amber time duration, and thus effective green time in this case would be equal to actual green time.

Another important traffic flow phenomenon is the diffusion of platoons as traffic leaves a traffic signal. Consider traffic passing through a pedestrian signalized crosswalk and moving downstream. A pedestrian signalized crosswalk is selected so that no vehicular traffic turns off or turns on to the road in question, and thus keeps the example simple. Such a situation is shown in Figure 3.16. The horizontal scale is time and is constructed in a fashion similar to Figure 3.15. The vertical scale is distance with the pedestrian signalized crosswalk at the beginning of the route and traffic moves up the diagram. The shaded area at the bottom of the figure is the discharge flow rate at the pedestrian signalized crosswalk and is similar to the shaded area in Figure 3.15.

As the traffic leaves the pedestrian signalized crosswalk and moves downstream, some vehicles travel a little faster and others travel a little slower than the average vehicles. By the time the platoon passes location A, there is a slight diffusion of the platoon, but it is still a very tight platoon. As traffic continues to move downstream past count locations B and C, there is further diffusion of the platoon. As the platoon moves even farther downstream, the platoons formed by individual green periods begin to overlap and the flow rates become more and more uniform. These phenomena are essential elements in the timing of traffic signal systems and constitute an integral part of network signal models such as TRANSYT [10]. Considerable research has been undertaken in platooning, bunching, and dispersion behavior and the reader is encouraged to refer to the rather extensive literature [19, 22–28].

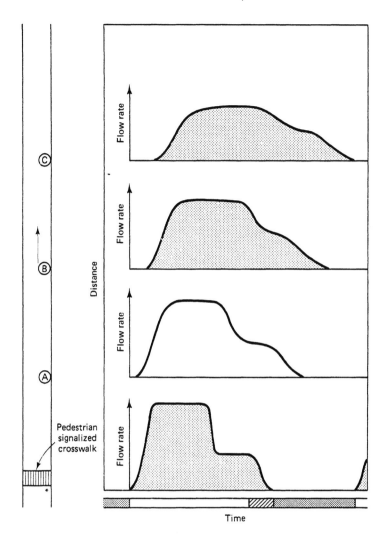

Figure 3.16   Hypothetical Example of Platoon Diffusion at a Traffic Signal

3.8 SELECTED PROBLEMS

1. Conduct a library search of field-measured monthly flow patterns, and plot the patterns in a fashion similar to Figure 3.1. Attempt to develop a classification scheme for grouping sites with similar monthly flow patterns.

2. Conduct a library search of field-measured daily flow patterns, and plot the patterns in a fashion similar to Figure 3.2. Attempt to develop a classification scheme for grouping sites with similar daily flow patterns.

3. Conduct a library search and locate traffic flow for one site in which hourly flow rates are available by direction and for several days. Plot the hourly flow patterns in a fashion

similar to Figures 3.3 and 3.4. Calculate directional splits during the peak hour (D) and the ratio of the peak-hour flow to the 24-hour flow (K).

4. Undertake a field study of turning movements of a single approach at a signalized intersection. Record counts by 15-minute periods for 1 hour. Plot the turning movement diagram, and calculate the peak-hour factor for each movement and for the total approach. How do your results compare with the discussion near the end of Section 3.1?

5. Conduct a library search and locate several data sets that give the distribution of traffic between lanes on two-lane directional roadways. Plot lane volume distributions in a fashion similar to Figure 3.6. Note similarities and differences between data sets, and attempt to explain.

6. Repeat Problem 5 but consider roadways with three lanes in one direction and compare results with Figure 3.6.

7. Conduct a library search and locate traffic flow data for a facility that is multimodal and has hourly flows for both vehicles and persons for several modes. Plot results in a fashion similar to Figures 3.7 through 3.9. Compare your results with those of the San Francisco–Oakland Bay Bridge given in Section 3.3.

8. N directional lanes pass through a toll plaza at the entrance to a bridge. Assume that the total vehicular demand at the toll plaza is 9000 vehicles per hour, consisting of 85 percent passenger vehicles (vehicles carrying one or two persons with average occupancy of 1.25 persons per vehicle), 10 percent priority passenger vehicles (vehicles carrying three or more persons with an average occupancy of 3.50 persons per vehicle), 3 percent buses (with an average occupancies of 40 persons per bus), and 2 percent trucks (with an average occupancy of 1.1 persons per truck). Design a lane configuration at the toll plaza so that n lanes of the N lanes are used only by priority passenger vehicles and buses and $N - n$ lanes of the N lanes are used only by passenger vehicles and trucks. Assume that trucks and buses are equivalent to 2.0 passenger car units. The passenger vehicles and trucks must stop at the toll plaza and pay a toll, so these lanes have a capacity of only 600 passenger car units per hour. The priority passenger cars and buses go through the toll plaza toll-free and only slow down, so that these lanes have a capacity of 1200 passenger car units per hour. Determine the minimum number of lanes possible so that traffic demands do not exceed toll plaza capacities. What is the average vehicle occupancy and number of person-trips on a lane basis for the two types of toll plaza lanes?

9. Conduct a field study at a site carrying 200 to 500 vehicles per hour and at which you feel the arrival flow pattern is fairly random. Record 1-minute flows for at least 1 hour. Plot your measured count distribution in a manner similar to Figure 3.11. Apply a Poisson count distribution to your data set and plot with your measured count distribution. Perform a chi-square test comparing your measured count distribution to the Poisson count distribution.

10. Conduct a field study at a site that you feel is operating at capacity, such as at a bottleneck on a freeway. Record 1-minute flows for at least 1 hour. Plot your measured count distribution in a manner similar to Figure 3.11. Apply a Poisson count distribution to your data set, and plot with your measured count distribution. Consider plotting a normal distribution to your data set. Estimate the capacity at the site and comment on the count distributions. (It would be desirable to have two-person teams for this exercise.)

11. Repeat Problem 10 for different volume–capacity ratio levels. Calculate the standard deviation for the measured count distribution for each v/c level, and compare with Figure 3.12.

12. Conduct a library study of a mathematical count distribution such as the negative-binomial, generalized Poisson, or binomial count distribution. Apply one of these distributions to the

example problem given in Section 3.4 and plot the results in a form similar to Figure 3.11. Discuss similarities and differences as compared to the Poisson count distribution.

13. Conduct a library search of K and D factors and attempt to develop a classification scheme for grouping sites with similar K and D factors. Reconstruct a graph similar to Figure 3.14 on which your classification scheme is identified.

14. A three-lane directional freeway carries 4500 vehicles per hour. The roadway and traffic conditions are as follows:
   - 60-mile per hour design speed
   - 11-foot lanes
   - 4-foot shoulders on both sides
   - Rolling terrain
   - 10 percent trucks, 5 percent recreational vehicles, and 1 percent buses
   - Commuter drivers

   Calculate the directional freeway capacity, the level of service, and estimate the resulting speed and density. Refer to the 1985 *Highway Capacity Manual*, Chapter 3 [1].

15. Conduct a field study on one approach to a fixed-time signal during a period of time that it is operating near capacity. It would be desirable that turning movements either be prohibited or handled by special turn lanes. Determine the length of the green period for your approach and the signal cycle time. Select a short time interval on the order of 4 to 8 seconds which divides the cycle time into equal intervals. The start of the green period will be recorded as time zero. During each short time interval, count the number of vehicles that discharge across the stop line in the straight-through vehicle lane (or lanes). Continue this counting process for at least 20 cycles. Construct a diagram like Figure 3.15 and estimate the saturation flow and capacity. (It would be desirable to have two-person teams for this exercise.)

16. Undertake a library study of literature pertaining to platoon diffusion with special attention to mathematical expressions and field-measured flow profiles such as those shown in Figure 3.16.

17. Left-turning vehicles randomly arrive at a signalized intersection that has a separate left-turn phase. The number of left-turning vehicles arriving during the design hour is 120 vehicles. The length of the signal cycle is 90 seconds. Determine the following:
   (a) The probabilities of 0, 1, 2, ... 10 vehicles arriving during the cycle length in the design hour.
   (b) The minimum-length left-turn storage lane so that during only one cycle in the design hour will the left-turning traffic block the through traffic. (Assume that stored left-turn vehicles occupy 30 feet of space and that there is no overflow from one cycle to the next.)
   (c) The probability of the through vehicles being blocked, assuming that the left-turn storage lane designed above is built but the hourly rate has doubled.

3.9 SELECTED REFERENCES

1. Transportation Research Board, *Highway Capacity Manual*, Special Report 209, TRB, Washington, D.C., 1985, 474 pages.

2. Institute of Transportation Studies, *Traffic Survey Series A-46 Bay Bridge*, ITS, Berkeley, Calif., Spring 1976, 10 pages.

3. F. Busch, Spurbelastungen und Häufigkeit von Spurwechseln auf einer dreispurigen BAB-Richtungsfahrbahn, *Strassenverkehrstechnik*, Vol. 28, No. 6, 1984, pages 228–231.

4. Adolf D. May, Traffic Characteristics and Phenomena on High Density Controlled Access Facilities, *Traffic Engineering*, Vol. 31, No. 6, March 1961.

5. Daniel L. Gerlough and Frank C. Barnes, *Poisson and Other Distributions in Traffic*, Eno Foundation, Saugatuck, Conn., 1971, pages 10–34.

6. Daniel L. Gerlough and Matthew J. Huber, *Statistics with Applications to Highway Traffic Analysis*, Eno Foundation, Saugatuck, Conn., 1978, pages 84–91.

7. Daniel L. Gerlough and Matthew J. Huber, *Traffic Flow Theory—A Monograph*, Transportation Research Board, Special Report 165, TRB, Washington, D.C., 1975, pages 31–37.

8. F. A. Haight, *Mathematical Theories of Traffic Flow*, Academic Press, Inc., New York, 1963, pages 97–98.

9. F. V. Webster and B. M. Cobbe, *Traffic Signals*, Her Majesty's Stationery Office, London, 1966, page 111.

10. Dennis I. Robertson, *TRANSYT: Traffic Network Study Tool*, Transport and Road Research Laboratory, Report 243, TRRL, Crowthorne, England, 1969, 37 pages.

11. Frank A. Haight, Bertram F. Whisler, and Walter W. Mosher, Jr., New Statistical Methods for Describing Highway Distribution of Cars, *Highway Research Board, Proceedings*, Vol. 40, 1961, pages 557–563.

12. Donald R. Drew, *Traffic Flow Theory and Control*, McGraw-Hill Book Company, New York, 1968, pages 120–146.

13. Institute of Transportation Engineering, *Transportation and Traffic Engineering Handbook*, 2nd Edition, Prentice-Hall, Inc., Englewood Cliffs, N.J., 1982.

14. Martin Wohl and Brian V. Martin, *Traffic Systems Analysis*, McGraw-Hill Book Company, New York, 1967, pages 164–174.

15. R. Akcelik, *Traffic Signals: Capacity and Timing Analysis*, Australian Road Research Board, Reprint 123, ARRB, Victoria, Australia, 1981.

16. Louis J. Pignataro, *Traffic Engineering—Theory and Practice*, Prentice-Hall Inc., Englewood Cliffs, N.J., 1973, pages 143–163.

17. Wolfgang S. Homburger and James H. Kell, *Fundamentals of Traffic Engineering*, Institute of Transportation Studies, Berkeley, Calif., 1984, pages 5-1 to 5-10.

18. D. J. Buckley, Road Traffic Counting Distributions, *Transportation Research*, Vol. 1, No. 2, 1967, pages 105–116.

19. A. Daou, On Flows within Platoons, *Australian Road Research Board*, Vol. 2, No. 7, 1966, pages 4–13.

20. P. G. Pak-Poy, The Use and Limitation of the Poisson Distribution in Road Traffic, *Australian Road Research Board Proceedings*, Vol. 2, 1964, pages 223–247.

21. R. T. Underwood, Discussion of Pak-Poy Paper, *Australian Road Research Board Proceedings*, Vol. 2, 1964, page 243.

22. M. J. Grace and R. B. Potts, Diffusion of Traffic Platoons, *Australian Road Research Board Proceedings*, Vol. 1, 1962, pages 260–267.

23. M. J. Grace and R. B. Potts, A Theory of the Diffusion of Traffic Platoons, *Operations Research*, Vol. 12, No. 2, 1964, pages 255–275.

24. R. Herman, R. B. Potts, and R. W. Rothery, Behavior of Traffic Leaving a Signalized Intersection, *Traffic Engineering and Control*, Vol. 5, No. 9, 1964, pages 539–533.

25. B. J. Lewis, *Platoon Movement of Traffic from an Isolated Signalized Intersection*, Highway Research Board Bulletin 178, HRB, Washington, D.C., 1958, pages 1–11.

26. G. M. Pacey, *The Progress of a Bunch of Vehicles Released from a Traffic Signal*, Road Research Laboratory, Research Notes RN/2665/GMP, RRL, Crowthorne, England, January 1956.

27. R. T. Underwood, Traffic Flow and Bunching, *Australian Road Research Board Proceedings*, Vol. 1, No. 6, 1963, pages 8–24.

28. C. C. Wright, Some Characteristics of Traffic Leaving a Signalized Intersection, *Transportation Science*, Vol. 4, No. 4, 1970, pages 331–346.

29. G. F. Newell, Stochastic Properties of Peak Short-Time Traffic Counts, *Transportation Science*, Vol. 1, No. 3, 1967, pages 167–183.

30. R. M. Oliver, A Traffic Counting Distribution, *Operations Research*, Vol. 9, No. 6, 1961, pages 807–810.

31. R. M. Oliver, Note on a Traffic Counting Distribution, *Operations Research Quarterly*, Vol. 13, No. 2, 1962, pages 171–178.

32. R. J. Vaughan, The Distribution of Traffic Volumes, *Transportation Science*, Vol. 4, No. 1, 1970, pages 97–112.

33. J. R. B. Whittlesey and F. A. Haight, Counting Distributions for an Erlang Process, *Annual Statistical Mathematics*, Tokyo, Vol. 13, No. 2, 1961.

34. F. A. Haight, Counting Distributions for Renewal Processes, *Biometrika*, Vol. 53, Nos. 3 and 4, 1965, pages 395–403.

35. F. A. Haight, *Handbook of the Poisson Distribution*, John Wiley & Sons, Inc., New York, 1967.

36. Bo E. Peterson, Arne Hansson, and Karl-Lennart Bang, *Swedish Capacity Manual*, Transportation Research Board, Record 667, TRB, Washington, D.C., 1978.

37. Organization for Economic Cooperation and Development, *Traffic Capacity of Major Routes*, OECD, Paris, July 1983, 120 pages.

38. J. Craus, A. Polus, and I. Grinberg, A Revised Method for the Determination of Passenger Car Equivalencies, *Transportation Research*, Vol. 14A, No. 4, August 1980, pages 241–246.

39. W. D. Cunagin and C. J. Messer, *Passenger Car Equivalents for Rural Highways*, Federal Highway Administration, Washington, D.C., 1982, 51 pages.

40. B. G. Heydecker, Vehicles, PCV's, and TCV's in Traffic Signal Calculations, *Traffic Engineering and Control*, Vol. 24, No. 3, March 1983, pages 111–114.

41. American Association of State Highway and Transportation Officials, *A Policy on Geometric Design of Highways and Streets*, AASHTO, Washington, D.C., 1984, 1087 pages.

4

Microscopic Speed Characteristics

Microscopic speed characteristics are those speed characteristics of *individual* vehicles passing a point or short segment during a specified period of time. Speeds and travel times over *longer* sections of roadways and statistical analysis between *groups* of vehicles will be considered as macroscopic speed characteristics and discussed in Chapter 5.

Speed is a fundamental measurement of the traffic performance on the highway system. Most analytical and simulation models of traffic predict speed as the measure of performance given the design, demand, and control on the highway system. More extensive models will then use speed as an input for the estimation of fuel consumption, vehicle emissions, and traffic noise. Speed is also used as an indication of level of service, in accident analysis, and in economic studies. Therefore, the traffic analyst must be familiar with speed characteristics and associated statistical analysis techniques.

This chapter contains five sections plus selected problems and references. The first section is devoted to speed trajectories of individual vehicles. The next section presents speed characteristics under uninterrupted flow conditions. The third and fourth sections are concerned with mathematical distributions and their evaluation. The final section describes procedures for estimating population means and sample size requirements.

4.1 VEHICULAR SPEED TRAJECTORIES

This section is about the trajectories of individual vehicles over space and time as influenced by interrupted flow and highway grade situations. Interrupted flow situations include sign- and signal-controlled intersections as well as railroad and pedestrian crossings. Highway grade situations will include the interactions between various types of vehicles and the length and steepness of grades.

First, equations are provided for determining vehicle speed trajectories over space and time based on specified acceleration and deceleration rates. Maximum and normal acceleration and deceleration rates are presented for various types of vehicles under various grade situations. Then the equations and rates of acceleration and deceleration are applied to several highway traffic situations.

4.1.1 Equations of Motion and Acceleration/Deceleration Rates

Two equations of motion that can be used to calculate distance traveled and elapsed time given the speed and acceleration (or deceleration) rates are*

$$t = \frac{\mu_e - \mu_b}{a} \tag{4.1}$$

$$d = 1.47\mu_b t + 0.733at^2 \tag{4.2}$$

where μ_b = speed (miles per hour) at the beginning of the acceleration (or deceleration) cycle

μ_e = speed (miles per hour) at the end of the acceleration (or deceleration) cycle

a = acceleration (or deceleration) rate (miles per hour/second)

t = time for vehicle to accelerate (or decelerate) at rate a from beginning speed (μ_b) to ending speed (μ_e)(seconds)

d = distance for vehicle to accelerate (or decelerate) at rate a from beginning speed μ_b to ending speed (μ_e) (feet)

As vehicles proceed along a highway, drivers may desire or be required to accelerate and/or decelerate their vehicles because of other vehicles in the traffic stream, interrupted flow situations, or highway design features. Acceleration and deceleration rates vary considerably between drivers, vehicles, traffic situations, roadway situations, and for different speed levels. One of the most comprehensive summaries of previous research and field studies of maximum and normal acceleration and deceleration rates is contained in the *Transportation and Traffic Engineering Handbook* [23] and is a starting point for materials covered in this chapter. The following two paragraphs are devoted to acceleration rates and deceleration rates.

Maximum and normal acceleration rates for various vehicle types, speed changes, and grade situations are summarized in Table 4.1. Maximum acceleration rates decrease with increased weight-to-horse-power ratios, with steeper grades, and with higher running speeds. Normal acceleration rates observed under typical driving conditions are considerably less than maximum acceleration rates. For example, normal acceleration rates for passenger vehicles in level terrain and in nonemergency situations are on the order of one-half to two-thirds of the maximum acceleration rates as shown in Table 4.1. Another important consideration is the acceleration capabilities (or the

---

*Coefficients are needed in equation (4.2) because speeds and acceleration rates are expressed in terms of miles per hour, while distances and time are expressed in feet and seconds.

TABLE 4.1   Maximum and Normal Acceleration Rates
(Miles per Hour/Second)[a]

COMPOSITE PASSENGER VEHICLE

| | Speed Change | | | | |
Grade	0–15	0–30	30–40	40–50	50–60
Level	8.0 (3.3)	5.0 (3.3)	4.7 (3.3)	3.8 (2.6)	2.8 (2.0)
+2%	7.8	4.6	4.2	3.4	2.4
+6%	6.7	3.7	3.4	2.5	1.5
+10%	5.8	2.8	2.5	1.6	0.6

PICKUP TRUCKS

	0–15	0–30	30–40	40–50	50–60
Level	8.0	5.0	2.0	1.8	1.5
+2%	7.8	4.6	1.6	1.4	1.0
+6%	6.7	3.7	0.7	0.5	0.2
+10%	5.8	2.8	[30]	—	—

TWO-AXLE, SIX-TIRE TRUCK

	0–15	0–30	30–40	40–50	50–60
Level	2.0	1.0	1.0	0.6	0.2
+2%	1.6	0.6	0.6	0.2	[50]
+6%	0.7	0.1	[30]	—	—
+10%	[14]	—	—	—	—

TRACTOR-SEMITRAILER TRUCK

	0–15	0–30	30–40	40–50	50–60
Level	2.0	1.0	0.8	0.4	0.1
+2%	1.6	0.6	0.3	[45]	—
+6%	0.7	[23]	—	—	—
+10%	[4]	—	—	—	—

[a]Normal acceleration rates are shown in parentheses. Some vehicle types on steeper grades cannot exceed a performance-limiting speed. These speeds, shown in brackets, are called crawl speeds and cannot be exceeded.

Source: Reference 23.

lack thereof) of vehicles with higher weight-to-horsepower ratios on steeper grades. For example, a tractor-semitrailer truck on a sustained 6 percent upgrade will be unable to accelerate once a speed on the order of 23 miles per hour is reached. Thereafter, the speed will remain constant on the upgrade, and this speed is referred to as the crawl speed. Table 4.2, which summarizes acceleration rates, distances traveled, and elapsed times for passenger vehicles on level terrain and under normal operating conditions, has been prepared for later analysis. The distances traveled and the elapsed times were calculated using equations (4.1) and (4.2).

TABLE 4.2 Normal Acceleration Rates with Associated
Distances Traveled and Elapsed Times
for Passenger Vehicles in Level Terrain[a]

Initial Speed (miles/hr)	Final Speed (miles/hr)				
	15	30	40	50	60
0	3.3 4.5 49	3.3 9.1 200	3.3 12.1 354	3.1 15.9 574	2.9 20.9 929
30	—	—	⁻ 3.3 3.0 154	2.9 6.8 374	2.5 11.8 729
40	—	—	—	2.6 3.8 220	2.3 8.8 575
50	—	—	—	—	2.0 5.0 355

[a]The first value given in each case is for acceleration rate in
miles per hour/second, the second for elapsed time in
seconds, and the third for distance traveled in feet.

Source: Reference 23.

Minimum safe stopping distances rather than maximum deceleration rates are considered because of the dependency on coefficient of friction rather than vehicle type and because most design issues are concerned with the distance requirements to stop a vehicle from some specified running speed. The equation for minimum stopping distance in terms of running speed, grade situation, and coefficient of friction between the tires and the pavement is

$$S = \frac{(\mu_b)^2}{30(f \pm g)} \qquad (4.3)$$

where S = minimum stopping distance in (feet)

μ_b = running speed at beginning of the deceleration (miles per hour)

f = coefficient of friction between tires and pavement

g = grade situation expressed as a decimal (i.e., a 3 percent upgrade is expressed as +0.03)

A table of minimum stopping distances to bring a vehicle to a complete stop as a function of initial running speed, grade situation, and coefficient of friction is presented as Table 4.3. The three subtables are for various tire–pavement situations: dry pavement–good tires, dry pavement–poor tires, and wet pavement. In each table, minimum stopping distances are given for selected running speeds and for three grade situations: 6 percent downgrade, level, and 6 percent upgrade. Note that coefficient of

TABLE 4.3 Minimum Stopping Distances (feet)

DRY PAVEMENT, GOOD TIRES

Grade	Running Speed (miles/hr)						
	20	30	40	50	60	70	80
f Value →	0.75	0.75	0.75				
−6%	19	43	77	—	—	—	—
Level	18	40	71	—	—	—	—
+6%	16	37	66	—	—	—	—

DRY PAVEMENT, POOR TIRES

f Value →	0.58	0.53	0.43				
−6%	26	64	148	—	—	—	—
Level	23	57	127	—	—	—	—
+6%	21	51	111	—	—	—	—

WET PAVEMENT

f Value →	0.40	0.36	0.33	0.31	0.30	0.29	0.27
−6%	39	100	198	333	500	710	1016
Level	33	83	162	269	400	563	790
+6%	29	71	137	225	333	467	646

Source: Reference 29.

friction values are not only a function of the tire–pavement situation but also depend on running speeds (i.e., lower coefficient of friction values are encountered at higher speeds). Also note that the size and type of vehicle are not considered directly. Table 4.4 has been prepared for later analysis which summarizes deceleration rates, distances traveled, and elapsed times for passenger vehicles in level terrain and under normal operating conditions. The distances traveled and the elapsed times were calculated using equations (4.1) and (4.2).

4.1.2 Highway Traffic Applications

Three applications will be made to demonstrate how acceleration and deceleration performance of vehicles can be used to develop vehicle speed trajectories for travel through an intersection, during a passing maneuver, and for trucks on grades. Normal acceleration and deceleration performance will be used in these examples and are based on data provided in Tables 4.2 and 4.4. Special care should be exercised in selecting appropriate acceleration and deceleration performance in such types of analysis.

TABLE 4.4 Normal Deceleration Rates with Associated Distances Traveled and Elapsed Times for Passenger Vehicles in Level Terrain[a]

Initial Speed (miles/hr)	Final Speed (miles/hr)				
	0	30	40	50	60
15	5.3 2.8 30	—	—	—	—
30	4.6 6.5 143	—	—	—	—
40	4.2 9.5 297	3.3 3.0 154	—	—	—
50	4.0 12.5 495	3.3 6.0 352	3.3 3.0 198	—	—
60	3.9 15.5 737	3.3 9.0 594	3.3 6.0 440	3.3 3.0 242	—
70	3.8 18.5 1023	3.3 12.0 880	3.3 9.0 726	3.3 6.0 528	3.3 3.0 286

[a]The first value given in each case is for the deceleration rate in miles per hour per second, the second for elapsed time in seconds, and the third for distance traveled in feet.

Source: Reference 23.

A graphical illustration of vehicle trajectories in the vicinity of a signalized inter-section in a rural area is shown in Figure 4.1. In each case shown, the running speed some distance upstream and downstream of the signal is assumed to be 50 miles per hour. The speeds at the stop line at the intersection are assumed to be 50, 30, and 0 miles per hour. The vehicle trajectories during constant running speed are shown as heavy lines while the vehicle trajectories during deceleration and acceleration are shown as lighter lines. Note that the lost time due to slowing to 30 miles per hour is only a few seconds, whereas a required stop (even without stopped time) is considerably more.

A graph of vehicle trajectories during a passing maneuver is shown in Figure 4.2. In this example a vehicle traveling 60 miles per hour approaches a vehicle ahead in the same lane which is traveling at a constant 40 miles per hour. The higher-speed vehicle decelerates, maintains a distance headway of 100 feet for 4 seconds, and then passes while accelerating back to 60 miles per hour. Note that the passing vehicle encroaches

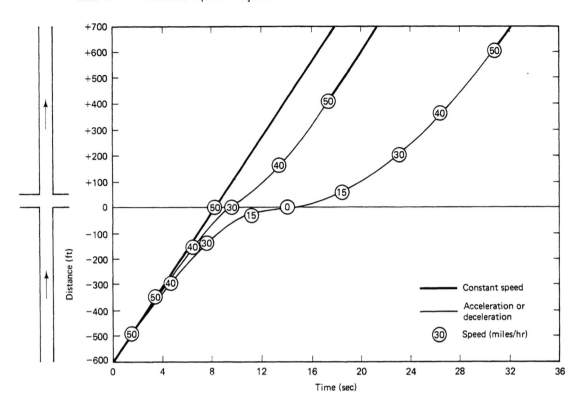

Figure 4.1   Typical Vehicle Trajectories in the Vicinity of a Signalized Intersection

in the adjacent lane for about 900 feet and 11 seconds. The vehicle trajectory of the second vehicle during encroachment is shown as a heavy line in the figure. The vehicle trajectory of the following vehicle is shown as a dashed line if passing was not required and a constant 60-mile per hour running speed was maintained.

The final example is concerned with the trajectory of trucks on grades. In this example maximum rather than normal acceleration and deceleration (due to limited vehicle performance not to braking) are considered. The performance of trucks are grouped on the basis of weight-to-horsepower ratios, and 100, 200, and 300 pounds per horsepower are used to represent light, typical, and heavy trucks, respectively. Special nomographs are prepared to aid in the analysis of truck projectories on grades. An example nomograph contained in the 1985 *Highway Capacity Manual* [20] for typical trucks [200 pounds per horsepower (lb/hp)] is reproduced here as Figure 4.3. Two situations are presented in this figure: acceleration on upgrades and downgrades, and deceleration on upgrades. In the acceleration situation, the initial speed is assumed to be zero and the vehicle accelerates to various speeds as a function of the length and steepness of grade. The dashed lines represent the acceleration performance of trucks on grades varying from +8 percent upgrade to -5 percent downgrade. Note the leveling off of speed with length of grade. For example, on a 6 percent upgrade after about

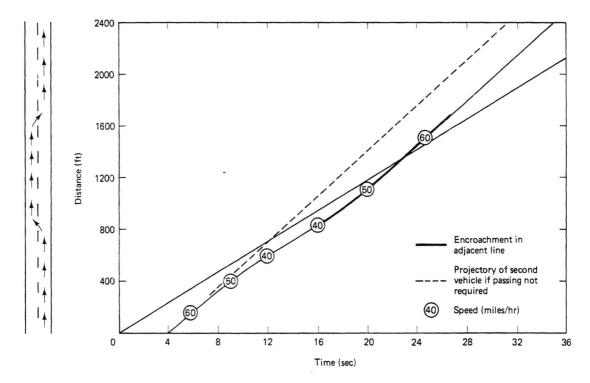

Figure 4.2   Vehicle Trajectories during a Passing Maneuver

3000 feet, the truck speed remains constant at about 23 miles per hour. This is because this particular type of truck cannot accelerate further once this speed is reached, and this speed is referred to as the crawl speed. In fact, in all cases there will be a leveling off of speed with length of grade because of performance limitations (upgrades) or because of drivers not using the full performance of the vehicle as the desired running speeds are approached (downgrades). In the deceleration situation, the initial speed is assumed to be 55 miles per hour, and the vehicle decelerates to various speeds as a function of length of grade and steepness of upgrade. The solid lines represent the deceleration performance for grades varying from +1 percent upgrade to +8 percent upgrade. Again, note the leveling off of speed with length of grade.

The nomograph is specially constructed so that initial speeds other than zero and 55 miles per hour can be considered. For example, if a 200-lb/hp truck approaches a +4 percent upgrade that is 2000 feet long and is followed by a +6 percent upgrade that is 1000 feet long, the anticipated speeds at the end of each individual grade can be predicted. Using the nomograph shown in Figure 4.3, the speed at the end of the 2000-foot-long +4 percent upgrade is found to be 39 miles per hour. Projecting this point horizontally to the left from the +4 percent to the +6 percent upgrade curve, and then moving along the +6 percent upgrade curve down and to the right for an additional distance of 1000 feet, the speed at the end of the second grade is found to be 26 miles per hour. Note that speeds above 55 miles per hour are not considered, and vehicle

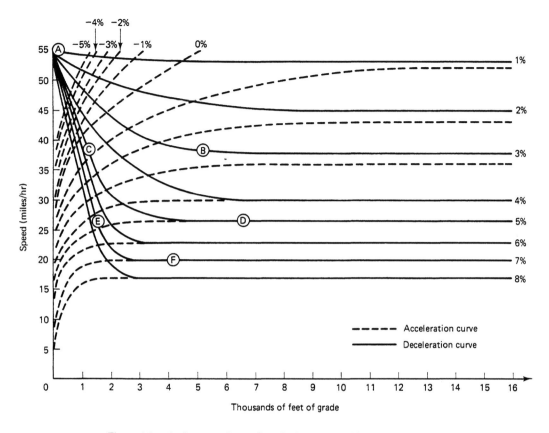

Figure 4.3   Performance Curves for a Typical Truck (200 lb/hp) (From Reference 20)

trajectories when speeds approach 55 miles per hour require some adjustments if the driver does not use the maximum acceleration performance of the truck.

For illustrative purposes, consider the speed trajectory of a 200-lb/hp truck along a highway that has a profile as shown in the top portion of Figure 4.4. The truck speeds at various points along the upgrades and downgrades are shown in the lower portion of Figure 4.4, and the results for the 200-lb/hp truck were obtained using the performance curves shown in Figure 4.3. The speed of the truck at the beginning of subsection 1 is assumed to be 55 miles per hour, and since the grade in subsection 1 is 0 percent, the truck speed of 55 miles per hour is maintained for the first 3 miles. The truck speed at the end of subsection 2, which is 1 mile long and has a grade of +3 percent, is reduced to 38 miles per hour (point B on Figure 4.3). The 1 mile, +5 percent grade reduced the truck speed further, to 26 miles per hour (point D on Figure 4.3), and the 1/2 mile, +7 percent grade reduced the truck speed even further, to 20 miles per hour (point F on Figure 4.3). At this point the summit is reached and a 1 mile, −3 percent downgrade is encountered. The truck is able to accelerate and has the performance capability to return to 55 miles per hour within approximately 2000 feet. However, in practice as the

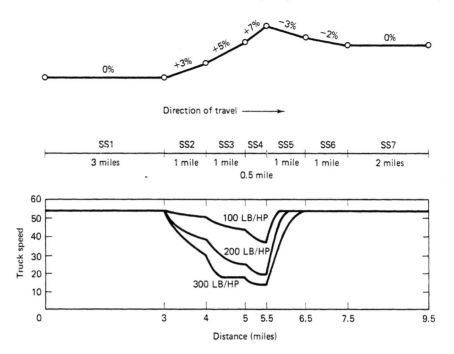

Figure 4.4   Truck Speeds as Affected by Grades

driver approaches his desired speed, the maximum acceleration is not maintained and so the truck will more likely reach a 55 mile-per hour running speed farther downstream as shown in Figure 4.4. For comparison purposes, truck speed profiles are also shown for light trucks (100 lb/hp) and heavy trucks (300 lb/hp) in the same example.

4.2 SPEED CHARACTERISTICS UNDER UNINTERRUPTED FLOW CONDITIONS

Consider standing at a point along a highway facility during a relatively short period of time under uninterrupted flow conditions: that is, a location away from intersections which has little or no roadside development. The speeds of individual vehicles are measured and recorded. This would result in a series of individual vehicular speeds such as 50, 46, 48, 55, 48, and so on, miles per hour. The sample mean and sample variance of these "ungrouped" speed observations would be

$$\bar{\mu} = \frac{\sum\limits_{i=1}^{N} \mu_i}{N} \tag{4.4}$$

$$s^2 = \frac{\sum\limits_{i=1}^{N} (\mu_i - \bar{\mu})^2}{N - 1} \tag{4.5}$$

$$s = \sqrt{s^2} \tag{4.6}$$

where $\bar{\mu}$ = sample mean speed (miles per hour)

μ_i = speed of vehicle i

N = total number of speed observations

s^2 = sample variance

s = sample standard deviation

In most cases the speed observations are "grouped." The frequencies of each speed level or speed interval are determined from the series of individual vehicular speeds observed. An example of grouped speed data is shown in the first two columns of Table 4.5 based on 200 individual speed observations. The cumulative frequency and percentile are shown in the next two columns and will be utilized a little later. The last two columns are calculations required in determining the sample mean and sample variance using the following equations:

$$\bar{\mu} = \frac{\sum_{i=1}^{g}\left(f_i\mu_i\right)}{N} \tag{4.7}$$

$$s^2 = \frac{\sum_{i=1}^{g} f_i\left(\mu_i\right)^2 - \frac{1}{N}\left(\sum_{i=1}^{g} f_i\mu_i\right)^2}{N-1} \tag{4.8}$$

$$s = \sqrt{s^2} \tag{4.9}$$

where $\bar{\mu}$ = mean speed of sample

g = number of speed groups

i = speed group i

f_i = number of observations in speed group i

μ_i = midpoint speed of group i

N = total number of speed observations

s^2 = sample variance

s = sample standard deviation

Note that in calculating the variance the term $N - 1$ rather than N is used in the demonstration because of dependency between the last $f_i\mu_i$ value and $\bar{\mu}$. However, for moderate sample sizes, using N or $N-1$ will have little effect on the numerical results. For the grouped speed data shown in Table 4.5, the sample mean and the sample variance are calculated as follows:

$$\bar{\mu} = \frac{\sum_{i=1}^{g}(f_i\mu_i)}{N} = \frac{10,460}{200} = 52.3 \text{ miles/hour} \tag{4.10}$$

$$s^2 = \frac{\sum_{i=1}^{g} f_i\left(\mu_i\right)^2 - 1/N\left(\sum_{i=1}^{g} f_i\mu_i\right)^2}{N-1} \tag{4.11}$$

$$= \frac{554,882 - 1/200(10,460)^2}{199}$$

$$= 39.3 \ (\text{miles/hour})^2$$

$$s = \sqrt{s^2} = 6.3 \ \text{miles/hour} \qquad (4.12)$$

TABLE 4.5 Grouped Speed Data

u_i	f_i	Cumulative Frequency	%	$f_i \mu_i$	$f_i (\mu_i)^2$
30	—	0	0	—	—
31	—	0	0	—	—
32	0	0	0	—	—
33	1	1	1	33	1,089
34	2	3	2	68	2,312
35	1	4	2	35	1,225
36	1	5	2	36	1,296
37	—	5	2	—	—
38	1	6	3	38	1,444
39	1	7	4	39	1,521
40	2	9	4	80	3,200
41	1	10	5	41	1,681
42	5	15	8	210	8,820
43	4	19	10	172	7,396
44	1	20	10	44	1,936
45	7	27	14	315	14,175
46	4	31	16	184	8,464
47	8	39	20	376	17,672
48	8	47	24	384	18,432
49	15	62	31	735	36,015
50	8	70	35	400	20,000
51	8	78	39	408	20,808
52	10	88	44	520	27,040
53	23	111	56	1,219	64,607
54	15	126	63	810	43,740
55	16	142	71	880	48,400
56	9	151	76	504	28,224
57	14	165	82	798	45,486
58	6	171	86	348	20,184
59	3	174	87	177	10,443
60	9	183	92	540	32,400
61	3	186	93	183	11,163
62	6	192	96	372	23,064
63	3	195	98	189	11,907
64	3	198	99	192	12,288
65	2	200	100	130	8,450
66	—	200	100	—	—
67	—	200	100	—	—
68	—	200	100	—	—
69	—	200	100	—	—
70	—	200	100	—	—
	200			10,460	554,882

Two types of distributions are often plotted in regard to measured speed distributions as shown in Figure 4.5. Figure 4.5a is a straight frequency distribution in which the vertical scale represents the number of observations in each speed group and the horizontal scale is speed in miles per hour. Figure 4.5b is a cumulative percentile distribution in which the vertical scale represents the percent of vehicles traveling at or less than the indicated speed group and the horizontal scale is speed in miles per hour. These distributions are helpful in showing important measures of central tendencies and dispersion as well as qualitatively observing the shape of the speed distribution.

The sample mean was calculated in equation (4.10) to be 52.3 miles per hour. The sample mode can be observed in Figure 4.5a to be 53 miles per hour. The sample median can be observed in Figure 4.5b to be 52.5 miles per hour. Hence the mean, mode, and median of the sample are the three measures of central tendency of the sample and are 52.3, 53.0, and 52.5 miles per hour, respectively.

The standard deviation of the sample was calculated in equation (4.12) to be 6.3 miles per hour. Figure 4.5 can be used to determine the total range (0 to 100 percent) and is found to be 65 miles per hour minus 33 miles per hour or 32 miles per hour. Figure 4.5b can be used to determine the 15 to 85 percentile range and is found to be 57.8 miles per hour minus 45.5 miles per hour, or 12.3 miles per hour. Another measure of dispersion used uniquely in speed distributions is the 10-mile per hour pace. This measure is defined as the highest percentile of vehicles in a 10-mile per hour range. Inspection of Table 4.5 and Figure 4.5 indicates that the 10-mile per hour pace is 47 to 57 or 48 to 58 miles per hour, each of which includes 62 percent of all vehicles in the sample. In summary, the four measures of dispersion: standard deviation, total range, 15 to 85 percentile range, and 10-mile per hour pace were found to be 6.3 miles per hour, 32 miles per hour, 12.3 miles per hour, and 62 percent (47 to 57 or 48 to 58 miles per hour), respectively.

The distributions graphically shown in Figure 4.5, and the calculated measures of central tendency and dispersion provide insights into the type of mathematical distribution that may represent the speed distribution. Inspection of the top portion of Figure 4.5a indicates a fairly symmetrical bell-shaped distribution while the corresponding cumulative distribution in the bottom portion indicates an S-shaped distribution. The three calculated measures of central tendency are all almost numerically equal. The calculated ranges and the pace which are measures of dispersion are nearly symmetrical about the mean. These observations would lead the analyst to consider strongly the possibility of a normal distribution. As measured distributions became nonsymmetrical or bimodal, other distributions, such as the log-normal and composite distributions, might be considered. These three distributions are discussed in the following section.

4.3 MATHEMATICAL DISTRIBUTIONS

Mathematical distributions most commonly employed to represent measured speed distributions are normal distributions, log-normal distributions, and composite distributions. Initial and primary attention is given to normal distributions and near the end of this section log-normal and composite distributions are described briefly.

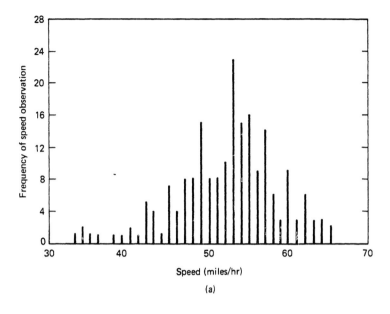

(a)

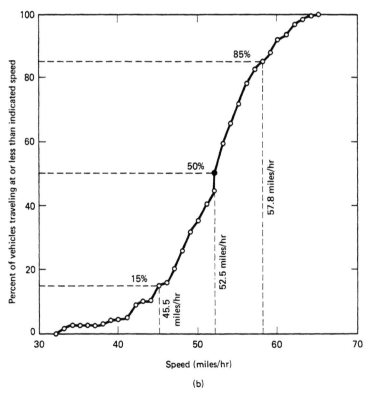

(b)

Figure 4.5   Measured Speed Distribution Example

It is desirable to find appropriate mathematical distributions to represent speed distributions for two major reasons. First, each mathematical distribution has unique attributes, and if a measured distribution can be represented by a mathematical distribution, the measured distributions can be said to have similar attributes and hence greater knowledge of the measured distribution can be inferred. It is also desirable to find appropriate distributions for purposes of computer simulation for which individual vehicle speeds are needed as input. Although measured speed distributions can be used within such models, it is easier and more flexible to use mathematical distributions.

A unique normal distribution is defined when the mean and standard deviation are specified. The normal distribution is symmetrical about the mean and the dispersion or spread is a function of the standard deviation. For example, three unique normal distributions are shown in Figure 4.6. In comparing normal distribution 1 with normal distribution 2, $s_1 = s_2$ but $\mu_1 < \mu_2$. In comparing normal distributions 2 and 3, $\mu_2 = \mu_3$ but $s_2 < s_3$.

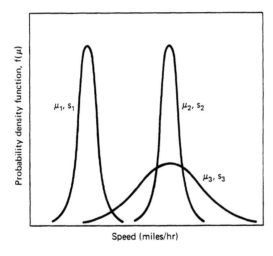

Figure 4.6   Examples of Normal Distributions

An additional attribute of the normal distribution is that the mean, mode, and median (the three measures of central tendency) are numerically equal. The dispersion is such that when the standard deviation is specified, 68.27 percent of the observations will be within 1 standard deviation of the mean, 95.45 percent within 2 standard deviations of the mean, and 99.73 percent within 3 standard deviations of the mean. These attributes are shown graphically in Figure 4.7.

The probability density function of the normal distribution is shown in Figure 4.7a and the general equation is

$$f(\mu_i) = \frac{1}{\sigma\sqrt{2\pi}}\, e^{-(\mu_i - \overline{U})^2/2\sigma^2} \tag{4.13}$$

where $f(\mu_i)$ = probability density function of individual speeds
  π = a constant, 3.1416
  e = a constant Napierian base of logarithms ($e = 2.71828$)

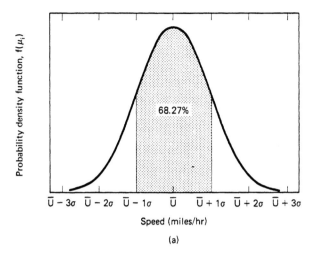

(a)

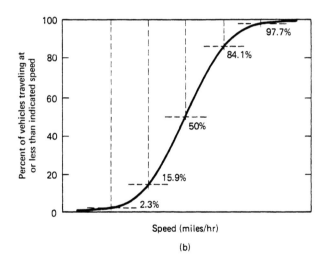

Speed (miles/hr)

(b)

Figure 4.7   Attributes of Normal Distributions

μ_i = speed value being investigated

$\overline{U}$ = population mean speed (miles per hour)

σ = population standard deviation

σ^2 = population variance

The total area under the probability density curve includes all possible outcomes and therefore is equal to unity (or 100 percent). The area under the curve between a speed value of, say, $\overline{U} - 1\sigma$ and $\overline{U} + 1\sigma$ (the shaded area in of Figure 4.7a) represents the probability of a speed between these two speed values, and in this case has a probability of 0.6827 (68.27 percent). In a similar way the area between $\overline{U} - 2\sigma$ and $\overline{U} + 2\sigma$ represents a probability of 0.9545 (95.45 percent) and the area between $\overline{U} - 3\sigma$ and

$\overline{U}$ + 3σ represents a probability of 0.9973 (99.73 percent). Note that one could ask What is the probability of an individual speed over $\overline{U}$ + 1σ? The total unshaded area represents a probability of 0.3173 (1.0000-0.6827), the normal distribution is symmetrical and therefore the unshaded areas representing the probability of $\mu_i > \overline{U}$ + 1σ is 0.3173/2, or 0.1586. That is, if individual speeds are normally distributed, 15 to 16 vehicles of every 100 vehicles would be expected to travel at speeds greater than $\overline{U}$ + 1σ. The cumulative form of the normal distribution is shown in Figure 4.7b. Note the S-shaped distribution and the indicated percentages of vehicles traveling at or less than indicated speeds.

Two issues must now be addressed in order to apply the normal distribution to a measured speed distribution. Equation (4.13) specifies that population mean ($\overline{U}$) and standard deviation of the population (σ) are required for the normal distribution. On the other hand, in most speed studies only the sample mean $\overline{\mu}$ and the standard deviation of the sample (s) are known. The sample mean is the best single estimate of the population mean and is used in the normal distribution. Estimating the standard deviation of the population from the standard deviation of the sample is more complex. The relationship between the two is a function of the sample size, N, and as N increases the difference between the two standard deviations becomes smaller and smaller. As N goes to infinity, the two standard deviations are identical. As a practical matter, if N > 30, the standard deviation of the sample is numerically substituted for the standard deviation of the population. For sample sizes less than 30, the t-distribution rather than the normal distribution is used.

The other issue involves the calculation procedures for the normal distribution. Inspection of equation (4.13) indicates that calculating the probability density function is rather tedious, and observing Figure 4.7a indicates that converting the probability density function calculations into probabilities for various speed ranges is very cumbersome. One solution is to "normalize" the normal distribution and integrate the probability density function between the mean speed value and other speed values. The resulting probabilities can be placed in a tabular form such as shown in Table 4.6. Then the analyst can calculate the $X/σ$ values and determine the probabilities from the table. The probabilities can then be determined as a function of $X/σ$.

$$P = f\left(\frac{X}{\sigma}\right) \tag{4.14}$$

where P = probability of an observation between some speed, μ_X, and the mean speed, $\overline{U}$

X = speed range or deviation between μ_x and $\overline{U}$ ($X = \mu_x - \overline{U}$) (absolute value)

$σ$ = standard deviation of the population (s is best estimate for large samples)

As an illustration, consider the example shown in Table 4.5. The sample mean is 52.3 miles per hour and the standard deviation of the speed sample is 6.3 miles per hour. The best estimate of the population mean is 52.3 miles per hour and since N > 30 (actually, N = 200), the standard deviation of the population is estimated to be 6.3 miles per hour and the normal distribution is used. Pose the question: What is the

TABLE 4.6   Calculating Probabilities from a Normal Distribution

x/s	\multicolumn{10}{c}{Second Decimal Place in x/s}									
	0.00	0.01	0.02	0.03	0.04	0.05	0.06	0.07	0.08	0.09
0.0	0.0000	0.0040	0.0080	0.0120	0.0160	0.0199	0.0239	0.0279	0.0319	0.0359
0.1	0.0398	0.0438	0.0478	0.0517	0.0557	0.0596	0.0636	0.0675	0.0714	0.0753
0.2	0.0793	0.0832	0.0871	0.0910	0.0948	0.0987	0.1026	0.1064	0.1103	0.1141
0.3	0.1179	0.1217	0.1255	0.1293	0.1331	0.1368	0.1406	0.1443	0.1480	0.1517
0.4	0.1554	0.1591	0.1628	0.1664	0.1700	0.1736	0.1772	0.1808	0.1844	0.1879
0.5	0.1915	0.1950	0.1985	0.2019	0.2054	0.2088	0.2123	0.2157	0.2190	0.2224
0.6	0.2257	0.2291	0.2324	0.2357	0.2389	0.2422	0.2454	0.2468	0.2517	0.2549
0.7	0.2580	0.2611	0.2642	0.2673	0.2704	0.2734	0.2764	0.2794	0.2823	0.2852
0.8	0.2881	0.2910	0.2939	0.2967	0.2995	0.3023	0.3051	0.3078	0.3106	0.3133
0.9	0.3159	0.3186	0.3212	0.3238	0.3264	0.3289	0.3315	0.3340	0.3365	0.3389
1.0	0.3413	0.3413	0.3461	0.3485	0.3508	0.3531	0.3554	0.3577	0.3599	0.3621
1.1	0.3643	0.3665	0.3686	0.3708	0.3729	0.3749	0.3770	0.3790	0.3810	0.3830
1.2	0.3849	0.3869	0.3888	0.3907	0.3925	0.3944	0.3962	0.3980	0.3997	0.4015
1.3	0.4031	0.4049	0.4066	0.4082	0.4099	0.4115	0.4131	0.4147	0.4162	0.4177
1.4	0.4192	0.4207	0.4222	0.4236	0.4251	0.4265	0.4279	0.4292	0.4306	0.4319
1.5	0.4332	0.4345	0.4357	0.4370	0.4382	0.4394	0.4406	0.4418	0.4429	0.4441
1.6	0.4452	0.4463	0.4474	0.4484	0.4495	0.4505	0.4515	0.4525	0.4535	0.4545
1.7	0.4554	0.4564	0.4573	0.4582	0.4591	0.4599	0.4608	0.4616	0.4625	0.4633
1.8	0.4641	0.4649	0.4656	0.4664	0.4671	0.4678	0.4686	0.4693	0.4699	0.4706
1.9	0.4713	0.4719	0.4726	0.4732	0.4738	0.4744	0.4750	0.4756	0.4761	0.4767
2.0	0.4772	0.4778	0.4783	0.4788	0.4793	0.4798	0.4803	0.4808	0.4812	0.4817
2.1	0.4821	0.4826	0.4830	0.4834	0.4838	0.4842	0.4846	0.4850	0.4854	0.4857
2.2	0.4861	0.4864	0.4868	0.4871	0.4875	0.4878	0.4881	0.4884	0.4887	0.4890
2.3	0.4893	0.4896	0.4898	0.4901	0.4904	0.4906	0.4909	0.4911	0.4913	0.4916
2.4	0.4918	0.4920	0.4922	0.4925	0.4927	0.4929	0.4931	0.4932	0.4934	0.4936
2.5	0.4938	0.4940	0.4941	0.4943	0.4945	0.4946	0.4948	0.4949	0.4951	0.4952
2.6	0.4953	0.4955	0.4956	0.4957	0.4959	0.4960	0.4961	0.4962	0.4963	0.4964
2.7	0.4965	0.4966	0.4967	0.4968	0.4969	0.4970	0.4971	0.4972	0.4973	0.4974
2.8	0.4974	0.4975	0.4976	0.4977	0.4977	0.4978	0.4979	0.4979	0.4980	0.4981
2.9	0.4981	0.4982	0.4982	0.4983	0.4984	0.4984	0.4985	0.4985	0.4986	0.4986
3.0	0.4987	0.4987	0.4987	0.4988	0.4988	0.4989	0.4989	0.4989	0.4990	0.4990
3.1	0.4990	0.4991	0.4991	0.4991	0.4992	0.4992	0.4992	0.4992	0.4993	0.4993
3.2	0.4993	0.4993	0.4994	0.4994	0.4994	0.4994	0.4994	0.4995	0.4995	0.4995
3.3	0.4995	0.4995	0.4995	0.4996	0.4996	0.4996	0.4996	0.4996	0.4996	0.4997
3.4	0.4997	0.4997	0.4997	0.4997	0.4997	0.4997	0.4997	0.4997	0.4997	0.4998
3.5	0.4998									
4.0	0.49997									
4.5	0.499997									
5.0	0.4999997									

Source: Standard Mathematical Tables, 15th ed. Reprinted by permission of CRC Press, Boca Raton, Fla.

probability of individual speeds between 35 and 40 miles per hour? Since the mean lies above both 35 and 40 miles per hour, two calculations are required: X between 35.0 and 52.3, and X between 40.0 and 52.3.

$$\left(\frac{X}{\sigma}\right)_{35\rightarrow52.3} = \frac{52.3 - 35.0}{6.3} = 2.75 \qquad (4.15)$$

$$\left(\frac{X}{\sigma}\right)_{40\rightarrow52.3} = \frac{52.3 - 40.0}{6.3} = 1.95 \qquad (4.16)$$

Entering Table 4.6 with X/σ values of 2.75 and 1.95, the corresponding probability values are 0.4970 and 0.4744. That is, 49.70 percent of the vehicles are traveling at speeds between 35 and 52.3 miles per hour and 47.44 percent are traveling at speeds between 40 and 52.3 miles per hour. Therefore, the probability of a speed between 35 and 40 miles per hour is the difference between these two probabilities and numerically equal to 0.0226. With a sample size of 200, the expected frequency would be 4 or 5. Note in Table 4.5 that the measured frequency was 4.

Two other distributions have been suggested to represent measured speed distributions: log-normal and composite distributions. The log-normal distribution is presented by Gerlough and Huber [1] and the earlier work of Haight and Mosher [2] is identified. It is particularly appropriate for unimodal-shaped distributions like the normal distribution, except the distribution is skewed with a larger tail of the distribution extending to the right. In the next section, graphical techniques are discussed that provide insights as to when the log-normal distribution might be considered. Several statistical references are available for detailed analysis of log-normal distributions [3, 4, 5].

The composite distribution has been proposed when the traffic stream consists of two classes of vehicles (or drivers), and there is little interference between these two classes. An example would be a relatively lightly traveled rural freeway with passenger cars and trucks each having different speed limits. There are several possible combinations of distributions that could be used in a composite distribution, that is, normal or log-normal for one or the other subpopulations. Subpopulations on a per-lane basis might be considered. The traffic analyst may consider some form of a composite distribution when the measured speed distribution is bimodal; that is, two modes some distance apart are clearly identified.

4.4 EVALUATION AND SELECTION OF MATHEMATICAL DISTRIBUTIONS

The suggested procedure is to assume initially that the measured speed distribution can be represented by a normal distribution. Numerical checks are made of the measures of central tendency and dispersion from the measured speed distribution. If the numerical checks appear to support the assumption of a normal distribution, it is suggested that a graphical plot be made of the measured speed distribution on normal probability paper. If the distribution appears on the normal probability paper as a straight line, then the measured speed distribution is evaluated using the chi-square test. If the hypothesis is accepted, then the search for a mathematical distribution is completed and the normal distribution is selected. On the other hand, if there is significant evidence to eliminate the normal distribution, the process is repeated assuming a log-normal and/or a composite distribution.

The procedure described above will be applied to the measured speed distribution shown in Table 4.5 and Figure 4.5. Numerical checks are made of the measures of central tendency. The mean, median, and mode are 52.3, 52.5, and 53.0 miles per hour, respectively, which are in very close agreement. The frequency distribution shown in Figure 4.5a displays a fairly strong single mode. Numerical checks are made of the measures of dispersion: standard deviation, total range, 15 to 85 percentile range, and 10 miles per hour pace are 6.3 miles per hour, 32 miles per hour, 12.3 miles per hour, and 62 percent (47 to 57 or 48 to 58 miles per hour), respectively. Some of the more important checks are shown below.

- The variance of a measured speed distribution normally should be less than the variance of a random distribution ($s_R^2 = m$):

$$s^2 = (6.3)^2 = 39.3 \text{ miles/hour}$$

$$s_R^2 = m = 52.3 \text{ miles/hour}$$

$$s^2 < s_R^2$$

- The standard deviation should be approximately one-sixth of the total range since the mean plus and minus three standard deviations encompasses 99.73 percent of the observations of a normal distribution:

$$s_{EST} = \frac{\text{total range}}{6} \qquad (4.17)$$

$$= \frac{32}{6} = 5.3 \text{ miles/hour}$$

$$s \approx s_{EST}$$

- The standard deviation should be approximately one-half of the 15 to 85 percent range since the mean plus and minus 1 standard deviation encompasses 68.27 percent of the observation of a normal distribution

$$s_{EST} = \frac{15\text{--}85 \text{ percentile range}}{2} \qquad (4.18)$$

$$= \frac{12.3}{2} = 6.15$$

$$s \approx s_{EST}$$

- The 10-mile per hour pace should be approximately straddling the sample mean

$$\bar{\mu} = 52.3 \text{ miles/hour}$$

$$10\text{-mile per hour pace} = 47 \text{ to } 57 \text{ or } 48 \text{ to } 58$$

$$\text{midpoint of pace} = 52 \text{ or } 53$$

$$\text{midpoint of pace} \sim \bar{\mu}$$

- The normal distribution has little skewness and the coefficient of skewness calculated below should be close to zero.

$$\text{coefficient of skewness} = \frac{\text{mean - mode}}{s} \tag{4.19}$$

$$= \frac{52.3 - 53.0}{6.3} = 0.1$$

or

$$\text{coefficient of skewness} = 3\left(\frac{\text{mean - median}}{s}\right) \tag{4.20}$$

$$= 3\left(\frac{52.3 - 52.5}{6.3}\right) = 0.1$$

The numerical checks appear to support the assumption of a normal distribution, so a graphical plot is made of the measured speed distribution on normal probability paper. The graphical plot is shown in Figure 4.8. The solid straight line represents a normal distribution in which the population mean $(\overline{U})$ is 52.3 miles per hour, and the population standard deviation of speed (σ) is 6.3 miles per hour. The plotted data points are taken from the measured speed distribution. There appears to be a very good fit in the 10 to 90 percentile range and as might be expected, a poor fit at the extreme tails of the distribution (less than 2 percent and more than 98 percentile ranges). Note that the slope of the line on the graph represents the standard deviation, with flatter slopes representing larger standard deviation values and steeper slopes representing smaller standard deviation values. The data points reveal that below the mean speed, the distribution is a little more spread out (a larger indicated standard deviation), while above the mean speed, the distribution is a little more dense (a smaller indicated standard deviation). Overall the graphical fit appears reasonable, so the next and last step is to perform a chi-square test.

The chi-square test was described in Section 2.6, so only the calculations and the conclusions will be presented here. Inspection of the observed frequencies in Table 4.5 reveals that the 1-mile per hour class interval is too small considering the number of observations ($N = 200$). Neiswanger [21] has proposed that there normally should be between 10 and 25 class intervals, depending on the range of the observation values and the number of observations. Sturges [22] proposed that the following equation be used to estimate the size of the class interval:

$$I = \frac{\text{Range}}{1 + (3.322) \log N} \tag{4.21}$$

where I = size of the class interval
  Range = total range (largest observed value minus smallest observed value)
  N = number of observations

Substituting 32 miles per hour for Range and 200 for number of observations, the size of the class can be determined.

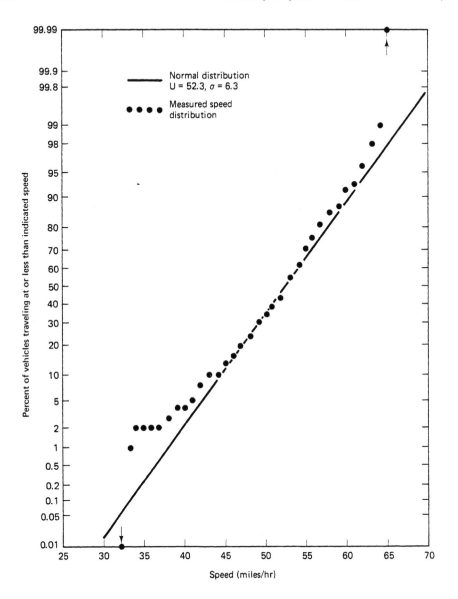

Figure 4.8   Measure Speed Distribution on Normal Probability Paper

$$I = \frac{32}{1 + (3.322) \log 200}$$

$$= \frac{32}{1 + 7.64} = 3.7$$

Considering this estimate of size of the class interval and the desire to have at least 10 class intervals, a class interval of 3 miles per hour is selected. Before performing the chi-square test, it is necessary to calculate the theoretical frequencies for the normal distribution having a mean of 52.3 miles per hour and a standard deviation of 6.3 miles per hour. This assumes that the population mean and standard deviation are numerically equal to the sample mean and standard deviation. The normal distribution calculations are shown in Table 4.7. The class interval limits are selected based on a class interval size of 3 miles per hour. The deviation (X) from each class interval limit to the mean is calculated and shown in the second column. The deviation is "normalized" by dividing it by the standard deviation. The probabilities of occurrence (P) between each class interval limit speed and the mean speed is obtained from Table 4.6 and entered in the fourth column. The theoretical probabilities for each class interval is then calculated. Finally, the theoretical frequency is determined by multiplying the theoretical probabilities by the total frequency $(N = 200)$.

TABLE 4.7   Normal Distribution Calculations

Class Interval Limit	X	$\dfrac{X}{\sigma}$	P	P_t	f_t
				0.0038	0.76
< 35.5	16.8	2.67	0.4962		
				0.0105	2.10
38.5	13.8	2.19	0.4857		
				0.0293	5.86
41.5	10.8	1.71	0.4564		
				0.0639	12.78
44.5	7.8	1.24	0.3925		
				0.1161	23.22
47.5	4.8	0.76	0.2764		
				0.1623	32.46
50.5	1.8	0.29	0.1141		
				0.1894	37.88
53.5	1.2	0.19	0.0753		
				0.1733	34.66
56.5	4.2	0.67	0.2486		
				0.1243	24.86
59.5	7.2	1.14	0.3729		
				0.0745	14.90
> 62.5	10.2	1.62	0.4474		
				0.0526	10.52

Table 4.8 summarizes the chi-square test calculations. The observed frequencies are taken from Table 4.5 and the theoretical frequencies are taken from Table 4.7. The resulting calculated value of chi-square is found to be 4.09. The degrees of freedom is equal to the number of class intervals minus three degrees of freedom. The three degrees of freedom lost are due to the normal distribution assuming the same mean,

TABLE 4.8   Chi-Square Test Calculations

Class Interval	f_o	f_t	$f_o - f_t$	$(f_o - f_t)^2$	$\dfrac{(f_o - f_t)^2}{f_t}$
< 35.5	4	0.76			
35.5–38.5	2	2.10	+1.28	1.64	0.19
38.5–41.5	4	5.86			
41.5–44.5	10	12.78	−2.78	7.73	0.60
44.5–47.5	19	23.22	−4.22	17.81	0.77
47.5–50.5	31	32.46	−1.46	2.13	0.07
50.5–53.5	41	37.88	+3.12	9.73	0.26
53.5–56.5	40	34.66	+5.34	28.52	0.82
56.5–59.5	23	24.86	−1.86	3.46	0.14
59.5–62.5	18	14.90	+3.10	9.61	0.64
> 62.5	8	10.52	−2.52	6.35	0.60
	200	200			$\chi^2_{CALC} = 4.09$

standard deviation, and frequency of the measured speed distribution. Note that it was necessary to combine the first three class intervals to obtain a minimum theoretical frequency of 5 or more. The resulting degrees of freedom is 6, and selecting an α value of 0.05, the table value of chi-square was found to be 12.59 from Appendix G. Since the calculated value is less than the table value of chi-square, the hypothesis is accepted and the concluding statement would be "There is no evidence of a statistical difference between the two distributions and the measured speed distribution could be identical to the normal distribution."

The measured speed distribution provided numerical checks, graphical plot, and chi-square test results which support the use of the normal distribution. If on the other hand, contrary results were found, attention would be directed to the log-normal and/or composite distributions. If there was no strong evidence of a bimodal distribution by inspection of a frequency distribution, as shown in the top portion of Figure 4.5, the log-normal distribution would be considered. The first step would be to plot the cumulative distribution of the speed measurements on logarithm-probability paper, as shown in Figure 4.9. The previously analyzed normal distribution is shown as a solid line, and the measured speed distribution is shown as a series of data points. A log-normal distribution would appear as a straight line in the figure. Therefore by inspection, the measured speed distribution is closer to a normal distribution than any log-normal distribution.

If, on the other hand, a measured speed distribution appears as a sloping straight line on Figure 4.9, the log-normal distribution would appear promising. For more quantitative evaluation, the theoretical frequency of the log-normal distribution could be determined and a chi-square test performed.

The composite distribution could be considered but usually after both the normal and log-normal distributions are rejected. Again the key characteristic for the composite

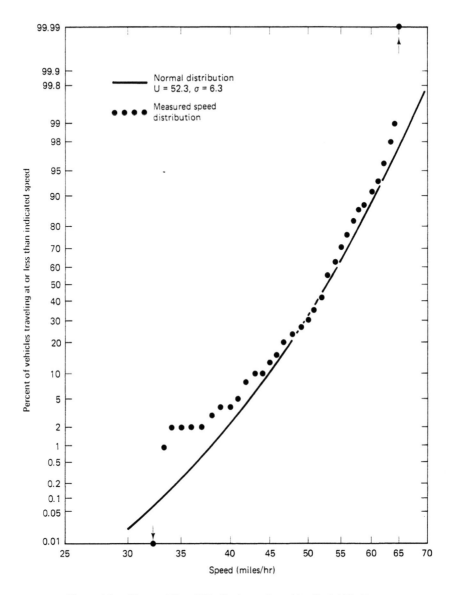

Figure 4.9   Measured Speed Distribution on Logarithm Probability Paper

distribution is a bimodal-appearing distribution. Two approaches may be considered. The speed measurements and field site could be restudied to attempt to identify the two subpopulations: that is, driver types, vehicle types, lane usage, and so on. If subpopulations can be identified, a second speed study should be undertaken and individual speeds of vehicles in the two subpopulations measured separately. Then the normal or log-normal distributions could be considered for each subpopulation. In simple, clear cases it might be possible to develop a composite distribution based on the original set of speed measurements if the mean, standard deviation, and frequency of each subpopulation can be determined.

4.5 ESTIMATION OF POPULATION MEANS AND SAMPLE SIZES

The best single estimate of the population mean is the sample mean. However, if several samples were taken under similar conditions and the sample means computed, there would be some numerical differences between sample mean values and hence differences in estimating the population mean. As more and more samples were taken and sample means computed, a distribution of sample means about the population mean would emerge in a fashion similar to a distribution of individual speeds about its sample mean. Of course, the distribution of sample means would be much more compact than the distribution of individual speeds, and standard statistics references [4, 5, 21] would show that the dispersion measure called the "standard error of the mean" would be

$$s_{\bar{x}} = \frac{s}{\sqrt{N}} \tag{4.22}$$

where $s_{\bar{x}}$ = standard error of the mean (miles per hour)
     s = standard deviation of the sample of individual speeds
    N = number of individual speeds observed

Further, the distribution of sample means would be normally distributed about the population mean even if the distribution of individual speed measurements were not normally distributed. It is significant to observe from equation (4.22) that only one sample of observations is required in order to calculate the standard error of the mean ($s_{\bar{x}}$). Using equation (4.22) and the speed observations presented in Table 4.5 and Figure 4.5, the standard error of the mean is

$$s_{\bar{x}} = \frac{s}{\sqrt{N}} = \frac{6.3}{\sqrt{200}} = 0.45 \text{ miles per hour}$$

Assuming the population mean equal to the sample mean (52.3 miles per hour) and using the calculated value of the standard error of the mean, a distribution of sample means can be calculated and the results are shown in Figure 4.10. Figure 4.10 is plotted in a similar fashion as Figure 4.5a except that the vertical scale is the probability density function of sample mean (μ's) rather than individual speeds (μ_i's). As mentioned before, the best single estimate of the population mean is the sample mean (52.3 miles per hour). However now with Figure 4.10, probability statements can be made about the population mean such as shown at the bottom of Figure 4.10. For example, with a probability of 0.9973, the population mean is expected to lie between 50.95 and 53.65 miles per hour. Other probability statements can be made as indicated on the figure.

    This concept can be carried further in determining sample size requirements for speed studies. In speed studies a set of observations are made, the sample mean is calculated, and the population mean which is the real objective is inferred. Although larger samples may result in better estimates of the population mean, larger samples

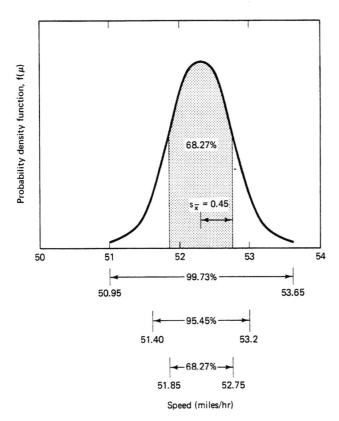

Figure 4.10   Distribution of Sample Means

require more time and effort in collection and analysis. Consequently, it is desirable to develop a technique that will provide the analyst with a means of selecting the smallest sample size possible while providing a limit on a prespecified probability that the population mean will be within a specified allowable error.

Rearranging equation (4.22) in order to solve for the required sample size (n), the equation becomes

$$n = \left(\frac{s}{s_{\bar{x}}}\right)^2 \tag{4.23}$$

Equation (4.23) has four variables, three that are clearly specified and one that is implied. The required sample size that is to be determined is represented by n. The standard deviation is a measure of dispersion of individually observed speeds that is obtained from a speed study or is known historically. The standard error of the mean, is a measure of the dispersion of sample means about the population mean, but in addition, probabilities are connected with different levels of the standard error of the mean. For example, in Figure 4.10, the population mean is expected to lie between 51.85 and 52.75 ($\pm 1s_{\bar{x}}$) with a probability of 0.6827. Or putting it another way, the error in the population mean is expected to be within ± 0.45 miles per hour ($\pm 1s_{\bar{x}}$) with a probability of 0.6827. In most speed studies, probabilities or confidence levels closer to 0.95 or

0.99 are selected. In these cases instead of using $\pm 1s_{\bar{x}}$, a coefficient on the order of 2 or 3 is used for the standard error of the mean. More specifically, if a probability level of 0.95 is selected, the coefficient can be obtained from Table 4.6. Since the distribution is symmetrical, a value of 0.4750 (one-half of 0.95) is entered and the coefficient is found to be 1.96. Equation (4.20) is now modified in two ways: the coefficient t is entered on the right side of the equation, and ε, the allowable error, is substituted for $ts_{\bar{x}}$. Equation (4.23) becomes

$$n = \left(\frac{ts}{ts_{\bar{x}}}\right)^2 = \left(\frac{ts}{\varepsilon}\right)^2 \tag{4.24}$$

where n = required sample size
$\quad s$ = standard deviation
$\quad \varepsilon$ = user-specified allowable error
$\quad t$ = coefficient of the standard error of the mean that represents user specified probability level

Figure 4.11 contains nomographs that can be used for the determination of sample size requirements based on equation (4.24). Figure 4.11a is for a 0.95 probability level, Figure 4.11b is for a 0.99 probability level. The analyst can enter the desired nomograph with the expected standard deviation value and the user-selected allowable error, and the minimum sample size can be determined. For example, the analyst may want to estimate the population mean within ±1 mile per hour with 99 percent confidence, and based on previous speed studies the standard deviation is expected to be 4 miles per hour. Figure 4.11b would indicate a minimum sample size of 100 observations. Another example would be the situation where the standard deviation is unknown and a pilot speed study is undertaken to estimate the standard deviation. The analyst then wishes to check to see if the pilot speed study sample is adequate or if a further speed study is required. Consider the example given in Table 4.5, in which 200 speeds are observed and the standard deviation is found to be 6.3 miles per hour. Assuming that the population mean is to be estimated within ±1 mile per hour with 95 percent confidence, Figure 4.11a would indicate a minimum sample size of approximately 152 observations. Since the number of observations in the original sample ($n = 200$) was larger than the determined minimum sample size ($N = 152$), no further observations are required.

4.6 SELECTED PROBLEMS

1. Undertake a library study of microscopic simulation models that require individual vehicular speeds to be generated within the model. What type(s) of mathematical distributions are employed?

2. Estimate the lost time due to a vehicle stopping (but with stopped delay being equal to zero) assuming cruise or running speeds upstream and downstream of the stop location of 30, 40, 50, and 60 miles per hour. Plot on graph similar to Figure 4.1.

3. There is considerable concern about traffic safety on high-speed approaches to rural signalized intersections. That is, when the signal changes to amber, the approaching vehicle may

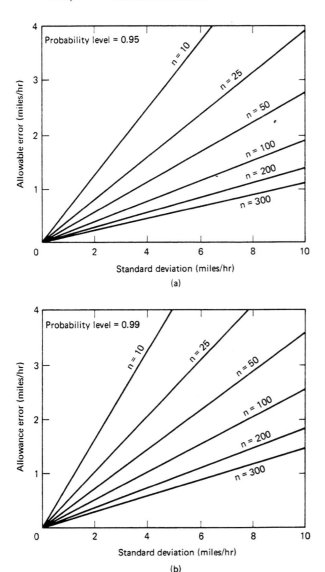

Figure 4.11   Sample Size Requirement Nomographs

be too far away from the stop line to accelerate and enter the intersection before the ending of the amber phase but too close to the intersection to decelerate and stop at the stop line. Consider the case of passenger vehicles in level terrain under wet pavement conditions. Assume approach speeds of 30, 40, 50, and 60 miles per hour and use maximum acceleration and deceleration rates. Determine the minimum amber phase to eliminate the so-called "dilemma" zone. (*Hint:* Plot a diagram in which the vertical scale is speed and the horizontal scale is distance to the stop line. Plot one curve for the minimum stopping distance and another for the maximum distance that can be traversed to the stop line as a function of approach speed and the length of the amber phase.) (See Reference 30 for more details.)

4. Some states have regulations that the approaching vehicle must clear the intersection before the end of the amber phase rather than to just enter the intersection. Solve Problem 3 assuming this regulation and an intersection width of 60 feet.

5. A truck approaches the foot of a 2-mile upgrade at a speed of 50 miles per hour and decelerates on the upgrade at a rate of 2 miles per hour/second until a crawl speed of 30 miles per hour is reached. Halfway up the grade a 2000 foot auxiliary lane is added, which the truck uses. A passenger vehicle travelling at 50 miles per hour is 1000 feet behind the truck at the foot of the grade but has the performance capability of maintaining 50 miles per hour if uninhibited by other vehicles. The passenger vehicle can pass the truck only in the auxiliary lane section, due to sight distance requirements. Plot the trajectories of the two vehicles and estimate the loss time to the passenger vehicle and the truck. Comment on the length of the auxiliary lane. Could it be shortened or should it be lengthened? Assume a minimum headway of 100 feet between vehicles at all times.

6. Plot a speed trajectory of a typical truck for the example shown in Figure 4.4 but modify the lengths of all subsections to $\frac{1}{2}$ mile.

7. The vertical and horizontal alignment of a new highway is to be selected to minimize the travel time for a typical truck to travel from a valley to a summit whose difference in elevation is 800 feet. Single or composite grades may be employed ranging from 0 percent grade to +8 percent upgrade.

8. Using the grouped speed data shown in Table 4.9, calculate various measures of central tendency and dispersion, and plot speed distributions similar to Figure 4.5. Qualitatively compare these results with the results obtained from the example problem presented in Section 4.2.

9. Demonstrate mathematically that equation (4.5) is identical to equation (4.8).

10. Conduct a field study at a location that exhibits uninterrupted flow conditions. Measure the individual speeds of 100 vehicles under fairly uniform low-flow conditions. Calculate various measures of central tendency and dispersion, and plot speed distributions similar to Figure 4.5.

11. Calculate the probability density function for a normal distribution that has parameter values identical to the example problem shown in Table 4.5. Plot diagrams similar to Figure 4.7.

12. Using a normal distribution that has parameter values identical to the example problem shown in Table 4.5, calculate the following probabilities and associated frequencies: $P(\mu_i < 30)$; $P(\mu_i > 55)$, $P(50 < \mu_i < 55)$, and $P(60 < \mu_i < 65)$.

13. Develop a normal distribution based on the parameter values of the example problem shown in Table 4.9. Compare various measures of central tendency and dispersion of the measured and normal distribution.

14. Undertake a library study of mathematical distributions used to represent measured speed distributions other than the normal distribution. Attempt to formulate guidelines as to their appropriate use.

15. Select a mathematical distribution for the example problem shown in Table 4.9 by following the procedures described in Section 4.4. Limit investigation to the normal distribution.

16. Extend the investigation described in Problem 15 to include other distributions in addition to the normal distribution. Include a chi-square test of the log-normal distribution.

17. Analyze the example problem shown in Table 4.9 to make statistical statements about its population mean.

TABLE 4.9 Grouped Speed Data for Problem 8

μ_i	f_i	Cumulative Frequency	%	$f_i\mu_i$	$f_i(\mu_i)^2$
30	—	0	0	—	—
31	—	0	0	—	—
32	1	1	1	32	1,024
33	—	1	1	—	—
34	—	1	1	—	—
35	—	1	1	—	—
36	1	2	1	36	1,296
37	—	2	1	—	—
38	—	2	1	—	—
39	—	2	1	—	—
40	3	5	3	120	4,800
41	2	7	4	82	3,362
42	4	11	6	168	7,056
43	6	17	9	258	11,094
44	4	21	11	176	7,744
45	5	26	14	225	10,125
46	8	34	18	368	16,928
47	6	40	21	282	13,254
48	18	58	31	864	41,472
49	9	67	36	441	21,609
50	18	85	45	900	45,000
51	8	93	50	408	20,808
52	16	109	58	832	43,264
53	10	119	64	530	28,090
54	13	132	71	702	37,908
55	11	143	76	605	33,275
56	8	151	81	448	25,088
57	13	164	88	741	42,237
58	6	170	91	348	20,184
59	4	174	93	236	13,924
60	3	177	95	180	10,800
61	3	180	96	183	11,163
62	3	183	98	186	11,532
63	1	184	98	63	3,969
64	1	185	99	64	4,096
65	—	185	99	—	—
66	1	186	99	66	4,356
67	1	187	100	67	4,489
68	—	187	100	—	—
69	—	187	100	—	—
70	—	187	100	—	—
	187			9,611	499,947

18. Is the sample size of the example problem shown in Table 4.9 adequate assuming that the population mean is to be estimated within ±1 mile per hour with 95 percent confidence?

4.7 SELECTED REFERENCES

1. Daniel G. Gerlough and Matthew J. Huber, *Traffic Flow Theory—A Monograph*, Transportation Research Board, Special Report 165, TRB, Washington D.C., 1975, pages 37 and 205–206.

2. F. A. Haight and W. W. Mosher, Jr., *A Practical Method for Improving the Accuracy of Vehicular Speed Distribution Measurements*, Highway Research Board Bulletin 341, HRB, Washington D.C., 1962, pages 92–116.

3. J. Aitchison and J. A. C. Brown, *The Lognormal Distribution*, Cambridge University Press, Cambridge, 1963.

4. A. Hald, *Statistical Theory With Engineering Applications*, John Wiley & Sons, Inc., New York, 1952.

5. M. G. Kendall and A. Stuart, *The Advanced Theory of Statistics*, Volume 1, Hafner Publishing Co., Inc., New York, 1958, pages 168–173.

6. H. J. W. Leong, The Distribution and Trend of Free Speeds on Two-Lane Rural Highways in New South Wales, *Australian Road Research Board Proceedings,* Vol. 4, 1968, pages 791–814.

7. R. L. Bleyl, Skewness of Spot Speed Studies, *Traffic Engineering*, Vol. 40, No. 10, 1970, pages 36–37 and 40.

8. D. S. Berry and D. M. Belmont, Distribution of Vehicle Speeds and Travel Times, *Second Berkeley Symposium on Mathematical Statistics and Probability Proceedings,* 1951, pages 589–602.

9. W. C. Taylor, Speed Zonings—A Theory and Its Proof, *Institute of Traffic Engineering Proceedings,* Vol. 34, 1964, pages 11–23.

10. N. C. Duncan, *A Method of Estimating the Distribution of Speeds of Cars on a Motorway,* Transport and Road Research Labortory, Report 598, TRRL, Crowthorne, England, 1973.

11. M. Mori, H. Takata, and T. Kisi, Fundamental Considerations on Speed Distributions of Road Traffic Flow, *Transportation Research*, Vol. 2, No. 1, 1968, pages 31–39.

12. J. C. Oppenlander, W. F. Bunte, and P. L. Kadakia, *Sample Size Requirements for Vehicular Speed Studies*, Highway Research Board, Bulletin 281, HRB, Washington D.C., 1961, pages 68–86.

13. R. P. Schumate and R. F. Crowthers, *Variability of Fixed-point Speed Measurements*, Highway Research Board, Bulletin 281, HRB, Washington D.C., 1961, pages 87–96.

14. Martin Wohl and Brian V. Martin, *Traffic System Analysis for Engineers and Planners,* McGraw-Hill Book Company, New York, 1967, pages 46–51.

15. Daniel L. Gerlough and Matthew J. Huber, *Statistics with Applications to Highway Traffic Analyses,* Eno Foundation, Saugatuck, Conn., 1978.

16. J. J. Leeming, *Statistical Methods for Engineers*, Blackie & Son Ltd, Glasgow, 1963.

17. Wolfgang S. Homburger and James H. Kell, *Fundamentals of Traffic Engineering,* Institute of Transportation Studies, Berkeley, Calif., 1988, pages 6–1 to 6–8.

18. Louis J. Pignataro, *Traffic Engineering—Theory and Practice,* Prentice-Hall, Inc., Englewood Cliffs, N. J., 1973, pages 116–142.

19. Donald R. Drew, *Traffic Flow Theory and Control,* McGraw-Hill Book Company, New York, 1968, pages 150–152.

20. Transportation Research Board, *Highway Capacity Manual,* Special Report 209, TRB Washington D.C., 1985, Chapter 2.

21. W. A. Neiswanger, *Elementary Statistical Methods,* Macmillan Publishing Co., Inc., New York, 1943, page 218.

22. Herbert A. Sturges, The Choice of a Class Interval, *Journal of the American Statistical Association,* March 1926, pages 65–66.

23. Institute of Transportation Engineers, *Transportation and Traffic Engineering Handbook,* 2nd Edition, Prentice-Hall, Inc., Englewood Cliffs, N., J., 1982, 883 pages.

24. V. Chrissikopoulos, J. Darzentas and M. R. C. McDowell, Deceleration Major-Road Vehicles Approaching a T-Junction, *Traffic Engineering and Control,* Vol. 24, No. 9, September 1983, pages 433–436.

25. D. Branston, Method of Estimating the Free Speed Distribution for a Road, *Transportation Science,* Vol. 13, No. 2, May 1979, pages 130–145.

26. J. R. McLean, Observed Speed Distributions and Rural Road Traffic Operations, *Australian Road Research Board Conference Proceeding,* Vol. 9, No. 5, 1979, pages 235–244.

27. J. Oppenlander, *Sample Size Determination for Spot Speed Studies at Rural, Intermediate, and Urban Locations,* Highway Research Board, Record 35, HRB, Washington D.C., 1962, pages 78–80.

28. *The Amount of Vehicle Operation over 50 mph,* Michigan Environmental Protection Agency, Ann Arbor, Michigan, July 1978, 23 pages.

29. American Association of State Highways and Transportation Officials, *A Policy on Geometric Design of Highways and Streets,* AASHTO, Washington D.C., 1984, 1087 pages.

30. Adolf D. May, *Clearance Interval at Signalized Intersections,* Highway Research Board, Record 221, HRB, Washington D.C., 1968, pages 41–71.

5

Macroscopic Speed Characteristics*

Macroscopic speed characteristics are those speed characteristics of vehicle groups passing a point or short segment during a specified period of time or traveling over longer sections of highway. Individual vehicle speed trajectories and individual vehicle speeds at a point or over a short segment were covered in Chapter 4 and serve as a beginning point for this chapter.

Speed and travel time are fundamental measurements of the traffic performance of the existing highway system, and speed is a key variable in the redesign or design of new facilities. Most analytical and simulation models of traffic predict speed and travel time as the measure of performance given the design, demand, and control along the highway system. More extensive models then use speed or travel time as an input for the estimation of fuel consumption, vehicle emissions, and traffic noise. Speed is also used as an indication of level of service, in accident analysis, in economic studies, and in most traffic engineering studies. Thus the traffic analyst must be familiar with microscopic and macroscopic speed characteristics, and associated statistical analysis techniques.

This chapter contains four major sections plus selected problems and references. The first section is concerned with speed and travel time variations over time, space, and between modes. The next section is directed toward mean–variance relationships, with attention given to coefficient of variation, time-mean-speed and space-mean-speed, and testing of significant differences. The third section is concerned with travel time characteristics, with emphasis on travel time study techniques and intersection delay

*If limited time is available, Section 5.4 can be scanned or omitted.

studies. The final section describes a comprehensive study of speeds, travel times, and delays in the operational evaluation of an arterial traffic signal system.

5.1 SPEED AND TRAVEL TIME VARIATIONS

Speeds and travel times vary over time, space, and between modes. Temporal variations may be due to changing traffic flows, mix of vehicle types and driver groups, lighting, weather, and traffic incidents. Spatial variations may be transversal (between lanes and/or directions) and/or longitudinal, due to differences in traffic flow, geometric design, and/or traffic control. Modal variations may be due to differences in driver desires combined with vehicle performance capabilities. Each of these variations will now be presented.

5.1.1 Temporal Variations

Speeds vary over time at specific locations in the highway system. Many factors influence these speed variations, and numerous field studies have been undertaken to attempt to explain and quantify temporal variations. The primary influencing factor is traffic flow intensity expressed as a flow rate or as a flow–capacity ratio. Secondary influencing factors include vehicle mix, driver mix, lighting, weather, and traffic incidents. Of course, design speed and speed limits influence the upper limits on speed.

The 1985 *Highway Capacity Manual* [8] provides procedures for predicting speed as a function of flow, vehicle mix, and driver mix for specific locations along freeways, multilane highways, and two-lane highways. Selected speed–flow relationships are shown in Figure 5.1. Figure 5.1a depicts speed–flow relationships for two-lane directional freeways, two-lane directional highways, and two-lane two-way highways under ideal conditions. Figure 5.1b displays speed–flow relationships for two-lane directional freeways with varying mixes of vehicle types and driver population.

Note that the capacity of two-lane two-way highways under ideal conditions is only 2800 vehicles per hour, while the capacity of two-lane directional freeways and highways under ideal conditions is 4000 vehicles per hour. In Figure 5.1b note that introducing trucks and noncommuter drivers into freeway traffic not only reduces the capacity but also lowers the upper limb of the speed–flow curve.

Design speed and speed limits influence the upper limits on traffic speeds. These effects are depicted in Figure 5.2. Figure 5.2a displays the effect of design speed on the speed–flow relationships for two-lane directional freeways. These relationships are based on procedures contained in the 1985 *Highway Capacity Manual* [8] and are representative for near-ideal conditions. Figure 5.2b illustrates the effect of speed limits on the speed–flow relationships for two-lane directional freeways having design speeds of 70 miles per hour. These relationships assume 50 percentile driver compliance and speed limits of 55, 60, and 65 miles per hour.

Note that lowering the design speed lowers the entire upper limb of the speed–flow curve while it only reduces the capacity slightly when the design speed is lowered to 50 miles per hour. On the other hand, reducing the speed limit only lowers

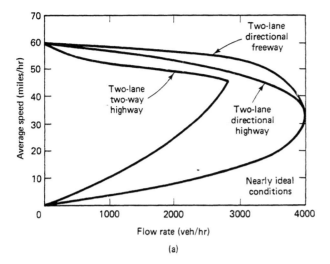

(a)

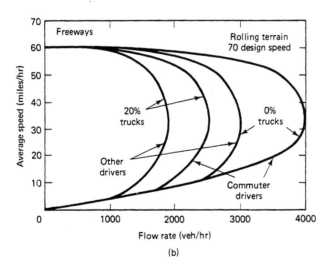

(b)

Figure 5.1   Effect of Flow Level, Vehicle Mix and Driver Population on Traffic Speeds (From Reference 8)

a portion of the upper limb of the speed–flow curve and has no effect on the capacity (unless speed limits less than approximately 30 miles per hour are imposed).

In Michigan, a study was undertaken at seven locations, representing different types of urban arterials and freeways for the purpose of studying the variation of traffic speeds by time of day [19, 27, 36]. The speed of every vehicle passing each location in a given direction for a 7 day period was measured and 15-minute average speeds calculated. The results for four locations (two freeway and two major arterial sites) for the period 6 A.M. to 6 P.M. are shown in Figure 5.3. The upper two curves represent the two freeways, and the lower two curves represent the two major arterials. One of the freeways is congested during the morning peak period, while one of the major arterials is congested during the afternoon peak period. The 24-hour average speed for each of the four sites is indicated to the right on the figure. It is of particular interest to note that

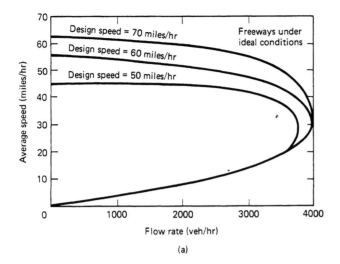

(a)

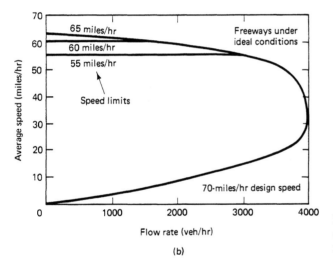

(b)

Figure 5.2   Influence of Design Speeds and Speed Limits on Traffic Speeds (From reference 8)

any 15-minute average speed for any site from 11 A.M. to 3 P.M. is within ±1 mile per hour of its 24-hour average speed.

Speed studies have also been undertaken on an annual basis at selected locations to determine trends in average speeds and speed distributions [12]. For example, Figure 5.4 depicts speed trends on rural interstate highways. Figure 5.4a shows average speed trends from 1965 to 1984. As can be seen, there are three apparent segments: 1965 to 1973 shows a steady increase from 60 miles per hour to 65 miles per hour; 1973 to 1974 shows an abrupt decrease (due to introduction of the 55-miles per hour speed limit) from 65 miles per hour to 57 miles per hour; and 1974 to 1984 again shows a steady but small increase from 57 miles per hour to 59 miles per hour. Figure 5.4b depicts percentage of vehicles exceeding 55-, 60-, and 65-miles per hour trends from 1965 to 1984 and their trends are quite similar to average speed trends.

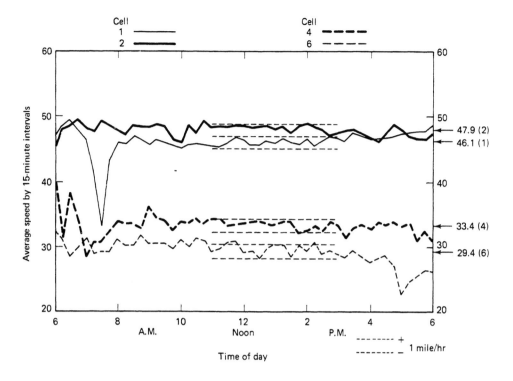

Figure 5.3   Variation of Average Speed by Time of Day (From Reference 19)

5.1.2 Spatial Variations

Traffic speeds may vary transversely across the highway between lanes and direction of travel, and longitudinally along the highway or street. Although such variations have not been precisely modeled, sufficient field studies have been undertaken to provide empirical evidence of such variations. These variations and results from selected field studies will now be presented.

Drivers on multilane directional freeways, highways, and streets select specific lanes for travel because of driver desires, vehicle performance capabilities, origin–destination of trips, traffic control measures, and downstream traffic conditions. For example, drivers may desire to travel in the rightmost lane if they are unsure of their route, wish to travel at a lower speed, or want to reduce lane changing. Vehicles with lower performance capabilities generally use the rightmost lane particularly on upgrades and in unsteady flow conditions. Origins and destinations of individual trips also influence which lanes vehicles will enter and leave the system and whether the trip along the route is long enough to warrant lane changing. Traffic control measures such as exclusive turn lanes and lane use regulations (i.e., "slow vehicles keep right," "trucks use two right lanes," etc.) influence lane usage. Finally, downstream traffic conditions either seen or experienced on previous trips may influence lane use. All of these factors

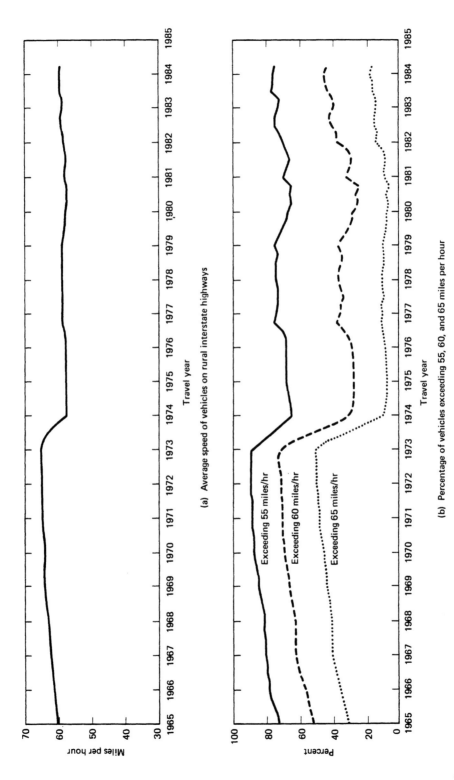

(a) Average speed of vehicles on rural interstate highways

(b) Percentage of vehicles exceeding 55, 60, and 65 miles per hour

Figure 5.4   Speed Trends on Rural Interstate Highways (From Reference 12)

work collectively to stratify the traffic flow by lane with the result that flow and speed vary between lanes. Figure 5.5 provides a comparison between lane speeds at six locations along the outbound Hollywood Freeway in Los Angeles, California. When free-flow conditions exist, the lane speeds vary considerably. For example, when average freeway speeds are over 40 miles per hour, the median lane speeds are on the order of 0 to 8 miles per hour higher, the middle lane is slightly higher, and the shoulder lane speeds are 2 to 5 miles per hour lower than the average freeway speed. However, under congested flow conditions the lane speeds are approximately equal.

Traffic speeds may vary between directions of travel at a location at which the directions of travel are separated or undivided. With separated or divided roadways, the speeds may be different because of differences in flows, designs, or controls. In addition, if the drivers in one direction can see the traffic conditions (accident, incident, etc.) in the other direction, the directional speeds may be influenced. On two-lane two-way highways not only the foregoing situations may exist, but also the traffic flow in one direction reduces the opportunity for passing in the other. Reduced passing opportunities increases the proportion of vehicles that are restricted by the vehicles ahead, and, of course, this reduces traffic speeds.

Traffic speeds may also vary longitudinally along the highway or street due to changes in traffic demands and/or geometric design as well as the introduction of traffic control measures. Two situations can be considered: uninterrupted and interrupted traffic situations. For the uninterrupted-flow situation, consider a single direction of a two-lane two-way rural highway that encounters a fairly steep and continuous upgrade followed by a downgrade. A profile of the route is shown in the upper portion of Figure 5.6 and immediately below is the speed profile and average subsection speeds of vehicles traveling along the route from left to right. These results are taken from an application of a macroscopic simulation model called RURAL2 [30]. The model is based on the 1985 *Highway Capacity Manual* [8] but includes subsection dependency factors with speed calculated on a directional basis.

For the interrupted-flow situation, consider a single direction of flow along a major arterial that encounters a number of signalized intersections. A profile of the route is shown in the upper portion of Figure 5.7, and immediately below is the speed profile of vehicles traveling along the route from left to right. Note that the horizontal scale is time rather than distance, which permits the display of stopped time, and the slopes of the speed profile depict acceleration (or deceleration) rates. Running time, stopped time, and total time are indicated for each link of the arterial.

5.1.3 Modal Variations

Most vehicle streams contain a mix of vehicle types that may have different speed characteristics and patterns. In level terrain and under uninterrupted flow situations, these speed differences are usually very small, on the order of 3 to 5 miles per hour. However, because of differing vehicle performance capabilities, different vehicle types travel at different speeds on long, sustained grades. Figure 5.8 is an attempt to depict graphically the effect of long, sustained grades on average speeds and the low $12^{1}/_{2}$

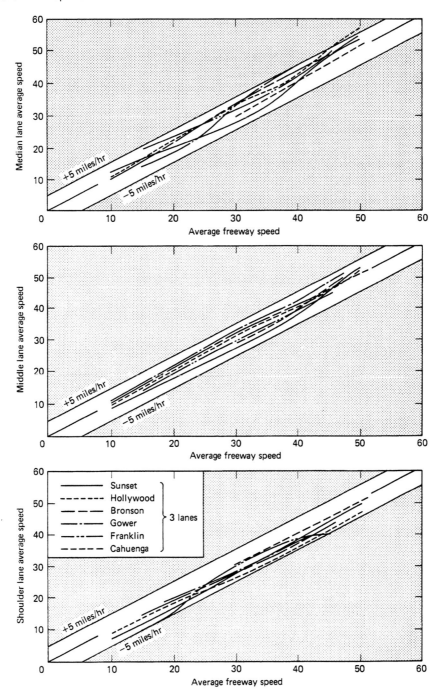

Figure 5.5   Comparison of Lane Speeds on Hollywood Freeway (From Reference 37)

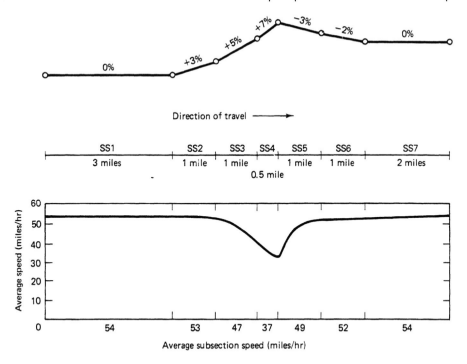

Figure 5.6   Average Speeds along an Uninterrupted Two-Lane Two-Way Rural Highway

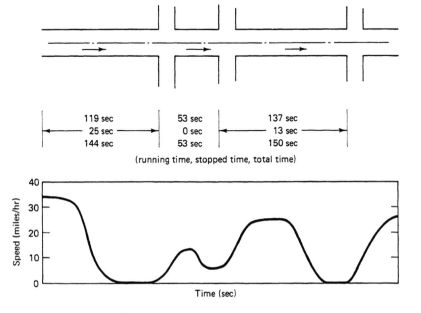

Figure 5.7   Average Speeds along a Signalized Arterial

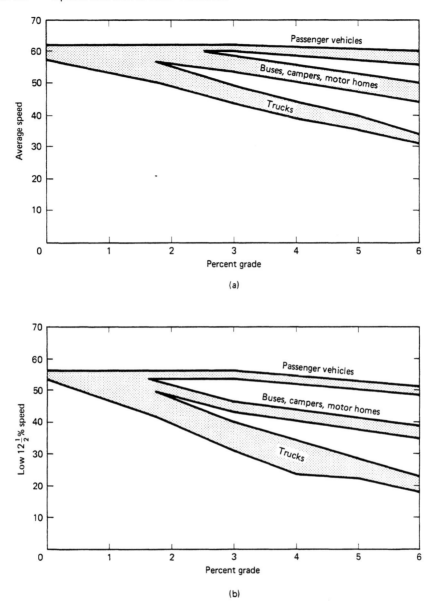

Figure 5.8   Speeds on Upgrades by Various Vehicle Types (From Reference 31)

percentile speeds. Three vehicle groups are identified: passenger vehicles; intermediate types of vehicles, such as buses, campers, and motor homes; and trucks, which are defined as freight-carrying vehicles with six or more tires touching the pavement. At grades flatter than 1 to 2 percent the drivers' characteristics and desires begin to overshadow the less limiting vehicle performance capabilities.

5.2 IMPORTANCE OF MEAN–VARIANCE RELATIONSHIPS

Measures of central tendency and dispersion for data sets of individual speed measurements were presented in Chapter 4. Further study of means, variances, and their relationships are now required as space-measured speeds as well as time-measured speeds are introduced and as means and variances of sample groups are compared and tested for significant differences. First, coefficient of variation calculations and empirical results will be presented followed by a discussion of time-mean-speed and space-mean-speed and their relationship. This section concludes with the presentation of techniques for testing of significant differences between sample means and sample variances.

5.2.1 Coefficient of Variation

The coefficient of variation of a distribution of individually measured speeds is defined as the standard deviation of the speed sample divided by the sample mean speed.

$$\text{C.V.} = \frac{s}{\bar{\mu}} \qquad (5.1)$$

where C.V. = coefficient of variation
  s = standard deviation of the speed sample (miles per hour)
  $\bar{\mu}$ = sample mean speed (miles per hour)

The coefficient of variation is a measure of dispersion and could be considered as a means of "normalizing" the standard deviation. For example, a standard deviation of 3 miles per hour might be considered small if the sample mean was 60 miles per hour (coefficient of variation = 0.05) but reasonably larger if the mean speed was 15 miles per hour (coefficient of variation = 0.20).

The coefficient of variation might range from approximately zero (all vehicles traveling at the same speed) to something on the order of the reciprocal of the square root of the mean speed (no interactions between vehicles and no vehicle performance limitations). However, empirical study results indicate that coefficients of variation are normally on the order of 0.08 to 0.17. These results are shown in Figure 5.9, where the standard deviation is plotted against the mean speed with coefficients of variations being represented by radial straight lines extending from the origin up and to the right. The portion of the curves at mean speeds below 20 miles per hour are shown as dashed lines since such low mean speeds are not obtained unless interrupted flow conditions are encountered. The coefficient-of-variation curves representing the empirical study results are generally less than the so-called "enclosure" curve because of design speed and speed limits, and because of interactions between vehicles. These interactions become more apparent as the mean speed decreases. Thus the analyst has a crude, first approach to estimating speed distributions by selecting an expected mean speed, using Figure 5.9 to select a value for the standard deviation, and assuming a normal distribution.

A very significant study undertaken in 1983 to 1984 of the 55-mile per hour speed limit included speed study results with particular attention given to average speeds and

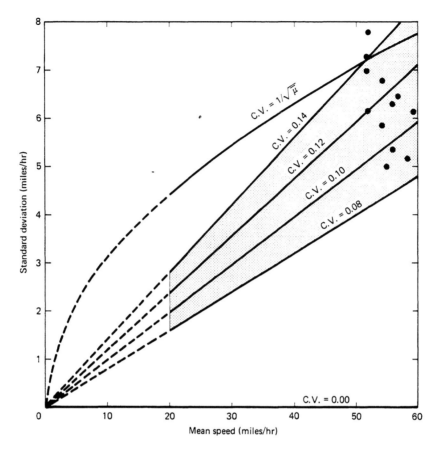

Figure 5.9   Standard Deviation–Mean Speed Relationships

speed variations [12]. Speed studies were for 1981 and 1982 for rural and urban portions of the highway system. The results obtained for 12 situations are superimposed on Figure 5.9. All sites were studied under free-flow conditions with average speeds between 51 and 59 miles per hour and standard deviations between 5.0 and 7.8 miles per hour.

In the same 55-mile per hour speed limit study, the annual trends from 1972 to 1983 in average speeds and standard deviations on interstate highways were analyzed, and results are shown in Figure 5.10. Since shortly after the establishment of the 55-mile per hour speed limit, the coefficient of variation has gradually been increasing from 0.08 to 0.10, with the urban values being slightly higher than the rural values.

5.2.2 Time-Mean-Speed and Space-Mean-Speed

In Chapter 4 and in the earlier portions of this chapter, the term "mean" speed or "average" speed has implied that a set of individual speed measurements have been obtained and the mean or average speed has been calculated using the following equation:

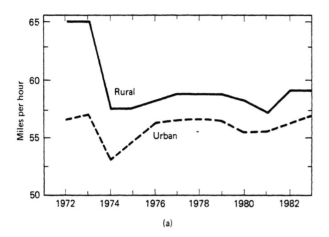

(a)

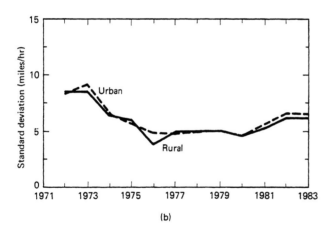

(b)

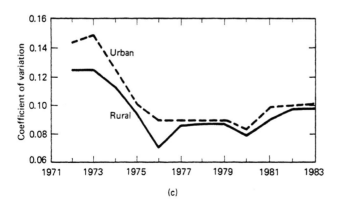

(c)

Figure 5.10   Annual Trends in Average Speeds, Standard Deviations, and Coefficient of Variations (From Reference 12)

$$\bar{\mu}_{TMS} = \frac{\sum\limits_{i=1}^{n} \mu_i}{n} \tag{5.2}$$

This mean or average speed should be referred to as a time-mean-speed since individual speeds were recorded for vehicles passing a particular point or short segment over a selected *time* period. If, however, the individual speeds were first converted to individual travel time rates, then an average travel time rate calculated, and finally an average speed calculated, the average speed would not be time-mean-speed but space-mean-speed.

$$\bar{\mu}_{SMS} = \frac{1}{\sum\limits_{i=1}^{n} t_i/n} = \frac{1}{1/n \sum\limits_{i=1}^{n} t_i} \tag{5.3}$$

where $\bar{\mu}_{SMS}$ = space-mean-speed (miles per hour)

t_i = travel time rate of vehicle i, in this equation expressed in hours per mile

n = number of observations in sample

For example, if three vehicles pass a point in a study period traveling at 30, 60, and 60 miles per hour, the time-mean-speed would be 50 miles per hour while the space-mean-speed would be

$$\bar{\mu}_{SMS} = \frac{1}{\sum\limits_{i=1}^{n} t_i/n} = \frac{1}{[0.0333 + 2(0.01667)]/3} = 45 \text{ miles per hour} \tag{5.4}$$

Another example of the difference between time-mean-speed and space-mean-speed is to place two vehicles (one traveling 30 miles per hour and the other traveling 60 miles per hour) on a 1-mile-long circular track. If an observer stood at a point over a period of *time* for 1 hour, and recorded the speed of each vehicle crossing the study point, the observer would record 60 vehicles traveling at 60 miles per hour and 30 vehicles traveling at 30 miles per hour and the time-mean-speed would be

$$\bar{\mu}_{TMS} = \frac{\sum\limits_{i}^{n} \mu_i}{n} = \frac{60 \cdot 60 + 30 \cdot 30}{90} = 50 \text{ miles per hour} \tag{5.5}$$

If, on the other hand, two consecutive aerial photographs were taken a few seconds apart of the circular track, two vehicle speeds would be recorded over *space*, and the space-mean-speed would be

$$\bar{\mu}_{SMS} = \frac{30 + 60}{2} = 45 \text{ miles per hour} \tag{5.6}$$

Wardrop in 1952 derived the equation relating time-mean-speed and space-mean-speed as shown in the following equation:

$$\bar{\mu}_{TMS} = \bar{\mu}_{SMS} + \frac{S^2_{SMS}}{\bar{\mu}_{SMS}} \tag{5.7}$$

In the circular track example, the space-mean-speed was found to be 45 miles per hour and the variance of the two sampled vehicles can be calculated to be 225. Substituting these values into equation (5.7), time-mean-speed can be found to be

$$\bar{\mu}_{TMS} = 45 + \frac{225}{45} = 50 \text{ miles per hour} \tag{5.8}$$

This checks the previous calculation for time-mean-speed.

It is important to note that space-mean-speed is equal to time-mean-speed only when the variance of the space-mean-speed is equal to zero, that is, only when all vehicles in the traffic stream all travel at exactly the same speed. If the variance of the space-mean-speed is greater than zero, the space-mean-speed is always *less* than the time-space-speed. Space-mean-speed can never be greater than the corresponding time-mean-speed.

The analyst may be interested in quantifying the numerical difference between time-mean-speed and space-mean-speed. The earlier section on coefficient of variation, and particularly Figure 5.10, can assist in this analysis. For example, if the mean speed in Figure 5.10 is assumed to be the space-mean-speed and the coefficient of variation for the space-mean-speed is assumed to be $1/\sqrt{\bar{\mu}_{SMS}}$, 0.14, 0.12, 0.10, and 0.08, the resulting $\bar{\mu}_{TMS}$ can be calculated using equation (5.7). The numerical and percentage difference can be calculated for each value of space-mean-speed and the results plotted as shown in Figure 5.11. Drake et al. conducted an investigation of the relationship between time-mean-speed and space-mean-speed based on an extensive set of speed measurements taken on the Eisenhower Expressway in Chicago, and their results are also shown in Figure 5.11. While a number of assumptions have been made in this analysis, the difference between time-mean-speed and space-mean-speed appears to be on the order of 1 mile per hour or 1 to 5 percent. Greater differences would occur when the coefficient of variation is large and the mean speed is small.

5.2.3 Testing Significant Differences between Means

It is unlikely that any two samples of speed measurements will have exactly the same mean even when both are taken from the same population. Hence some numerical differences between sample means may be due to chance, while in other situations the differences are significant. It is the purpose of this section to present an analytical procedure to distinguish between differences in sample mean values due to chance from those due to true statistical differences.

Two speed studies are taken under situations that are considered identical and the following results are obtained:

	Study 1	Study 2
Mean	30.8	32.0
Standard deviation	6.2	5.4
Sample size	100	200

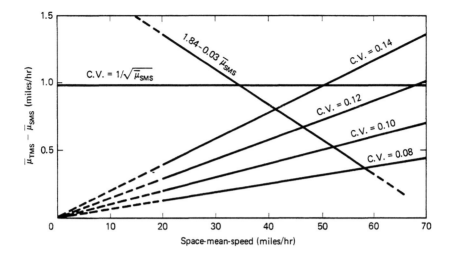

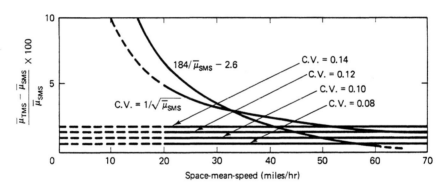

Figure 5.11   Estimated Numerical and Percentage Differences between Time-Mean-Speed (From Reference 38)

The initial assumption is that although the means are different (i.e., 30.8 and 32.0 miles per hour), the analyst assumes that the difference is due to chance since in the field study, the two situations were considered identical. The question is whether this assumption is in fact true (or at least statistically true) or if in fact the situations were not identical (i.e., schoolchildren were walking along the shoulder in study 1 and not in study 2) and the means are significantly different.

The key comparison is between the numerical differences in the means (1.2 miles per hour) and the variability of the speed measurements in the two studies, which is characterized as the standard deviation of the difference of the means and can be calculated as

$$\hat{s} = \sqrt{\frac{s_1^2}{n_1} + \frac{s_2^2}{n_2}} = \sqrt{(s_{\bar{\mu}})_1^2 + (s_{\bar{\mu}})_2^2} \tag{5.9}$$

where $\hat{s}$ = standard deviation of the difference of the means
            s_1, n_1 = standard deviation and number of observations in study 1
            s_2, n_2 = standard deviation and number of observations in study 2
      $(s_{\bar{\mu}})_1, (s_{\bar{\mu}})_2$ = standard error of the mean for study 1 and study 2

Substituting the numerical values into equation (5.9), the standard deviation of the difference of the means is found to be 0.73 miles per hour. Figure 5.12 is constructed as a normal distribution of $\bar{\mu}_1 - \bar{\mu}_2$ values with a standard deviation of $\hat{s}$. In this particular example, the difference between means of samples taken from the same population due to chance only would be expected to be within ±0.73 miles per hour, ±1.46 miles per hour, and ±2.19 miles per hour in 68.26, 95.46, and 99.73 percent of the cases. Thus a measured difference between means greater than ±1.46 miles per hour is highly suspect and is unlikely to be due to chance alone. Since the measured difference between means was actually only −1.2 miles per hour, and this is within ±1.46 miles per hour, the hypothesis is accepted and a statement can be made that there is no evidence that there is a significant difference between the two sample means.

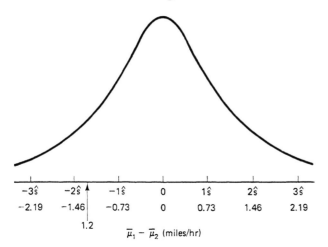

Figure 5.12   Distribution of Differences between Sample Means

5.2.4 Testing Significant Differences between Variances

In a fashion similar to differences between sample means, it is unlikely that any two samples of speed measurements will have exactly the same variance even when both are taken from the same population. Hence some numerical differences between sample variances may be due to chance while in other situations the differences are significant. It is the purpose of this section to present an analytical procedure to distinguish between differences in sample variance values due to chance from those due to true statistical differences.

The hypothesis to be tested is that there is no difference between two sample variances. If the hypothesis is accepted, the statement is made that there is no evidence that the sample variances are different. If the hypothesis is rejected, the statement is made that it is unlikely that the two sample variances are identical.

The F test is used to compare the ratio of the two sample variances with values taken from the F distribution. The F distribution is a distribution of sample variance ratios where the samples are independent and taken from the same population. If the ratio of the measured variance is less than the Table F value, the hypothesis is accepted, while in the reverse case, the hypothesis is rejected. The larger of the two measured variances is placed in the numerator of the ratio, so that the ratio is always equal to or greater than one which simplifies the testing procedure. The F distribution table is given in Table 5.1.

The two speed studies utilized in the preceding section will again be used in this example, and the results are summarized below.

	Study 1	Study 2
Mean	30.8	32.0
Standard deviation	6.2	5.4
Sample size	100	200

The calculated ratio of sample variances is found to be 1.318. The table F value is obtained from Table 5.1 and is found to be 1.32 at the 0.05 level. Since the calculated value is less than the F-table value, the hypothesis is accepted and the statement can be made that there is no evidence that the sample variances are statistically different. Note the effect of selecting the significance level on the outcome of this example.

5.3 TRAVEL TIME AND DELAY STUDY TECHNIQUES

Travel time and travel speed information is needed to identify and assess operational problems along highway facilities. Travel time information is also necessary in traffic signal control coordination, as input to traffic assignment algorithms, in economic studies, and in "before" and "after" studies.

In this section travel time studies and intersection delay studies are presented. The travel time study techniques include license plate studies and a variety of test vehicle studies. The intersection delay studies include elapsed-time studies and stopped vehicle studies.

5.3.1 Travel Time Study Techniques

The most direct way of obtaining the travel time for several vehicles between two points in the highway system is by recording the time of entry and the time of exit for individual vehicles traversing the study section. The labeling of individual vehicles is usually accomplished by recording the license plate number with the entry and exit times. The license plate numbers are then matched, and the elapsed travel times are determined assuming that the vehicles did not make an intermediate stop. Tape

TABLE 5.1 *F* Distribution

n_1	n_1 Degrees of Freedom (for Greater Mean Square)[a]											
n_2	1	2	3	4	5	6	7	8	9	10	11	12
1	161	200	216	225	230	234	237	230	241	242	243	244
	4,052	4999	5,403	5,625	5,764	5,859	5,928	5,981	6,022	6,056	6,082	6,106
2	18.51	29.00	19.16	19.25	19.30	19.33	19.36	19.37	19.38	19.30	19.40	19.41
	98.49	99.00	99.17	99.25	99.30	99.33	99.34	99.36	99.38	99.40	99.41	99.42
3	10.13	9.155	9.28	9.12	9.01	8.94	8.88	8.84	8.81	8.78	8.76	8.74
	34.12	30.82	29.46	28.71	28.24	27.91	27.67	27.49	27.34	27.23	27.13	27.05
4	7.71	6.94	6.50	6.30	6.26	6.16	6.00	6.04	6.00	5.96	5.93	5.91
	21.20	18.00	16.69	15.98	15.52	15.21	14.98	14.80	14.66	14.54	14.45	14.37
5	6.61	5.79	5.41	5.19	5.06	4.95	4.88	4.82	4.78	4.74	4.70	4.68
	16.26	13.26	12.06	11.39	10.97	10.67	10.45	10.27	10.15	10.05	9.96	9.89
6	5.00	5.14	4.76	4.53	4.39	4.28	4.21	4.15	4.10	4.06	4.03	4.00
	13.74	10.92	9.78	9.15	8.75	8.47	8.26	8.10	7.98	7.87	7.79	7.72
7	5.50	4.74	4.35	4.12	3.97	3.87	3.79	3.73	3.68	3.63	3.60	3.57
	12.25	9.55	8.45	7.85	7.46	7.19	7.00	6.84	6.71	6.62	6.54	6.47
8	5.32	4.46	4.07	3.84	3.60	3.58	3.50	3.44	3.39	3.34	3.31	3.28
	11.26	8.65	7.59	7.01	6.63	6.37	6.19	6.03	5.91	5.82	5.74	5.67
9	5.12	4.26	3.86	3.63	3.48	3.37	3.20	3.23	3.18	3.13	3.10	3.07
	10.56	8.02	6.99	6.42	6.06	5.80	5.62	5.47	5.35	5.26	5.18	5.11
10	4.96	4.10	3.71	3.48	3.33	3.22	3.14	3.07	3.02	2.97	2.94	2.91
	10.04	7.56	6.55	5.99	5.64	5.39	5.21	5.06	4.95	4.85	4.78	4.71
11	4.84	3.08	3.69	3.36	3.20	3.09	3.01	2.05	2.90	2.86	2.82	2.79
	9.65	7.20	6.22	5.67	5.32	5.07	4.88	4.74	4.63	4.54	4.46	4.40
12	4.75	3.68	3.49	3.26	3.11	3.00	2.92	2.85	2.80	2.76	2.72	2.69
	9.33	6.93	5.95	5.41	5.06	4.82	4.65	4.50	4.39	4.30	4.22	4.16
13	4.67	3.80	3.41	3.18	3.02	2.92	2.81	2.77	2.72	2.67	2.68	2.60
	9.07	6.70	5.74	5.20	4.86	4.62	4.44	4.30	4.19	4.10	4.02	3.96
14	4.60	3.74	3.34	3.11	2.98	2.86	2.77	2.70	2.65	2.60	2.56	2.53
	8.86	6.81	5.56	5.03	4.69	4.46	4.28	4.14	4.93	3.94	3.86	3.80
15	4.54	3.08	3.20	3.06	2.00	2.79	2.70	2.64	2.59	2.56	2.51	2.48
	8.68	6.36	5.42	4.89	4.56	5.32	4.14	4.00	3.89	3.80	3.73	3.67
16	4.40	3.83	3.21	3.01	2.85	2.74	2.66	2.50	2.54	2.40	2.45	2.42
	8.53	6.23	5.29	4.77	4.44	4.20	4.03	3.89	3.78	3.69	3.61	3.55
17	4.45	3.60	3.20	2.06	2.81	2.70	2.02	2.55	2.50	2.45	2.41	2.38
	8.40	6.11	5.18	4.67	4.34	4.10	3.93	3.79	3.68	3.59	3.52	3.45
18	4.41	3.55	3.16	2.03	2.77	2.66	2.58	2.51	2.46	2.41	2.37	2.34
	8.28	6.01	5.09	4.58	4.25	4.01	3.85	3.71	3.60	3.51	3.44	3.77
19	4.38	3.62	3.13	2.00	2.74	2.63	2.65	2.48	2.43	2.38	2.34	2.31
	8.18	5.93	5.01	4.50	4.17	3.94	3.77	3.63	3.52	3.43	3.36	3.30
20	4.35	3.40	3.10	2.87	2.71	2.60	2.52	2.45	2.40	2.35	2.31	2.28
	8.10	5.85	4.94	4.43	4.10	3.87	3.71	3.56	3.45	3.37	3.30	3.23
21	4.32	3.47	3.07	2.84	2.68	2.57	2.40	2.42	2.37	2.32	2.28	2.25
	8.02	5.78	4.87	4.37	4.04	3.81	3.65	3.51	3.40	3.31	3.24	3.17
22	4.30	3.44	3.05	2.82	2.06	2.55	2.47	2.40	2.36	2.30	2.26	2.23
	7.94	5.72	4.82	4.31	3.99	3.76	3.59	3.45	3.35	3.26	3.18	3.12
23	4.28	3.42	3.03	2.80	2.64	2.53	2.45	2.38	2.32	2.28	2.24	2.20
	7.88	5.66	4.76	4.26	3.94	3.71	3.54	3.41	3.30	3.21	3.14	3.07
24	4.26	3.40	3.01	2.78	2.62	2.51	2.43	2.36	2.30	2.26	2.22	2.18
	4.72	5.61	4.72	4.22	3.90	3.67	3.50	3.36	3.25	3.17	3.09	3.03
25	4.24	3.38	2.99	2.76	2.60	2.40	2.41	2.34	2.28	2.24	2.20	2.16
	7.77	5.57	4.68	4.18	3.86	3.63	3.46	3.32	3.21	3.13	3.05	2.99
26	4.22	3.37	2.08	2.74	2.50	2.47	2.30	2.32	2.27	2.22	2.18	2.15
	7.72	5.53	4.64	4.14	3.82	3.59	3.42	3.29	3.17	3.09	3.02	2.96

[a]The first value in each pair is for 5% points for distribution of F, the second is for 1%.

TABLE 5.1 (cont'd.) F Distribution

				n_1 Degrees of Freedom (for Greater Mean Square)[a]							
14	16	20	24	30	40	50	75	100	200	500	∞
245	246	248	249	250	251	252	153	253	254	254	254
6,142	6.269	6,208	6.234	6,258	6.286	6,302	6,323	6,334	6,352	6,361	6,366
19.42	19.43	19.44	19.45	19.46	19.47	19.47	19.48	19.49	19.49	19.50	19.50
99.43	99.44	99.45	99.46	99.47	99.48	99.48	99.49	99.49	99.49	99.50	99.50
8.71	8.69	8.00	8.04	8.62	8.50	8.58	8.57	8.56	8.54	8.54	8.53
26.92	26.83	26.69	26.60	26.50	26.41	26.35	26.27	26.28	26.18	26.14	26.12
5.87	5.84	5.80	5.77	5.74	5.71	5.70	5.08	5.66	5.65	5.64	5.63
14.24	14.15	14.02	13.93	13.83	13.74	13.69	13.61	13.57	13.52	13.48	13.46
4.64	4.60	4.50	4.53	.4.50	4.46	4.44	4.42	4.40	4.38	4.37	4.36
9.77	9.68	9.55	9.47	9.38	9.29	9.24	9.17	9.13	9.07	9.04	9.02
3.96	3.92	3.87	3.84	3.81	3.77	3.76	3.72	3.71	3.69	3.68	3.76
7.60	7.52	7.39	7.31	7.23	7.14	7.03	7.02	6.99	6.94	6.90	6.88
3.52	3.40	3.44	3.41	3.38	3.34	3.32	3.29	3.28	3.25	3.24	
6.35	6.27	6.15	6.07	5.98	5.90	5.85	5.78	5.75	5.70	5.67	5.65
3.23	3.20	3.15	3.12	3.08	3.05	3.03	3.00	2.98	2.98	2.94	2.93
5.56	5.48	5.36	5.28	5.20	5.11	5.06	5.00	4.96	4.91	4.88	4.86
3.02	2.98	2.93	2.90	2.88	2.82	2.80	2.77	2.76	2.73	2.72	2.71
5.00	4.97	4.80	4.73	4.64	4.56	4.51	4.45	4.41	4.36	4.33	4.31
2.86	2.82	2.77	2.74	2.70	2.67	2.64	2.61	2.59	2.56	2.55	2.54
4.60	4.52	4.41	4.33	4.25	4.17	4.12	4.05	4.01	3.96	3.93	3.91
2.74	2.70	2.65	2.61	2.57	2.53	2.50	2.47	2.45	2.42	2.41	2.40
4.29	4.21	4.10	4.02	3.94	3.86	3.80	3.74	3.70	3.66	3.62	3.60
3.64	2.60	2.54	2.50	2.46	2.42	2.40	2.36	2.35	2.32	2.31	2.30
4.05	3.98	3.86	3.78	3.70	3.61	3.56	3.49	3.46	3.41	3.38	3.36
2.55	2.51	2.46	2.42	2.38	2.34	2.32	2.28	2.26	2.24	2.22	2.21
3.85	3.78	3.67	3.59	3.51	3.42	3.37	3.30	3.27	3.21	3.18	3.16
2.48	2.44	2.39	2.35	2.31	2.27	2.24	2.21	2.19	2.16	2.14	2.13
3.70	3.62	3.51	3.43	3.34	3.26	3.21	3.14	3.11	3.06	3.02	3.00
2.43	2.30	2.33	2.20	2.25	2.21	2.18	2.15	2.12	2.10	2.08	2.07
3.56	3.48	3.36	3.29	3.20	3.12	3.07	3.00	2.97	2.92	2.89	2.87
2.37	2.33	2.28	2.24	2.20	2.16	2.13	2.09	2.07	2.04	2.02	2.01
3.45	3.37	3.25	3.18	3.10	3.01	2.96	2.89	2.86	2.80	2.77	2.75
2.33	2.29	2.23	2.10	2.15	2.10	2.08	2.04	2.02	1.90	1.97	1.95
3.35	3.27	3.16	3.08	3.00	2.92	2.86	2.79	2.76	2.70	2.67	2.65
2.29	2.25	2.19	2.15	2.11	2.07	2.04	2.00	1.98	1.95	1.93	1.93
3.27	3.19	3.07	3.00	2.91	2.83	2.78	2.71	2.68	2.62	2.59	2.57
2.26	2.21	2.15	2.11	2.07	2.02	2.00	1.98	1.94	1.91	1.90	1.88
3.19	3.12	3.00	2.92	2.84	2.76	2.70	2.63	2.60	2.54	2.51	2.49
2.23	2.18	2.12	2.08	2.04	1.99	1.96	1.92	1.90	1.87	1.85	1.84
3.13	3.05	2.94	2.86	2.77	2.69	2.63	2.56	2.53	2.47	2.44	2.42
2.20	2.15	2.00	2.05	2.00	1.96	1.93	1.89	1.87	1.84	1.82	1.81
3.07	2.99	2.88	2.80	2.72	2.63	2.58	2.51	2.47	2.42	2.38	2.36
2.18	2.13	2.07	2.03	1.98	1.93	1.91	1.87	1.84	1.81	1.80	1.78
3.02	2.94	2.83	2.75	2.67	2.58	2.53	2.46	2.42	2.37	2.33	2.31
2.14	2.10	2.04	2.00	1.96	1.91	1.88	1.84	1.82	1.79	1.77	1.76
2.97	2.89	2.78	2.70	2.62	2.53	2.48	2.41	2.37	2.32	2.28	2.26
2.13	2.09	2.01	1.98	1.94	1.89	1.86	1.82	1.80	1.76	1.74	1.73
2.93	2.85	2.74	2.66	2.58	2.49	2.44	2.36	2.33	2.27	2.23	2.21
2.11	2.06	2.00	1.96	1.92	1.87	1.84	1.80	1.77	1.74	1.72	1.71
2.89	2.81	2.70	2.62	2.54	2.45	2.40	2.32	2.29	2.23	2.19	2.17
2.10	2.05	1.09	1.95	1.90	1.86	1.82	1.78	1.76	1.72	1.70	1.69
2.86	2.77	2.66	2.58	2.50	2.41	2.36	2.28	2.25	2.19	2.15	2.13

Source: Reference 41.

TABLE 5.1 (cont'd.) *F* Distribution

n_2 \ n_1	n_1 Degrees of Freedom (for Greater Mean Square)[a]											
	1	2	3	4	5	6	7	8	9	10	11	12
27	4.21	3.35	2.96	2.73	2.57	2.46	2.37	2.30	2.25	2.20	2.10	2.13
	7.68	5.49	4.60	4.11	3.79	3.56	3.39	3.26	3.14	3.06	2.98	2.93
28	4.20	3.34	2.95	2.71	2.50	2.44	2.36	2.20	2.24	2.10	2.15	2.12
	7.64	5.45	4.57	4.07	3.76	3.53	3.36	3.23	3.11	3.03	2.95	2.90
29	4.18	3.33	2.93	2.70	2.64	2.43	2.36	2.28	2.22	2.18	2.14	2.10
	7.60	5.42	4.54	4.04	3.73	3.50	3.33	3.20	3.08	3.09	2.92	2.87
30	4.17	3.32	2.02	2.09	2.53	2.42	2.34	2.27	2.21	2.16	2.12	2.09
	7.56	5.39	4.51	4.02	3.70	3.47	3.30	3.17	3.06	2.98	2.90	2.84
32	4.15	3.30	2.00	2.67	2.51	2.40	2.32	2.25	2.19	2.14	2.10	2.07
	7.50	5.34	4.46	3.97	3.66	3.42	3.25	3.12	3.01	2.94	2.86	2.80
34	4.13	3.28	2.88	2.05	2.49	2.38	2.30	2.23	2.17	2.12	2.08	2.05
	7.44	5.29	4.42	3.93	3.61	3.38	3.21	3.08	2.97	2.89	2.82	2.76
36	4.11	3.26	2.86	2.63	2.48	2.36	2.28	2.21	2.15	2.10	2.06	2.03
	7.39	5.25	4.38	3.89	3.58	3.35	3.18	3.04	2.94	2.86	2.78	2.72
38	4.10	3.26	2.85	2.02	2.46	2.35	2.26	2.19	2.14	2.09	2.05	2.01
	7.35	5.21	4.34	3.86	3.54	3.32	3.15	3.02	2.91	2.82	2.75	2.69
40	4.08	3.23	2.84	2.61	2.45	2.34	2.25	2.18	2.12	2.07	2.04	2.00
	7.31	5.18	4.31	3.83	3.51	3.29	3.12	2.99	2.88	2.80	2.73	2.66
42	4.07	3.22	2.83	2.59	2.44	2.32	2.24	2.17	2.11	2.06	2.02	1.09
	7.27	5.15	4.29	3.80	3.49	3.26	3.10	2.96	2.86	2.77	2.70	2.64
44	4.00	3.21	2.82	2.58	2.43	2.31	2.23	2.16	2.10	2.05	2.01	1.98
	7.24	5.12	4.26	3.78	3.46	3.24	3.07	2.94	2.84	2.75	2.68	2.62
46	4.05	3.20	2.81	2.57	2.42	2.30	2.23	2.14	2.09	2.04	2.00	1.97
	7.21	5.10	4.24	3.76	3.44	3.22	3.05	2.92	2.82	2.73	2.66	2.60
48	4.04	3.10	2.80	2.50	2.41	2.30	2.21	2.14	2.08	2.03	1.99	1.96
	7.19	5.08	4.22	3.74	3.42	3.20	3.04	2.90	2.80	2.71	2.64	2.58
50	4.03	3.18	2.79	2.58	2.40	2.20	2.20	2.13	2.07	2.02	1.08	1.05
	7.17	5.06	4.20	3.72	3.41	3.18	2.02	2.88	2.78	2.70	2.62	2.56
55	4.02	3.17	2.78	2.54	2.38	2.27	2.18	2.11	2.05	2.00	1.97	1.93
	7.12	5.01	4.16	3.68	3.37	3.15	2.98	2.85	2.75	2.66	2.59	2.53
60	4.00	3.15	2.76	2.52	2.37	2.25	2.17	2.10	2.04	1.99	1.95	1.92
	7.08	4.98	4.13	3.65	3.34	3.12	2.95	2.82	2.72	2.63	2.56	2.50
65	3.99	3.14	2.76	2.51	2.36	2.24	2.15	2.08	2.02	1.98	1.94	1.90
	7.04	4.95	4.10	3.62	3.31	3.09	2.93	2.79	2.70	2.61	2.54	2.47
70	3.98	3.13	2.74	2.50	2.35	2.23	2.14	2.07	2.01	1.97	1.03	1.89
	7.01	4.92	4.08	3.60	3.29	3.07	2.91	2.77	2.67	2.59	2.51	2.45
80	3.98	3.11	2.72	2.48	2.33	2.21	2.12	2.06	1.99	1.95	1.91	1.88
	6.96	4.88	4.04	3.56	3.25	3.04	2.87	2.74	2.64	2.55	2.48	2.41
100	2.91	3.00	2.70	2.46	2.30	2.19	2.10	2.03	1.97	1.92	1.88	1.85
	6.90	4.82	3.98	3.51	3.20	2.99	2.82	2.69	2.59	2.51	2.43	2.36
125	3.92	3.07	2.08	2.44	2.20	2.17	2.08	2.01	1.95	1.90	1.84	1.83
	6.84	4.78	3.94	3.47	3.17	3.98	2.79	2.65	2.56	2.47	2.40	2.33
160	3.91	3.06	2.67	2.43	2.27	2.16	2.07	2.00	1.94	1.89	1.85	1.82
	6.81	4.75	3.91	3.44	3.14	2.92	2.76	2.62	2.53	2.44	2.37	2.30
200	3.80	3.01	2.05	2.41	2.20	2.14	2.05	1.98	1.92	1.87	1.83	1.80
	6.76	4.71	3.88	3.41	3.11	2.90	2.73	2.60	2.50	2.41	2.34	2.28
400	2.86	3.02	2.02	2.30	2.23	2.12	2.03	1.06	1.90	1.85	1.81	1.78
	6.70	4.66	3.83	3.36	3.06	2.85	2.69	2.55	2.46	2.37	2./29	2.23
1000	3.85	3.00	2.61	2.38	2.22	2.10	2.02	1.05	1.89	1.84	1.80	1.76
	6.66	4.62	3.80	3.34	3.04	2.82	2.66	2.53	2.43	2.34	2.26	2.20
∞	3.84	2.00	2.60	2.37	2.21	2.09	2.01	1.94	1.88	1.83	1.70	1.75
	6.64	4.60	3.78	3.32	2.02	2.80	2.64	2.51	2.41	2.32	2.24	2.18

[a]The first value in each pair is for 5% points for distribution of *F*, the second is for 1%.

TABLE 5.1 (cont'd.) *F* Distribution

				n_1 Degrees of Freedom (for Greater Mean Square)[a]							
14	16	20	24	30	40	50	75	100	200	500	∞
2.08	2.03	1.97	1.93	1.88	1.84	1.80	1.76	1.74	1.71	1.68	1.67
2.83	2.74	2.63	2.55	2.47	2.38	2.33	2.25	2.21	2.16	2.12	2.10
2.06	2.02	1.96	1.91	1.87	1.81	1.78	1.75	1.72	1.69	1.67	1.65
2.80	2.71	2.60	2.52	2.44	2.35	2.30	2.22	2.18	2.13	2.09	2.06
2.05	2.00	1.94	1.90	1.85	1.80	1.77	1.73	1.71	1.68	1.65	1.64
2.77	1.68	1.57	1.49	2.41	2.32	1.27	2.19	2.15	2.10	2.06	2.03
2.04	1.99	1.93	1.89	1.84	1.79	1.76	1.72	1.69	1.66	1.64	1.62
2.74	2.66	2.55	2.47	2.38	2.29	2.24	2.16	2.13	2.07	2.03	2.01
2.02	1.97	1.91	1.86	¯1.82	1.76	1.74	1.69	1.67	1.64	1.61	1.59
2.70	2.62	2.51	2.42	2.34	2.25	2.20	2.12	2.08	2.02	1.98	1.96
2.00	1.95	1.80	1.84	1.80	1.74	1.71	1.67	1.61	1.61	1.59	1.57
2.66	2.58	2.47	2.38	2.30	2.21	2.15	2.08	2.04	1.98	1.94	1.91
1.98	1.93	1.87	1.82	1.78	1.72	1.69	1.65	1.62	1.59	1.58	1.56
2.62	2.54	2.43	2.35	2.26	2.17	2.12	2.04	2.00	1.94	1.90	1.87
1.96	1.92	1.85	1.80	1.76	1.71	1.67	1.63	1.60	1.57	1.54	1.53
2.59	2.51	2.40	2.32	2.22	2.14	2.08	2.00	1.97	1.90	1.86	1.84
1.95	1.90	1.84	1.79	1.74	1.69	1.66	1.61	1.59	1.55	1.53	1.51
2.56	2.49	2.37	2.29	2.20	2.11	2.05	1.97	1.94	1.88	1.84	1.81
1.94	1.89	1.82	1.78	1.73	1.68	1.64	1.60	1.57	1.54	1.51	1.49
2.54	2.46	2.35	2.26	2.17	2.08	2.02	1.94	1.91	1.85	1.80	1.78
1.92	1.88	1.81	1.76	1.72	1.66	1.63	1.58	1.56	1.52	1.50	1.48
2.52	2.44	2.32	2.24	2.15	2.06	2.00	1.92	1.88	1.82	1.78	1.75
1.91	1.87	1.80	1.76	1.71	1.65	1.62	1.57	1.54	1.51	1.48	1.46
2.50	2.42	2.30	2.22	2.13	2.04	1.98	1.90	1.86	1.80	1.76	1.72
1.90	1.86	1.79	1.74	1.70	1.64	1.61	1.56	1.53	1.50	1.47	1.45
2.48	2.40	2.28	2.20	2.11	2.02	1.96	1.88	1.84	1.78	1.73	1.70
1.90	1.85	1.78	1.74	1.69	1.63	1.60	1.55	1.52	1.48	1.46	1.44
2.46	2.39	2.26	2.18	2.10	2.00	1.94	1.86	1.82	1.76	1.71	1.68
1.88	1.83	1.76	1.72	1.67	1.61	1.58	1.52	1.50	1.46	1.43	1.41
2.43	1.35	2.23	2.15	2.06	1.96	1.90	1.82	1.78	1.71	1.66	1.64
1.86	1.81	1.75	1.70	1.65	1.59	1.56	1.50	1.48	1.41	1.41	1.39
2.40	2.32	2.20	2.12	2.03	1.92	1.87	1.79	1.74	1.68	1.63	1.60
1.85	1.80	1.73	1.68	1.63	1.57	1.54	1.49	1.40	1.42	1.39	1.37
2.37	2.30	1.18	2.09	1.00	1.90	1.84	1.76	1.71	1.64	1.60	1.56
1.84	1.70	1.72	1.67	1.62	1.56	1.53	1.47	1.45	1.40	1.37	1.35
2.25	2.28	2.15	2.07	1.98	1.88	1.82	1.74	1.69	1.62	1.56	1.53
1.82	1.77	1.70	1.65	1.60	1.54	1.51	1.45	1.42	1.38	1.35	1.32
2.32	2.24	2.11	2.03	1.94	1.84	1.78	1.70	1.65	1.57	1.52	1.49
1.79	1.75	1.68	1.63	1.57	1.51	1.48	1.42	1.30	1.31	1.30	1.28
2.26	2.19	2.06	1.98	1.89	1.79	1.73	1.64	1.59	1.51	1.46	1.43
1.77	1.72	1.65	1.60	1.55	1.49	1.45	1.30	1.36	1.31	1.27	1.25
2.23	2.15	2.03	1.94	1.85	1.75	1.68	1.59	1.54	1.46	1.40	1.37
1.76	1.71	1.64	1.59	1.54	1.47	1.44	1.37	1.34	1.29	1.25	1.22
2.20	2.12	2.00	1.91	1.83	1.72	1.66	1.56	1.51	1.43	1.37	1.33
1.74	1.69	1.62	1.57	1.52	1.45	1.42	1.35	1.32	1.26	1.22	1.10
2.17	2.09	1.97	1.88	1.79	1.69	1.62	1.53	1.48	1.39	1.33	1.28
1.72	1.67	1.60	1.54	1.49	1.42	1.38	1.32	1.28	1.22	1.16	1.13
2.12	2.04	1.92	1.84	1.74	1.64	1.57	1.47	1.42	1.32	1.24	1.19
1.70	1.65	1.58	1.53	1.47	1.41	1.36	1.30	1.26	1.19	1.13	1.08
2.09	2.01	1.89	1.81	1.71	1.61	1.54	1.44	1.38	1.28	1.19	1.11
1.09	1.04	1.57	1.52	1.46	1.40	1.35	1.28	1.24	1.17	1.11	1.00
2.07	1.99	1.87	1.79	1.69	1.69	1.52	1.41	1.36	1.25	1.15	1.00

Source: Reference 41.

recorders are available for recording and transcribing license plate data, and computer programs are available for the matching process. Fairly small sample sizes on the order of 50 to 100 license plate matches are required to predict overall travel time with a reasonable level of confidence [24]. Sample size requirements can be determined from equation (4.24) derived in Chapter 4 and which is as follows:

$$n = \left(\frac{ts}{\varepsilon}\right)^2 \tag{5.10}$$

where n = sample size

t = value for the selected confidence level using the t distribution (the t distribution is used instead of the normal distribution if sample sizes approach 30 or less)

s = standard deviation of the sample

ε = user-selected allowable error

The so-called license plate technique has some inherent disadvantages if additional trip information is required, such as stopped time, number of brake applications, fuel consumption, and so on, along the route. Another consideration is whether travel time information between more than two points is required. License plate surveys can handle almost any number of segments, but each added segment will require an additional license plate recording station, which of course increases the cost of such surveys. For an efficient matching process, a high proportion of vehicles should appear at each pair of stations. Note that some origin and destination as well as routing information can be obtained from license plate surveys.

A second type of travel time study technique is the test vehicle technique. This is the most popular method employed, and there are a number of variations depending on the purpose and scope of the study. All variations require that a vehicle is placed in the traffic stream and the driver instructed to travel as the "average" vehicle. Under moderate to heavy flow conditions this is relatively easy to do, while under low-flow conditions it can be much more difficult. The variations in the test vehicle method are due to what data are recorded as the vehicle proceeds along the study section as the "average" vehicle. Almost all test vehicle methods include the recording of travel time information, and usually travel time is considered the primary measurement. Equation (5.10) can be used to estimate the number of runs required. Fairly small sample sizes on the order of 5 to 50 test runs are normally required, depending on travel time variation and confidence level specified. The variation of test vehicle methods discussed in the following paragraphs are: travel time contour, speed contour, flow–travel time, and comprehensive performance variations.

The travel time contour variation of the test vehicle method was developed to estimate trip times from a central point to various locations in a metropolitan area. By repeating the experiment on an annual basis, trends in travel time changes could be assessed. For a number of years the Automobile Club of Southern California conducted such studies utilizing their employees during the afternoon peak period on their way from the downtown office of the Automobile Club to their individual homes in various locations in the metropolitan area [34]. The drivers left the downtown office at about the same time and each recorded their location on the trip home at 15-minute intervals. This information was processed and 15-minute trip contour times were constructed and superimposed over a map of the Los Angeles metropolitan area. Figure 5.13 is a reproduction of one such travel time contour map.

The speed contour map variation of the test vehicle method was developed primarily to identify bottlenecks and resulting queueing patterns on urban freeways. In addition to the normal travel time information, the speedometer reading is recorded at 0.2-mile intervals when speeds are over 30 miles per hour and at 0.1-mile intervals when speeds are reduced below 30 miles per hour. Test vehicles are launched into the traffic stream about every 10 to 15 minutes. The trajectory of each vehicle is plotted on a time–space diagram and its speedometer readings are recorded along the trajectory. Speed contour lines are then constructed with particular attention given to the speed contour line just above the speed at capacity flow (usually 40 miles per hour) and the speed contour line just below the speed at capacity flow (usually 30 miles per hour). The time–space areas above 40 miles per hour represent free-flow conditions, the areas between 30 and 40 miles per hour represent transitional conditions, and the areas below 30 miles per hour represent congested flow conditions. The three areas are often referred to as the green, amber, and red zones, respectively. A typical speed contour map is shown in Figure 5.14. The time–space areas in dark shading indicate zones of congestion. Bottlenecks can be identified on the downstream edge of the congested zones and located within the white areas. The lightly shaded areas indicate areas of free-flow conditions.

The flow–travel time variation of the test vehicle method, sometimes referred to as the "moving-vehicle method," is a technique whereby the test vehicle obtains not only travel time information but traffic flow information as well. It is of particular importance when attempting to estimate total vehicle-miles and vehicle-hours of travel in a sizable network of low- to moderate-volume highways. In addition to recording the travel time in each direction along a study section, three vehicle counts are taken: the number of opposing vehicles met, the number of vehicles overtaking the test vehicle, and the number of vehicles passed by the test vehicle. The hourly flow in one direction can be estimated by using the following equation:

$$V_N = \frac{60[M_S + (O_N - P_N)]}{T_N + T_S} \tag{5.11}$$

where V_N = northbound flow (vehicles per hour)
M_S = number of opposing vehicles met when test vehicle was traveling south
O_N = number of vehicles overtaking the test vehicle when test vehicle was traveling north
P_N = number of vehicles passed by test vehicle when test vehicle was traveling north
T_N, T_S = travel time in minutes for northbound and southbound test vehicle runs, respectively

Note that if the difference between O_N and P_N is small in comparison with M_S, the equation is simplified to become the ratio of the number of vehicles met divided by the

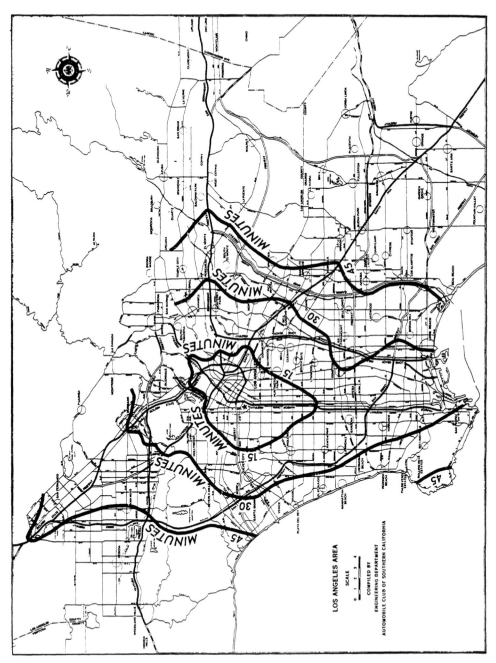

Figure 5.13 Travel Time Contour Map (From Reference 34)

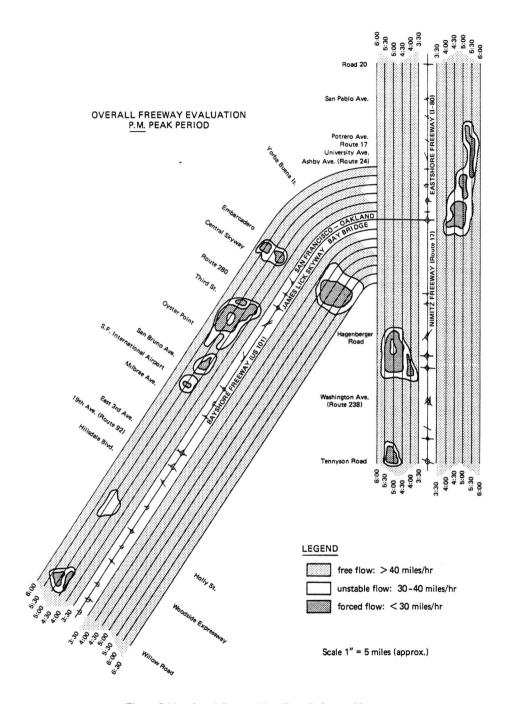

Figure 5.14   Speed Contour Map (From Reference 39)

sum of the travel times in both directions. This technique was originally developed in England in the early 1950s [23] and applied extensively in Illinois in the mid-1950s [32].

The comprehensive performance variation of the test vehicle method was developed to evaluate a significant number of measures of performance simultaneously, and by so doing, to develop relationships between the performance measures. A number of sensing devices are required, such as a fuel meter, brake light indicator, precise odometer, precise speedometer, timing device, and manual pushbuttons. An automatic data collection and recording device is required such as a microcomputer. These measurements are recorded, and later off-line, are processed for analysis. Figure 5.15 illustrates the types of results that are available from such a test vehicle study. These results were obtained from a study in Michigan of some 41 sections of major urban arterials in which over 1900 test runs were made utilizing a specially equipped research vehicle [27, 36]. One of the obvious limitations of such a technique is the significant equipment costs and the fact that normally only one or two such vehicles are available for a particular study.

5.3.2 Intersection Delay Study Techniques

A special type of travel time study is concerned with the measurement of delay at signal-controlled and sign-controlled intersections. Total delay is defined as the difference in travel time when a vehicle is unaffected by the controlled intersection and when a vehicle is affected by the controlled intersection. This delay, which is called total delay, includes lost time due to deceleration and acceleration as well as stopped delay. Thus intersection delay studies are directed toward estimating total delay or simply, stopped delay.

Total delay can be estimated using several different techniques. It consists of two steps. The travel time without delay is first estimated by establishing a distance of travel and specifying an unaffected cruise speed. The distance should be selected between an unaffected point upstream of the intersection to a similar point downstream of the intersection. The actual travel time is then measured by observing the total elapsed time over the selected distance. This actual travel time can be measured using the test vehicle technique or measuring the entrance and exit times of vehicles already in the travel stream.

It is more common to measure stopped delay and to use a correction factor if an estimate of total delay is desired. There are several methods available for measuring stopped delay, but the most common is the stopped vehicle count method. This method requires the counting of the number of stopped vehicles at regular time intervals. The time interval usually selected is 10 to 20 seconds and should not be a multiple of the green phase or cycle length duration at signalized intersections. The stopped delay is equal to the average number of stopped vehicles multiplied by the time interval. Recent research has shown that there is a bias in measuring stopped delay, and actual stopped delay is about 92 percent of measured stopped delay [25]. The total delay can be approximated by multiplying the stopped delay by a factor of 1.3 [25].

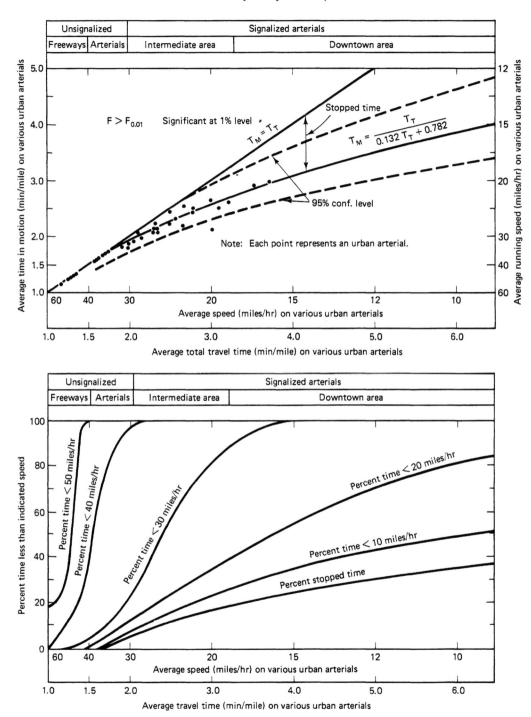

Figure 5.15   Comprehensive Performance Results with Test Vehicle (From Reference 27)

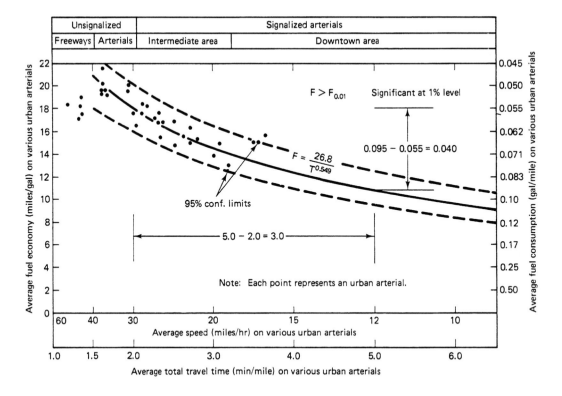

Figure 5.15   Cont'd

Several analytical techniques are available for mathematically calculating delay at signalized intersections. One method is the Webster method [33] and another is the 1985 *Highway Capacity Manual* method [8]. The Webster equation is as follows:

$$d_t = \frac{C(1-\lambda)^2}{2(1-\lambda x)} + \frac{x^2}{2q(1-x)} - 0.65\left(\frac{C}{q^2}\right)^{1/3} x^{2+5\lambda} \tag{5.12}$$

where d_t = average total delay per vehicle on the particular approach of the intersection (seconds/vehicle)

C = cycle time (seconds)

λ = proportion of the cycle that is effectively green for the phase under consideration (i.e., g/C)

q = flow (vehicles per second)

s = saturation flow (vehicles per second of green)

x = degree of saturation (i.e., $x = q/\lambda s$)

Note that the equation has three terms. The first term represents average delay for a particular approach assuming uniform arrivals at a fixed-time signal-controlled intersection and can be derived from deterministic queueing theory. The second term is added to account for random arrivals. The third term is subtracted from the first two terms and varies from a value equal to zero (in the event of random arrivals) to a value equal to the second term (in the event of uniform arrivals). Through simulation and field measurements, Webster has empirically developed the third term.

The 1985 *Highway Capacity Manual* contains a chapter devoted to signalized intersections, and this chapter includes a mathematical formulation for estimating stopped delays to an approach of a signalized intersection. The mathematical formulation is shown in the following equation:

$$d_s = 0.38C\frac{[1 - g/C]^2}{[1 - (g/C)(X)]} + 173X^2 \left[(X - 1) + \sqrt{(X - 1)^2 + (16X/c)}\right] \quad (5.13)$$

where d_s = average stopped delay per vehicle for the subject lane group (seconds per vehicle)

C = cycle length (seconds)

g/C = ratio of effective green time to cycle time

X = ratio of volume to capacity for subject lane group

c = capacity of the subject lane group (vehicles per hour)

Note that the equation has only two terms, with the first term accounting for uniform stopped delay and the second term accounting for incremental delay of random arrivals over uniform arrivals and for the additional delay due to cycle failures. Note also that the *Highway Capacity Manual* equation is for average stopped delay rather than average total delay.

5.4 TRAVEL TIME AND DELAY STUDY*

This portion of the chapter presents a comprehensive travel time and delay study that was conducted for the operational evaluation of an arterial traffic signal system. All 11 signals in the system were fixed-time two-phase signals. The project included before and after field studies as well as comparison with predicted before and after results obtained from the application of the TRANSYT model. The presentation and discussion will be limited to the before field study because of the extensiveness of the project. Primary attention will be given to the design of experiment, results, and statistical analysis.

The study was limited to the afternoon peak period during good weather conditions and data were collected by 15-minute intervals from 4 to 6 P.M.. Data were only collected on Mondays through Thursdays because of the higher flow situations on Fridays. The 2-hour period included pre-peak, peak and after-peak flow conditions, and recording by 15 minutes permitted analysis for these three subperiods. Travel time and stopped delay were selected as the prime measures of effectiveness. Secondary measures of effectiveness included percentage of vehicles stopped and maximum queue lengths. Since all the performance indices measured were very sensitive to flow changes, traffic flows were measured at the same time that stopped delays were measured and four semipermanent count stations were maintained. Travel time was obtained

---

*This section contains highlights of the study described in Reference 40.

from test vehicles traveling along the major arterials while stopped delay (and percentage of vehicles stopped and maximum queue lengths) were recorded manually for each lane group of each approach to the 11 intersections in the traffic signal system. A map showing the layout of the study section is presented in Figure 5.16. The test vehicle travel time substudy is discussed first, and then attention is directed to the intersection stopped delay substudy.

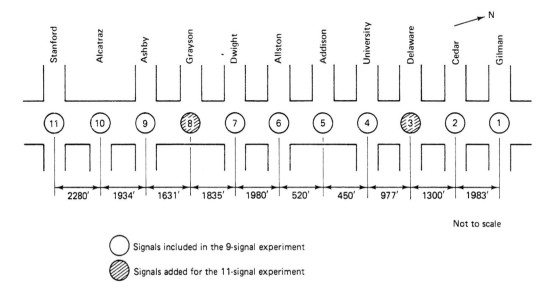

Figure 5.16 Layout of Study Section [14,989 ft (2.82 mi)]

5.4.1 Travel Time Study Results

Two test vehicles were employed, and at 4 P.M. one vehicle started to drive south from a point approximately 1 mile north of the study section, while the other vehicle started to drive north from a point approximately 1 mile south of the study section. The drivers were instructed to "float" with traffic and to attempt to represent the "average" vehicle. A one-way trip normally took about 15 minutes, so each vehicle could make four round trips per afternoon study period. The variations in travel time were unknown, so the number of required test runs could not be precisely estimated. However, the study period was scheduled to last for 2 weeks (eight afternoon peak periods) and a total of 60 plus test runs per direction seemed quite adequate based on previous studies.

Two persons occupied each vehicle: a driver and a data recorder. The following data were recorded:

- Identification of the vehicle
- Driver and observer names
- Date

- Starting time for each run
- Cumulative time at each node between adjacent links
- Stops due to traffic signals
- Time when headlights turned on (headlights not needed in before study)
- Time when windshield wipers were turned on (windshield wipers not needed in before study)
- Comments in regard to unusual roadway and traffic occurrences (no unusual occurrences encountered in before study)

The results of the test vehicle travel time study are arranged in Table 5.2 to permit testing of significant differences between test vehicles (drivers), days of the week, and direction of travel. The techniques presented in Sections 5.2.3 and 5.2.4 were applied to these data, and the results of the statistical tests are summarized in Table 5.3. A confidence level of 0.05 was selected. In the southbound direction in the tests between vehicles (drivers), there was no significant difference between means nor variances. The same was true for the tests between vehicles (drivers) in the northbound direction. In the tests between days of the week, the largest difference in means (Mondays versus Thursdays) and variances (Mondays versus Tuesdays) were selected. The tests revealed no significant differences of means nor variances for different days of the week. It should be noted that the differences in mean travel times between Mondays and Thursdays although quite large (27 seconds), was offset by a higher standard deviation of the

TABLE 5.2 Test Vehicle Travel Time Study Results by Vehicle, Day of Week, and Direction of Travel

Situation		Average Travel Time (sec)	Standard Deviation (sec)	Number of Runs
Southbound	Vehicle 1	491	39	35
	Vehicle 2	499	38	32
Northbound	Vehicle 1	486	36	36
	Vehicle 2	494	46	29
Combined directions both vehicles	Mondays	470	36	14
	Tuesdays	492	45	14
	Wednesdays	495	38	14
	Thursdays	497	41	18
Both vehicles all days	Southbound	495	38	67
	Northbound	490	41	65

TABLE 5.3   Statistical Test Results by Vehicle, Day of Week, and Direction of Travel

COMPARISON OF MEANS

Test Situation		ΔT	$\hat{s}$	Conclusion
Southbound	Test between vehicles (drivers)	8	9.4	8 < 18.8, no significant difference
Northbound	Test between vehicles (drivers)	8	10.3	8 < 20.6, no significant difference
Both directions	Test between days	27	13.6	27 < 27.2, no significant difference
Both vehicles	Test between directions	5	6.9	5 < 13.8, no significant difference

COMPARISON OF VARIANCES

Test Situation		Calculated F Value	Table F Value	Conclusion
Southbound	Test between vehicles (drivers)	1.05	2.10	1.05 < 2.10, no significant difference
Northbound	Test between vehicles (drivers)	1.60	2.10	1.60 < 2.10, no significant difference
Both directions	Test between days	1.56	1.95	1.56 < 1.95, no significant difference
Both vehicles	Test between directions	1.16	1.53	1.16 < 1.53, no significant difference

difference of the means ($\hat{s}$). This higher standard deviation of the difference of the means value was due to a smaller sample size (14 to 18). In the final comparison of means and variances between directions of travel, no significant differences were found.

The results of the test vehicle travel time study are rearranged in Table 5.4 to permit testing of significant differences between 15-minute time periods. The travel times varied from 463 to 525 seconds and the variances ranged from 14 to 61 seconds. Note that the number of runs in each time period is much smaller than the number of runs

TABLE 5.4   Test Vehicle Travel Time Study Results by Time of Day

Situation		Average Travel Time (sec)	Standard Deviation (sec)	Number of Runs
Southbound	4:00–4:15	525	38	8
	4:15–4:30	490	42	8
	4:30–4:45	492	32	10
	4:45–5:00	495	26	6
	5:00–5:15	502	43	14
	5:15–5:30	476	25	4
	5:30–5:45	502	36	9
	5:45–6:00	463	36	8
Northbound	4:00–4:15	485	31	10
	4:15–4:30	496	48	8
	4:30–4:45	492	14	7
	4:45–5:00	511	34	9
	5:00–5:15	485	47	8
	5:15–5:30	493	61	11
	5:30–5:45	466	30	7
	5:45–6:00	478	38	5

displayed in Table 5.2. This will cause the standard deviation of the differences of the means and the table F value to be considerably greater, and thereby make it more difficult to detect significant differences. The same statistical techniques were applied and the statistical statements are summarized in Table 5.5. In each cell the results are shown for the comparison of means and variances. A blank signifies no significant difference, and the symbol "SIG" indicates significant difference. In the southbound direction the 5:45 to 6:00 P.M. travel time was found to be significantly less than the 4:00 to 4:15 P.M., 5:00 to 5:15 P.M., and 5:30 to 5:45 P.M. mean travel times. The 5:15 to 5:30 mean travel time was found to be significantly less than the 4:00 to 4:15 P.M. mean travel time. There were no significant differences between time period variances. In the northbound direction the 5:30 to 5:45 P.M. mean travel time was found to be significantly less than the 4:30 to 4:45 P.M. and the 4:45 to 5:00 P.M. mean travel times. The 4:30 to 4:45 P.M. travel time variance was determined to be significantly less than the 5:15 to 5:30 travel time variance.

5.4.2 Intersection Delay Study Results

Stopped delay was measured for each lane group on each approach to the 11 intersections in the traffic signal system. The lane group numbering system used in the study is shown in Figure 5.17. Since there were no special signal turn phases for right-turning traffic, the right-turning traffic and right-turn lanes (if any existed) were always

TABLE 5.5 Statistical Test Results by Time of Day[a]

SOUTHBOUND

Time Period	4:00	4:15	4:30	4:45	5:00	5:15	5:30	5:45
4:00	-	-	-	-	-	SIG	-	SIG
	-	-	-	-	-	-	-	-
4:15	-	-	-	-	-	-	-	-
	-	-	-	-	-	-	-	-
4:30	-	-	-	-	-	-	-	-
	-	-	-	-	-	-	-	-
4:45	-	-	-	-	-	-	-	-
	-	-	-	-	-	-	-	-
5:00	-	-	-	-	-	-	-	SIG
	-	-	-	-	-	-	-	-
5:15	SIG	-	-	-	-	-	-	-
	-	-	-	-	-	-	-	-
5:30	-	-	-	-	-	-	-	SIG
	-	-	-	-	-	-	-	-
5:45	SIG	-	-	-	SIG	-	SIG	-
	-	-	-	-	-	-	-	-

NORTHBOUND

Time Period	4:00	4:15	4:30	4:45	5:00	5:15	5:30	5:45
4:00	-	-	-	-	-	-	-	-
	-	-	-	-	-	-	-	-
4:15	-	-	-	-	-	-	-	-
	-	-	-	-	-	-	-	-
4:30	-	-	-	-	-	-	SIG	-
	-	-	-	-	-	SIG	-	-
4:45	-	-	-	-	-	-	SIG	-
	-	-	-	-	-	-	-	-
5:00	-	-	-	-	-	-	-	-
	-	-	-	-	-	-	-	-
5:15	-	-	-	-	-	-	-	-
	-	-	SIG	-	-	-	-	-
5:30	-	-	SIG	SIG	-	-	-	-
	-	-	-	-	-	-	-	-
5:45	-	-	-	-	-	-	-	-
	-	-	-	-	-	-	-	-

[a]In each cell the results are shown for the comparison of means and variances. A dash signifies no significant difference and the symbol "SIG" indicates significant difference.

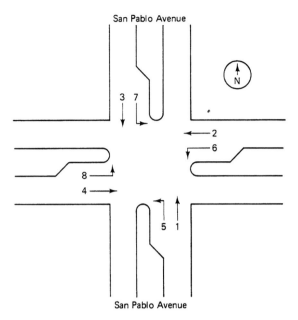

Figure 5.17   The Lane Group Numbering System

combined with the through-traffic-lane group. Left-turning traffic was handled as a separate lane group if a separate left-turn lane existed. Otherwise, the left-turning traffic was combined with the through-traffic-lane group. There were a total of 64 lane groups located on the 42 approaches to the 11 signalized intersections. Note in Figure 5.16 that three of the 11 signalized intersections are tee-type intersections.

The intersection delay study was conducted for eight lane groups each afternoon during the 2 week before study period (a total of 8 days, Mondays through Thursdays). This permitted each lane group to be studied for one afternoon peak period. Ten two-person teams were hired and trained for the intersection delay study. One afternoon was used to train the teams to ensure correct and consistent results. One team member counted the number of vehicles stopped in the queue at 13-second intervals for later calculations of stopped delay. The other team member counted the number of vehicles required to stop and the total number of vehicles passing through the intersection for later calculations of total flow and percent stopped vehicles.

Since only one set of measurements were recorded for each lane group, it was not possible to compare means and variances. However, since total delay in vehicle-hours and total flow in vehicles per hour were determined for each 15-minute period, it was possible to analyze the effect that the flow rate had on total delay for each lane group, at each intersection, and for the total 11-signal system. As an example, these relationships are depicted in Figure 5.18 for each of the individual seven lane groups at Ashby Avenue. Lane groups 1, 2, 3, and 4 are through-traffic-lane groups, while lane groups 5, 7, and 8 are left-turn traffic lane groups. Linear regressions were performed for each lane group and the resulting curves and equations are shown in each diagram. The data points represent 15-minute time periods for the 2-hour afternoon peak period. All equations have a positive slope as would be expected and except for one case, all y-

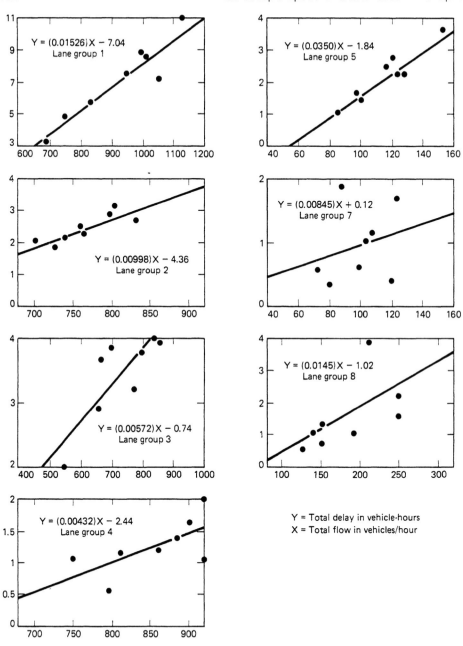

Figure 5.18   Effect of Flow on Total Delay for Individual Lane Groups

intercepts are negative. The curves obviously must pass through the origin, and hence the negative y-intercepts implies that the relationship particularly at the lower flow range is curvilinear.

Flow-delay relationships were developed for the total of all lane groups at Ashby and for the entire signal system as shown in Figure 5.19. Again, linear regressions were performed, and the resulting curves and equations are shown in each diagram. The 15-minute data points are shown. Again, the equations have positive slopes and negative y-intercepts.

This portion of the chapter has attempted to provide the reader with some appreciation of the planning, conducting, and analyzing of travel time and delay studies by covering one such study in some detail. Only selected portions of the study have been presented and the reader may wish to refer to the complete report for more detail [40].

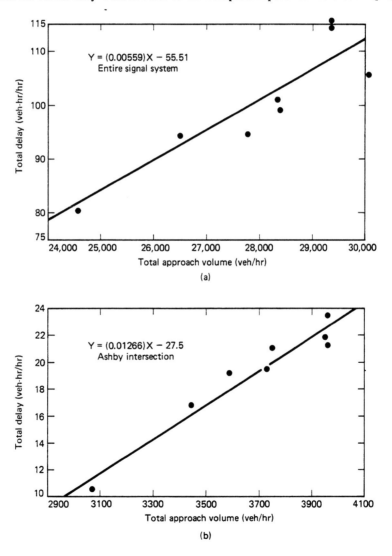

Figure 5.19   Effect of Flow on Total Delay for an Intersection and for the Signal System

5.5 SELECTED PROBLEMS

1. The traffic flow along an existing radial urban two-lane directional freeway in the peak direction is 3000 vehicles per hour, which includes 5 percent recreational vehicles, 2 percent buses, and 10 percent trucks. The existing freeway has the following characteristics: 60-mile per hour design speed, 12-foot lanes, 8-foot shoulders on each side, and rolling terrain. The annual growth in traffic is increasing at a rate of 5 percent per year, with the vehicle composition factors remaining about the same. Consideration is being given to redesigning the freeway by adding a third lane, but because of right-of-way constraints, the lane widths would be reduced to 11 feet, the left shoulder would be abandoned, and the right shoulder would be reduced to 6 feet. The reconstruction is expected to take 3 years. Estimate average speeds during this design hour under existing conditions, three years from now with the existing design, and 3 years from now with the new design.

2. Conduct a library study of annual trends in freeway speeds for the period 1960 to the present time. Assess the impact that the 55-mile per hour speed limit has had on average speeds and speed distributions. Attempt to predict average speeds and speed distributions when the speed limit was increased to 65 miles per hour on the rural freeway system.

3. Solve Problem 1 for the existing condition only. Estimate average speeds in each lane and the average speed in the other direction during the same hour assuming a 60:40 directional split in volume and the same vehicle mix.

4. Calculate the average two-way speed profile along the two-lane two-way rural highway shown in Figure 5.6 using the 1985 *Highway Capacity Manual*. The characteristics of the highway and traffic situation are given below.

Geometric Characteristics		Traffic Characteristics	
Road Type	Two-lane		
Grade	3%, 5%	Two-way volume (veh/hr)	600
Lane width (ft)	12	Percent Trucks	5
Shoulder width (ft)	6	Percent RVs	5
Design Speed (miles per hour)	60	Directional split	50:50
Percent no-passing zones	50		

Compare and assess differences between your calculations and the results shown in Figure 5.6.

5. The running or cruise speed of vehicles along an arterial is 30 miles per hour. There are six signalized intersections along the arterial and are spaced 1/4 mile apart. You are to make three estimates of the travel time from a point 1/4 mile upstream of the first signal to just downstream of the stop line at the last signal. The three situations to be analyzed are (a) theoretical minimum travel time at a constant speed of 30 miles per hour; (b) at nighttime the signals are placed in flashing yellow operations along the arterial and vehicles slow down to 20 miles per hour at each signalized intersection momentarily; and (c) during the time of day that signals are coordinated well for the opposite direction of traffic, the traffic in one study direction is required to stop at every signal for an average of 20 seconds. Assume acceleration and deceleration rates considering values shown in Tables 4.2 and 4.4.

6. Conduct a library study to assess expected average speeds by vehicle types under open-highway and level-terrain conditions. Has there been any noticeable change in the past 10 years? Compare with results shown in Figure 5.8.

7. Conduct a library study of empirical studies in which coefficients of variations were calculated and superimpose results on Figures 5.9 and 5.10. Identify similarities and differences and attempt to explain.

8. Derive equation (5.7) for time-mean-speed as a function of space-mean-speed.

9. Conduct a library study of the relationship of time-mean-speed and space-mean-speed. Can you add further insights to the text of Section 5.2.2 and Figure 5.11?

10. The walking speeds of men and women on two different days crossing a major arterial at a signalized intersection were measured and the results are shown in the following table. Test for significant differences in means and variances between the various possibilities. Note the rather small sample sizes.

		Wednesday	Friday	Both Days
$\bar{\mu}$ (ft/sec)	Men	5.5	5.3	5.4
	Women	4.9	4.8	4.8
	All	5.2	5.0	5.1
s^2 (ft/sec)^2	Men	0.48	0.43	0.62
	Women	0.25	1.50	0.91
	All	0.47	0.99	0.82
n (number)	Men	17	8	25
	Women	12	12	24
	All	29	20	49

11. Consider the example problem in Sections 5.2.3 and 5.3.4 and solve the following problems. (a) What minimum $\bar{\mu}_2$ value must occur for $\bar{\mu}_2$ to be significantly larger than $\bar{\mu}_1$ assuming a 95 percent and a 99 percent confidence level. (b) What maximum s_2 value must occur for s_2 to be significantly smaller than s_1 assuming a 95 percent and a 99 percent confidence level? (c) How large could n_2 become before there were a significant difference between means and variances at the 95 percent confidence level given the original means and standard deviation values?

12. Identify the assumptions and limitations in testing for significant differences between means and variances and possible ways of overcoming these limitations.

13. Inspection of the speed contour map shown in Figure 5.14 reveals several bottlenecks along the freeway. Study each bottleneck carefully and specify (a) bottleneck locations; (b) duration of congestion; (c) maximum length of congestion; and (d) minute–miles of congestion. Plot a time–distance vehicle projectory through one congested section starting every 30 minutes from just before the start of congestion until the end of congestion (suggest the northbound east shore freeway I-80).

14. A moving vehicle method study was undertaken on a 40-mile rural highway network to estimate travel time and flow, and the following table summarizes the results of the study. Estimate the vehicle-hours and vehicle-miles of travel for each direction of traffic. Assuming a 95 percent confidence level, what statement can be made about the maximum error of the travel time means?

Run Number		T (min)	M (number)	O (number)	P (number)
Eastbound	1	55.1	712	11	7
	2	61.7	677	3	4
	3	55.3	681	7	10
	4	52.4	733	4	5
	5	57.0	660	10	9
	6	56.6	685	9	8
	7	50.5	704	5	2
	8	51.8	698	11	13
Westbound	1	53.0	801	9	12
	2	58.9	930	13	7
	3	54.8	872	8	10
	4	61.2	903	9	15
	5	57.3	917	14	10
	6	60.1	861	11	13
	7	59.7	929	12	11
	8	60.4	935	10	10

15. Multivariate analysis techniques have been used to develop equations for the prediction of travel time as a function of roadway design, travel demand, and traffic control. Pignataro summarizes results of some of the earlier studies [7]. Perform a library study and report the highlights of more recent studies.

16. A stopped vehicle count study was undertaken on one approach to a signalized intersection and the field measurements are summarized in the following table. A total of 241 vehicles passed through the approach, of which 203 were required to stop. Calculate the following flow and performance measures: flow rate, stopped delay, total delay, percent stopped vehicles, stopped delay per vehicle, and total delay per vehicle. Plot the number of stopped vehicles versus time and attempt to superimpose the signal timing along the time scale, assuming a cycle length of 80 seconds and a 50:50 split in green time.

Minute	Number of Stopped Vehicles at:				
	0 sec	12 sec	24 sec	36 sec	48 sec
0	8	13	14	8	2
1	0	5	10	15	12
2	6	1	2	6	11
3	16	9	4	0	3
4	8	13	14	8	2
5	0	5	11	15	12
6	6	0	2	7	11
7	17	10	4	0	3
8	8	13	15	7	2
9	0	5	9	14	12

17. A FHWA study [25] reported that actual stopped delay was 92 percent of measured stopped delay and total delay was 130 percent of stopped delay. Attempt to understand and explain these relationships.

18. Calculate the average total delay per vehicle using the Webster equation for a single approach to a fixed-time signal-controlled intersection which has the following characteristics:

    • Signal timing of 36 seconds of green, 4 seconds of amber, and 20 seconds of red
    • Two-lane approach with lane saturation flow rates of 1600 vehicles per hour of green in the median lane and 1500 vehicles per hour of green in the shoulder lane
    • Demand flows equivalent from 10 to 110 percent of capacity in steps of 10 percent

19. Repeat Problem 18 above using the 1985 *Highway Capacity Manual* method. Check the saturation flow rate and compare the two calculated average total delay per vehicle values. Attempt to explain differences.

5.6 SELECTED REFERENCES

1. Daniel L. Gerlough and Matthew J. Huber, *Traffic Flow Theory—A Monograph*, Transportation Research Board, Special Report 165, TRB, Washington, D.C., 1975, pages 8–9, 12–13, and 199–201.

2. Daniel L. Gerlough and Matthew J. Huber, *Statistics with Applications to Highway Traffic Analyses*, Eno Foundation, Saugatuck, Conn., 1978, 179 pages.

3. J. J. Leeming, *Statistical Methods for Engineers*, Blackie & Son Ltd, Glasgow, 1963, 146 pages.

4. Donald R. Drew, *Traffic Flow Theory and Control*, McGraw-Hill Book Company, New York, 1968, pages 312–314, and 319–325.

5. Martin Wohl and Brian V. Martin, *Traffic System Analysis for Engineers and Planners*, McGraw-Hill Book Company, New York, 1967, pages 322–329.

6. Wolfgang S. Homburger and James H. Kell, *Fundamentals of Traffic Engineering*, Institute of Transportation Studies, Berkeley, Calif., 1988, pages 7–1 to 7–6.

7. Louis J. Pignataro, *Traffic Engineering—Theory and Practice*, Prentice-Hall Inc., Englewood Cliffs, N. J., 1973, pages 106–115, 128–131, and 139–142.

8. Transportation Research Board, *Highway Capacity Manual*, Special Report 209, TRB, Washington D. C., 1985.

9. Institute of Transportation Engineers, *Transportation and Traffic Engineering Handbook*, 2nd Edition, Prentice-Hall, Inc., Englewood Cliffs, N. J., 1982, 883 pages.

10. American Association of State Highways and Transportation Officials, *A Policy on Geometric Design of Highways and Streets*, AASHTO, Washington D. C., 1984, 1087 pages.

11. S. Yagar, Predicting Speeds for Rural 2-Lane Highways, *Transportation Research, Part A*, Vol. 18A, No. 1, January 1984, pages 61–70.

12. Transportation Research Board, *55 — A Decade of Experience*, Special Report 204, TRB, Washington D. C., 1984, 262 pages.

13. S. Yagar and S. E. H. Vanar, Geometric and Environmental Effects on Speeds of 2-Lane Highways, *Transportation Research, Part A*, Vol. 17A, No. 4, July 1983, pages 315–325.

14. B. Kolsrud, G. K. Nilsson, and S. Rigefalk, *Follow-up Study of Speeds on Rural Roads*, National Swedish Road and Traffic Research Institute, Linkoeping, Sweden, 1983, 19 pages.

15. D. B. Kamerud, The 55 mph Speed Limit: Costs, Benefits, and Implied Tradeoffs, *Transportation Research, Part A*, Vol. 17A, No. 1, January 1983, pages 51–64.

16. M. Armour, *Vehicle Speeds on Residential Streets*, Australian Road Research Board, Victoria, Australia, 1982, pages 190–205.

17. T. M. Klein, *An Evaluation of the 55 mph Speed Limit*, National Highway Traffic Safety Administration, Washington D. C., May 1981, 8 pages, and December 1981, 33 pages.

18. F. C. M. Wegman, *Speed Limits in the Netherlands*, Institute for Road Safety Research, Leidschendam, The Netherlands, 1981, 32 pages.

19. F. A. Wagner and A. D. May, *Volume and Speed Characteristics at Seven Study Locations*, Highway Research Board Bulletin 281, HRB, Washington D. C., 1960, pages 48–67.

20. J. C. Oppenlander, *Multivariate Analysis of Vehicular Speeds*, Highway Research Board, Record 35, 1961, pages 41–77.

21. J. C. Oppenlander, *Variables Influencing Spot Speed Characteristics*, Highway Research Board, Special Report 89, HRB, Washington D. C., 1966, 39 pages.

22. J. C. Estep and R. N. Smith, 55 mph—How Are We Doing? *Institute of Traffic Engineers Proceedings*, Vol. 45, August 1975, pages 42–50.

23. J. G. Wardrop and G. Charlesworth, A Method for Estimating Speed and Flow of Traffic from a Moving Vehicle, *Institute of Civil Engineers Proceedings, London*, Part II, No. 2, 1954, pages 158–171.

24. D. S. Berry and F. M. Green, Techniques for Measuring Overall Speeds in Urban Areas, *Highway Research Board, Proceedings*, Vol. 29, 1949, pages 311–318.

25. W. R. Reilly, C. C. Gardner, and J. H. Kell, *A Technique for Measurement of Delay at Intersections*, Federal Highway Adimnstration, Reports FHWA-RD-76-135, FHWA-RD-136, and FHWA-RD-137, FHWA, Washington D. C., September 1976.

26. R. H. Wortman, *A Multivariate Analysis of Vehicular Speeds on Four-Lane Rural Highways*, Highway Research Board, Record 72, HRB, Washington D. C., 1965, pages 1–18.

27. A. D. May, A Friction Concept of Traffic Flow, *Highway Research Board Proceedings*, Vol. 38, 1959, pages 493–510.

28. J. Treadway and J. C. Oppenlander, *Statistical Modeling of Travel Speeds and Delays on a High-Volume Highway*, Highway Research Board, Record 199, HRB, Washington D. C., 1967, pages 1–18.

29. J. W. Horn, P. D. Cribbins, J. D. Blackburn, and C. E. Vick Jr., *Effects of Commercial Roadside Development on Traffic Flow in North Carolina*, Highway Research Board Bulletin 303, HRB, Washington D. C., 1961, pages 76–93.

30. Juan C. Sananez, Laura Wingerd, and Adolf D. May, *A Macroscopic Model for the Analysis of Traffic Operations on Rural Highways—Final Report*, Institute of Transportation Studies, Berkeley, Calif., April 1985, 388 pages.

31. P. Y. Ching and F. D. Rooney, *Truck Speeds on Grades in California*, California Department of Transportation, Sacramento, Calif., June 1979.

32. J. Mortimer, *Moving Vehicle Method of Estimating Traffic Volumes and Speeds*, Highway Research Board Bulletin 156, HRB, Washington D. C., 1957, pages 14–26.

33. F. V. Webster and B. M. Cobbe, *Traffic Signals*, Road Research Laboratory, Technical Paper No. 56, Her Majesty's Stationery Office, London, 1956.

34. *Peak Hour Driving Study—Metropolitan Los Angeles*, Automobile Club of Southern California, Los Angeles, 1962, 18 pages.

35. U.S. Department of Transportation, *Highway Statistics 1983*, U. S. DOT, Washington D. C., 1983, 181 pages.

36. Adolf D. May and Frederick A. Wagner, Jr., *A Summary of Quality and Fundamental Characteristics of Traffic Flow*, Michigan State University, East Lansing, Mich., 1961, 64 pages.

37. Adolf D. May, *California Freeway Operations Study*, Thompson Ramo Wooldridge, Chatsworth, Calif., January 1962, 126 pages.

38. Joseph S. Drake, Joseph L. Schofer, and Adolf D. May, A Statistical Analyses of Speed Density Hypotheses, *Third International Symposium on the Theory of Traffic Flow Proceedings*, 1965, pages 112–117.

39. Adolf D. May and Brian Allen, *System Evaluation of Freeway Design and Operations*, Highway Research Board Special, Report 107, HRB, Washington D. C., 1968, pages 45–59.

40. Marthinus J. Vermeulen, Nicolas Lermant, and Adolf D. May, *Guidelines for Improving Traffic Signal Timing, UCB-ITS-RR-84-10*, University of California, Berkeley, Calif., June 1984, 239 pages.

41. E. W. Snedecor, *Statistical Methods*, 7th Edition, Iowa State University Press, Ames, Iowa, 1980, 507 pages.

6

Microscopic Density

Characteristics

The longitudinal space occupied by individual vehicles in the traffic stream is the subject covered in this chapter and can be considered as a microscopic representation of traffic density characteristics. Macroscopic density characteristics will be discussed in Chapter 7. The longitudinal spacing of vehicles in the traffic stream is of particular importance considering safety, capacity, and level of service. Minimum space must be available in front of every vehicle so that the driver can control his vehicle without colliding with the vehicle ahead or with fixed objects on the roadway.

The spacing of vehicles on the other hand, influences the capacity of the roadway and in turn affects the level of service to the users. Naively, the reader might first assume that higher flows can be obtained by spacing vehicles closer and closer together. This is true up to the critical spacing, but if the spacing becomes smaller than the critical spacing, the gain in potential capacity due to decreased spacing is more than offset by the loss in capacity due to lower speeds.

Vehicle spacing is also important from the point of view of level of service. If vehicle spacing is large, the driver has considerable freedom and a high level of service. As the spacing decreases drivers are required to give more attention to the driving task and may have to reduce their operating speed. This results in lower levels of service but increased productivity as long as the spacing remains greater than the critical spacing. If the spacing becomes less than the critical spacing, not only does the productivity drop below the capacity level but the level of service becomes lower and lower. The extreme case would be vehicles in a stopped queue where spacing would be minimum but level of service would be the lowest and flow would be zero.

In the following section attention will be given to distance headway characteristics with particular emphasis given to definitions and relationships. A major section will be devoted to a chronological development of the most pertinent car-following theories.

An application of car-following theory will be presented in which two vehicles are tracked through space and time as the lead vehicle proceeds on an acceleration-constant speed–deceleration driving cycle. The question of traffic stability is a logical extension of car-following theory, and limits and examples of local and asymptotic stability are given. The final section is devoted to measurements of microscopic and macroscopic density characteristics with single and dual presence-type detectors. Selected problems and references are provided at the end of the chapter.

6.1 DISTANCE HEADWAY CHARACTERISTICS

The longitudinal space occupied by individual vehicles in the traffic stream consists of space occupied by the physical vehicles and the gaps between vehicles. The two microscopic measures considered are distance headway and distance gap.

Distance headway is defined as the distance from a selected point on the lead vehicle to the same point on the following vehicle. Usually, the front edges or bumpers are selected since they are more often detected in automatic detection systems. Hence distance headway includes the length of the lead vehicle and the gap length between the lead and following vehicle as shown in Figure 6.1 and in the following equation:

$$d_{n+1}(t) = L_n + g_{n+1}(t) \tag{6.1}$$

where $d_{n+1}(t) =$ distance headway of vehicle $n + 1$ at time t (feet)
$L_n =$ physical length of vehicle n (feet)
$g_{n+1}(t) =$ gap length between vehicle n and $n + 1$ at time t (feet)

The distance gap is defined as the gap length between the rear edge of the lead vehicle and the front edge of the following vehicle and is shown in Figure 6.1 and as $g_{n+1}(t)$ in equation (6.1). As will be shown in the following paragraphs, distance headway rather than distance gap is normally used as the primary microscopic characteristic of density because of its more direct relationship to time headway and density.

Time headway rather than distance headway is more often encountered because of the greater ease of measuring time headway. Distance headway can be obtained photographically, however it is more often obtained by calculation based on time headway and individual speed measurements, as shown in the following equation:

$$d_{n+1} = h_{n+1}\dot{x}_n \tag{6.2}$$

where $d_{n+1} =$ distance headway of vehicle $n + 1$ (feet)
$h_{n+1} =$ time headway of vehicle $n + 1$ at point p (seconds)
$\dot{x}_{n+1} =$ speed of vehicle n during the time period h_{n+1} (ft/sec)

It is also important to introduce traffic density at this time briefly because of its relationship to distance headway. Traffic density, hereafter referred to as density, is the macroscopic density characteristic and will be presented in detail in Chapter 7. Density is defined as the number of vehicles occupying a length of roadway usually in a single lane over a length of 1 mile. If the average distance headway is known, the density can be calculated from the following equation:

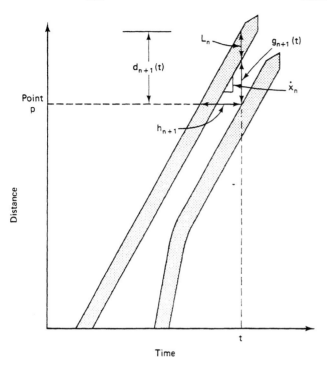

Figure 6.1   Distance Headway
Characteristics

$$k = \frac{5280}{\bar{d}} \tag{6.3}$$

where k = density (vehicles per mile per lane)

5280 = a constant representing the number of feet per mile

$\bar{d}$ = average distance headway (feet per vehicle)

Of course, the average distance headway can be obtained from individual distance headways from the following equation.

$$\bar{d} = \frac{\sum\limits_{n=1}^{N} d_n}{N} \tag{6.4}$$

where $\bar{d}$ = average distance headway (feet per vehicle)

d_n = individual distance headway (feet per vehicle)

N = number of observed distance headways

6.2 CAR-FOLLOWING THEORIES

Theories describing how one vehicle follows another vehicle were developed primarily in the 1950s and 1960s. Reuschel [8, 9] and Pipes [10] were pioneers in the development of car-following theories in the early 1950s. The work of Pipes will be described in a later subsection. Three parallel efforts were undertaken in the late 1950s and continued to the mid-1960s. Kometani and Sasaki [11–14] in Japan, Forbes [15–17] at the Institute for Research and Michigan State University, and a group of researchers associated with General Motors made significant contributions to car-following theory [18–23, 26–28]. The work of Forbes will be described later, and particular attention will be given to the comprehensive efforts and accomplishments of the General Motors researchers in a later section. The work at General Motors was of particular importance because of the accompanying field experiments and the discovery of the mathematical bridge between microscopic and macroscopic theories of traffic flow. Some limited research in car-following theory has been undertaken since the mid-1960s [24, 25].

Before describing the contributions of Pipes, Forbes, and the General Motors researchers, a comprehensive set of notations and definitions will be presented. These were uniquely developed by the General Motors researchers for use in car-following theories.

6.2.1 Notations and Definitions

The notations and definitions used below are summarized in Figure 6.2. Two vehicles are moving from left to right with vehicle n as the lead vehicle and having a length of

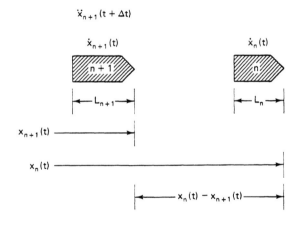

n = lead vehicle
n + 1 = following vehicle
L_n = length of lead vehicle (ft)
L_{n+1} = length of following vehicle (ft)
x_n = position of lead vehicle (ft)
x_{n+1} = position of following vehicle (ft)
$\dot{x}_n$ = speed of lead vehicle (ft/sec)
$\dot{x}_{n+1}$ = speed of following vehicle (ft/sec)
$\ddot{x}_{n+1}$ = acceleration rate (or deceleration rate) of the following vehicle (ft/sec^2)
t = at time t
t + Δt = Δt time after t time

Figure 6.2   Car Following Theory Notations and Definations

L_n and vehicle $n + 1$ as the following vehicle and having a length of L_{n+1}. The distance positions, speeds, and acceleration rates (or deceleration rates) are denoted as x (feet), $\dot{x}$ (feet per second), and $\ddot{x}$ feet per second^2, respectively. Since the vehicles change distance positions, speeds, and acceleration rates (or deceleration rates) over time, the subscript t is used to specify time.

Four other points are worth noting. The acceleration rate (or deceleration rate) of the following vehicle $(\ddot{x}_{n+1})$ is specified as occurring at time $t + \Delta t$, not t. The Δt represents an interval of time between the time a unique car-following situation occurs (t) and the time the driver of the following vehicle decides to apply a specified acceleration rate (or deceleration rate) at $(t + \Delta t)$. This interval of time is often referred to as the reaction time. The distance headway between the lead vehicle and the following vehicle is denoted as $[x_n(t) - x_{n+1}(t)]$. The relative velocity of the lead vehicle and the following vehicle is denoted as $[\dot{x}_n(t) - \dot{x}_{n+1}(t)]$. If the relative velocity is positive, the lead vehicle has a higher speed and the distance headway is increasing. A negative value implies that the following vehicle has a higher speed and the distance headway is decreasing. Finally, the acceleration rate (or deceleration rate) $[\ddot{x}_{n+1}(t + \Delta t)]$ can be positive or negative, with a positive value indicating that the following vehicle is accelerating and increasing its speed while a negative value indicates the reverse.

6.2.2 Pipes' Theory

Pipes [10] characterized the motion of vehicles in the traffic stream as following rules suggested in the California Motor Vehicle Code, namely: "A good rule for following another vehicle at a safe distance is to allow yourself at least the length of a car between your vehicle and the vehicle ahead for every ten miles per hour of speed at which you are traveling." The resulting equation for distance headway as a function of speed is shown in the following equation.

$$d_{\text{MIN}} = \left[x_n(t) - x_{n+1}(t)\right]_{\text{MIN}} = L_n \left[\frac{\dot{x}_{n+1}(t)}{(1.47)(10)}\right] + L_n \tag{6.5}$$

Assuming a vehicle length of 20 feet, equation (6.5) can be expressed as follows:

$$d_{\text{MIN}} = 1.36 \left[\dot{x}_{n+1}(t)\right] + 20 \tag{6.6}$$

Selecting speeds from 0 to 88 feet per second (60 miles per hour), the minimum safe distance headways can be computed. The results are shown in Figure 6.3a. Associated minimum safe time headways can be determined by combining equations (6.2) and (6.6) as shown in the following equation:

$$h_{\text{MIN}} = 1.36 + \frac{20}{\dot{x}_{n+1}(t)} \tag{6.7}$$

Again selecting speeds from 0 to 88 feet per second (60 miles per hour), the minimum safe time headways can be computed and the results are shown in Figure 6.3b.

According to Pipes' [1] car-following theory, the minimum safe distance headway increases linearly with speed. The associated minimum safe time headway continuously decreases with speed and theoretically reaches an absolute minimum time headway of

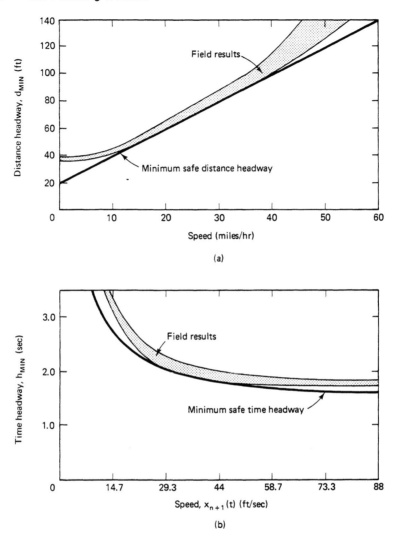

Figure 6.3   Minimum Headways according to Pipes' Theory

1.36 seconds at a speed of infinity. A shaded band is superimposed on both diagrams, which represents the results of field studies in which minimum distance headways observed for the average driver when following another vehicle were recorded as well as the speed of the two vehicles [3]. The minimum headways according to Pipes' theory are slightly less than corresponding field measurements in the speed range of 15 to 35 miles per hour but are considerably less at low speeds and high speeds. Selection of a slightly longer vehicle length on the order of 21 to 22 feet would provide a closer comparison of measured and theoretical minimum headways. Considering the simplicity of the car-following model, the agreement with calibration of the vehicle length is surprisingly close.

Another observation should be noted from equation (6.7) and from Figure 6.3b in terms of capacity. Note that as speed increases the minimum safe distance headway increases but the minimum safe time headway decreases. Since flow rate is the reciprocal of the time headway, the possible flow rate increases with increased speeds. This is true at lower speeds where almost all vehicles are in a car-following mode and are maintaining headways near the minimum headway. However, above some midrange speed all vehicles are not traveling at the same speed, nor are all vehicles traveling at minimum headways. Therefore, while theoretically the possible flow rate continues to increase with higher speeds, the traffic capacity is reached in the real world at midrange speeds on the order of 30 to 40 miles per hour. It is interesting to speculate what increases in capacity could be obtained from automatic vehicle control systems which could be designed to maintain minimum headway at high rates of speed.

6.2.3 Forbes' Theory

Forbes [15–17] approached car-following behavior by considering the reaction time needed for the following vehicle to perceive the need to decelerate and apply the brakes. That is, the time gap between the rear of the lead vehicle and the front of the following vehicle should always be equal to or greater than the reaction time. Therefore, the minimum time headway is equal to the reaction time (minimum time gap) and the time required for the lead vehicle to traverse a distance equivalent to its length. This relationship is shown mathematically in the following equation.

$$h_{\text{MIN}} = \Delta t + \frac{L_n}{\dot{x}_n(t)} \tag{6.8}$$

Forbes conducted many field studies of minimum time gaps and found considerable variations between drivers and sites. Minimum time gaps varied from 1 to 2 or 3 seconds. Assuming a reaction time of 1.5 seconds and a vehicle length of 20 feet, equation (6.8) can be rewritten as follows:

$$h_{\text{MIN}} = 1.50 + \frac{20}{\dot{x}_n(t)} \tag{6.9}$$

Selecting speeds from 0 to 88 feet per second (60 miles per hour) the minimum safe time headway can be computed, and the results are shown in Figure 6.4b. Associated minimum distance headways can be determined by combining equations (6.2) and (6.9) as shown in the following equation:

$$d_{\text{MIN}} = 1.50 \left[\dot{x}_n(t) \right] + 20 \tag{6.10}$$

Again selecting speeds from 0 to 88 feet per second (60 miles per hour), the minimum safe distance headways can be computed, and the results are shown in Figure 6.4a.

The results of Forbes' car-following theory is very similar to Pipes' results with minimum safe distance headway increasing linearly with speed while the minimum safe time headway continuously decreases with speed. Again a shaded band is superimposed on both diagrams which represents earlier described results of field studies [3]. There is

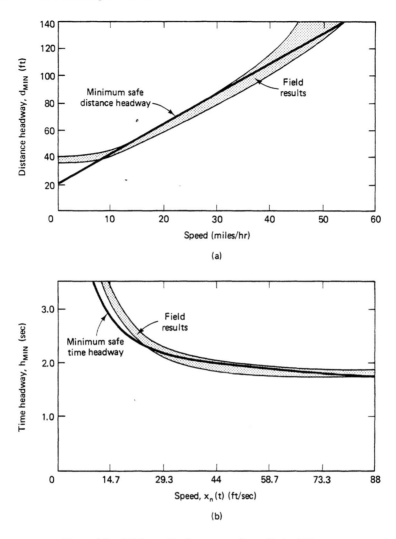

Figure 6.4   Minimum Headways according to Forbes' Theory

very close agreement between Forbes' model and the field study results in the midspeed range, but at lower and higher speeds there is considerable difference in a pattern similar to Pipe's model. The concluding comments about capacity discussed with Pipes' model are also applicable with Forbes' model.

6.2.4 General Motors' Theories

The car-following theories developed by researchers associated with the General Motors group were much more extensive and are of particular importance because of the accompanying comprehensive field experiments and the discovery of the mathematical bridge between microscopic and macroscopic theories of traffic flow. Another strength of this research effort was the integrated contributions of a large number of researchers both within and outside General Motors.

The research team developed five generations of car-following models, all of which took the form

$$\text{response} = \text{func(sensitivity, stimuli)}$$

The response was always represented by the acceleration (or deceleration) of the following vehicle, while the stimuli was always represented by the relative velocity of the lead and following vehicle. The difference in the different levels of models was the representation of the sensitivity.

The first model assumed that the sensitivity term was a constant and the model formulation is shown in the following equation.

$$\ddot{x}_{n+1}(t + \Delta t) = \alpha \left[\dot{x}_n(t) - \dot{x}_{n+1}(t) \right]$$
(6.11)

The stimuli term could be positive, negative, or zero, which could cause the response to be an acceleration, deceleration, or constant speed. This is shown in the following expressions.

$$\text{if } \dot{x}_n(t) > \dot{x}_{n+1}(t), \quad \text{then } \ddot{x}_{n+1}(t + \Delta t) \text{ is positive}$$
(6.12)

$$\text{if } \dot{x}_n(t) < \dot{x}_{n+1}(t), \quad \text{then } \ddot{x}_{n+1}(t + \Delta t) \text{ is negative}$$
(6.13)

$$\text{if } \dot{x}_n(t) = \dot{x}_{n+1}(t), \quad \text{then } \ddot{x}_{n+1}(t + \Delta t) = 0$$
(6.14)

Field experiments were conducted on the General Motors test track to quantify the parameter values for the reaction time (Δt) and the sensitivity parameter (α). The experiment consisted of two vehicles with a cable on a pulley attached between them. The driver of the lead vehicle was instructed to follow a prespecified speed pattern, while the driver in the following vehicle was unaware of the prespecified speed pattern and instructed to maintain a safe minimum distance behind the lead vehicle. The minimum, average, and maximum values for the Δt and α parameters of different test drivers are shown in Table 6.1. Note that the dimension of the sensitivity term is seconds^{-1}.

TABLE 6.1 Parameter Values for First
GM Model

Measured Value	Reaction Time, Δt (sec)	Sensitivity, α (sec^{-1})
Minimum	1.0	0.17
Average	1.55	0.37
Maximum	2.2	0.74

The significant range in the sensitivity value (0.17 to 0.74) alerted the investigators that the spacing between vehicles should be introduced into the sensitivity term. This led to the development of the second model, which proposed that the sensitivity term should have two states. That is, when the two vehicles were close together, a high sensitivity value (α_1) should be employed, while if the two vehicles were far apart, a lower sensitivity value (α_2) should be used. This formulation can be shown by the equation

$$\ddot{x}_{n+1}(t + \Delta t) = \begin{matrix} \alpha_1 \\ \text{or} \\ \alpha_2 \end{matrix} \left[\dot{x}_n(t) - \dot{x}_{n+1}(t) \right] \tag{6.15}$$

Very quickly the investigators saw the difficulty in selecting the α_1 and α_2 values and the difficulty associated with discontinous states. This led to further field experiments to determine means of incorporating the distance headway, into the sensitivity term. The experiments and the analysis of the results provided a significant breakthrough. The numerical values for the sensitivity term (α) were measured as an inverse function of the distance headway, and the data points closely followed a linear curve sloping upward to the right from the origin as shown in Figure 6.5. The slope of the line was designated as α_0, and for simplification d was substituted for $[x_n(t) - x_{n+1}(t)]$. The relationship was then determined for α_0, d, and α, as shown in the following equations:

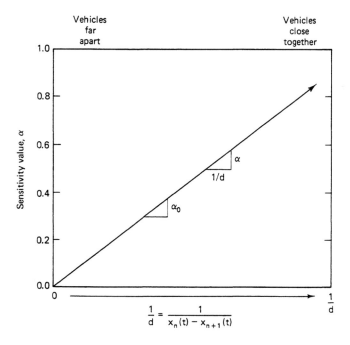

Figure 6.5   Sensitivity Value as a Function of Spacing in the General Motors Model

$$\alpha_0 = \frac{\alpha}{1/d} = \alpha d \tag{6.16}$$

$$\alpha = \frac{\alpha_0}{d} = \frac{\alpha_0}{x_n(t) - x_{n+1}(t)} \tag{6.17}$$

Then equation (6.17) for α was substituted into equation (6.11), and the third model resulted, as shown in the equation

$$\ddot{x}_{n+1}(t + \Delta t) = \frac{\alpha_0}{x_n(t) - x_{n+1}(t)} \left[\dot{x}_n(t) - \dot{x}_{n+1}(t) \right] \tag{6.18}$$

Note that the sensitivity term in this third model is a function of a constant, α_0, and distance headway. As the vehicles come closer and closer together, the sensitivity term becomes larger and larger. One other point worth noting is that the dimension of the α_0 constant in the sensitivity term is feet per second, or velocity. This was a clue used in later work that bridged between this particular microscopic model and the Greenberg macroscopic model, which will be discussed later. Field experiments were conducted to obtain the parameter values for α_0 and Δt for test drivers on the General Motors test track. Results are shown in Table 6.2.

TABLE 6.2   Parameter Values for Third GM
Model

Location	Reaction Time, Δt (sec)	Sensitivity Parameter, α_0 (ft/sec)
General Motors test track	1.5	40.3
Holland Tunnel	1.4	26.8
Lincoln Tunnel	1.2	29.8

The fourth model was a further development toward improving the sensitivity term by introducing the speed of the following vehicle. The concept was that as the speed of the traffic stream increased, the driver of the following vehicle would be more sensitive to the relative velocity between the lead and following vehicle. The formulation of this fourth model is shown in the following equation:

$$\ddot{x}_{n+1}(t + \Delta t) = \frac{\alpha'[\dot{x}_{n+1}(t + \Delta t)]}{x_n(t) - x_{n+1}(t)} \left[\dot{x}_n(t) - \dot{x}_{n+1}(t) \right] \tag{6.19}$$

In this formulation the sensitivity term has three components: a constant α', the speed of the following vehicle, and the distance headway. Note that the constant α' is dimensionless.

The fifth and final model was a continued effort to improve and generalize the sensitivity term. The question raised was whether the speed and distance headway components should be raised to the first power or whether an improved and more generalized approach could be accomplished by introducing generalized exponents. This was implemented by introducing m and l exponents as shown in the equation

$$\ddot{x}_{n+1}(t + \Delta t) = \frac{\alpha_{l,m}[\dot{x}_{n+1}(t + \Delta t)]^m}{[x_n(t) - x_{n+1}(t)]^l} \left[\dot{x}_n(t) - \dot{x}_{n+1}(t)\right] \tag{6.20}$$

This was the final car-following model developed and all previous General Motors car-following models were special cases of this generalized model as shown in Figure 6.6. In the figure, the horizontal scale is the speed exponent (m), while the vertical scale is the distance headway exponent l. The first and second models are represented by the matrix point $(m = 0, l = 0)$, while the third and fourth models are represented by the matrix points $(m = 0, l = 1)$ and $(m = 1, l = 1)$, respectively. The fifth and final model is represented by the total m,l matrix. Negative exponents are shown in Figure 6.6 but would not be expected to occur since the sensitivity component would be changed to its reciprocal; that is, the speed component would be transposed to the denominator and the distance headway term would be transposed to the numerator.

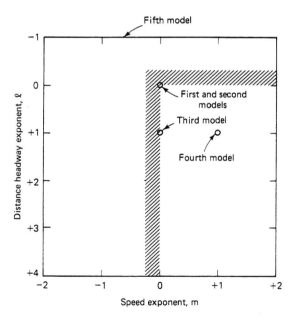

Figure 6.6 m,l Exponents of Sensitivity Components

At this point in the development further field experiments were undertaken in the tunnels and bridges of New York in cooperation with the Port of New York Authority. Again the two vehicles with a cable on a pully attached between them were introduced into the traffic stream, and the values of the α_0 and Δt parameters were determined for the third model as shown previously in equation (6.18) and Table 6.2. At the same time the Port of New York researchers were developing and evaluating a macroscopic flow model of speed as a function of density and included a parameter, μ_0, which was defined as optimum speed. Optimum speed was defined as that speed which existed when the traffic flow level was at capacity. The model was called the Greenberg model and its formulation is shown in the following equation.

$$\mu = \mu_0 \ln \left(\frac{k_j}{k}\right) \tag{6.21}$$

where μ = space mean speed (mph)
μ_0 = optimum speed (mph)
$\ln$ = natural log
k_j = jam density in vehicles per lane-mile (that density when vehicles are bumper to bumper and stopped)
k = density (vehicles per lane-mile)

The field experiments were also directed to quantify the optimum speed value (μ_0).

The surprising result was that the numerical values for α_0 (in the microscopic model) and μ_0 (in the macroscopic model) were almost identical for different types of highway facilities. Further, the dimensions of each (α_0 and μ_0) were the same (feet per second) [22]. This caused Gazis and others to suspect that there was a relationship between the third microscopic car-following model [equation (6.18)] and the Greenberg macroscopic model [equation (6.21)]. The derivation of this relationship will now be presented. Equation (6.18) is the starting point.

$$\ddot{x}_{n+1}(t + \Delta t) = \frac{\alpha_0}{x_n(t) - x_{n+1}(t)} \left[\dot{x}_n(t) - \dot{x}_{n+1}(t)\right]$$

Integrating with respect to t yields

$$\dot{x}_{n+1} = \alpha_0 \left[\ln x_n - x_{n+1}\right] + C_1 \tag{6.22}$$

If μ is substituted for $\dot{x}_{n+1}$ and $1/k$ is substituted for $x_n - x_{n+1}$, then

$$\mu = \alpha_0 \ln \left(\frac{1}{k}\right) + C_1 \tag{6.23}$$

Let $\alpha_0 \ln C_2$ be substituted for C_1; then

$$\mu = \alpha_0 \ln \left(\frac{1}{k}\right) + \alpha_0 \ln C_2 \tag{6.24}$$

$$\mu = \alpha_0 \ln \left(\frac{C_2}{k}\right) \tag{6.25}$$

When $k = k_j$, $\mu = 0$ (vehicles bumper to bumper but no movement):

$$0 = \alpha_0 \ln \left(\frac{C_2}{k_j}\right) \tag{6.26}$$

Solving for C_2 yields

$$\ln \left(\frac{C_2}{k_j}\right) = 0 \qquad \left(\frac{C_2}{k_j}\right) = 1 \quad \text{and} \quad C_2 = k_j \tag{6.27}$$

Substituting k_j for C_2 in equation (6.25) gives

$$\mu = \alpha_0 \ln \left(\frac{k_j}{k} \right) \tag{6.28}$$

Since $q = uk$, then multiplying each side of equation (6.28) by k gives

$$q = k\alpha_0 \ln \left(\frac{k_j}{k} \right) \tag{6.29}$$

Differentiating, $dq/dk = 0$ when $k = k_0$; then

$$\frac{dq}{dk} = 0 = \alpha_0 \ln \left(\frac{k_j}{k_0} \right) + \alpha_0 k_0 \left[\frac{k_0}{k_j} \cdot \left(-\frac{k_j}{k_0^2} \right) \right] \tag{6.30}$$

$$0 = \alpha_0 \ln \left(\frac{k_j}{k_0} \right) - \alpha_0 \tag{6.31}$$

Solving for k_0, we have

$$\ln \left(\frac{k_j}{k_0} \right) = 1 \qquad \left(\frac{k_j}{k_0} \right) = e \quad \text{and} \quad k_0 = \frac{e}{k_j} \tag{6.32}$$

Returning to equation (6.28) and substituting μ_0 for μ and e/k_j for k_0 gives

$$\mu_0 = \alpha_0 \ln \left(\frac{k_j}{e/k_j} \right) = \alpha_0 \ln e = \alpha_0 \tag{6.33}$$

Returning to equation (6.28) and substituting μ_0 for α_0 yields

$$\mu = \mu_0 \ln \left(\frac{k_j}{k} \right) \tag{6.34}$$

This is the Greenberg macroscopic model given in equation (6.21).
    This bridge between the third microscopic car-following model and the Greenberg macroscopic model was a very important discovery, and in Chapter 9 this bridge will be greatly expanded to provide a connection between the matrix of microscopic models shown in Figure 6.6 and most macroscopic theories of traffic flow.

6.3 CAR-FOLLOWING THEORY APPLICATION

An application of car-following theory will be presented in this section. Although the example will be rather simplistic, it is intended to provide a demonstration of such applications. This example is based on the work of the General Motors researchers [19, 21] and the applications developed by Gerlough and Huber [1].
    The example is concerned with two vehicles, a lead vehicle and a following vehicle, which are stopped and have a initial distance headway of 25 feet. The lead vehicle is automatically controlled to accelerate at a constant rate of 3.3 feet per second^2 until a speed of 44 feet per second (30 miles per hour) is reached at which time a constant speed of 44 feet per second is maintained for 10 seconds. Following the 10-second

period of constant speed the lead vehicle decelerates at a constant rate of 4.6 feet per second² until it is stopped. The driver of the following vehicle is instructed to follow the lead vehicle with a safe minimum distance headway. The objective of this example problem is to track over space and time the trajectories of these two vehicles through this acceleration, constant speed, and deceleration driving cycle. The first task is to track the lead vehicle over space and time since its trajectory has been prespecified and the following vehicle's trajectory is based on the lead vehicle's trajectory. Then the second task is to track the following vehicle over space and time as a function of the trajectory of the lead vehicle through car-following theory and equations of motion.

The trajectory of the lead vehicle is calculated using the equations below, and its speed and distance position for each second is tabulated in Table 6.3 in columns 3 and 4.

$$\dot{x}_1(t+T) = \dot{x}_1(t) + \left[\frac{\ddot{x}_1(t) + \ddot{x}_1(t+T)}{2}\right] T \tag{6.35}$$

Since T is selected as 1 second, and acceleration rate (and deceleration rate) is constant for the lead vehicle, equation (6.35) can be simplified to

$$\dot{x}_1(t+1) = \dot{x}_1(t) + \ddot{x}_1 \tag{6.36}$$

Then the distance position can be calculated as

$$x_1(t+T) = x_1(t) + \dot{x}_1(t)\, T + \left[\frac{\ddot{x}_1(t) + \ddot{x}_1(t+T)}{2}\right] \frac{T^2}{2} \tag{6.37}$$

and can be simplified in the same way to

$$x_1(t+1) = x_1(t) + \dot{x}_1(t) + \frac{\ddot{x}_1}{2} \tag{6.38}$$

The lead vehicle accelerates and after traveling 293.3 feet in 13.3 seconds reaches it cruise speed of 44 feet per second. After 10 seconds (time period 13.3 to 23.3) of a constant cruise speed of 44 feet per second, the lead vehicle decelerates and completely stops after traveling a total distance of 943.4 feet in an elapsed time of 32.9 seconds.

Now attention is turned to the second task of tracking the trajectory of the following vehicle. Three equations are needed. The first equation is the car-following model, and the first General Motors car-following model shown earlier as equation (6.11) is selected (for ease of calculation) and is shown again below:

$$\ddot{x}_{n+1}(t + \Delta t) = \alpha \left[\dot{x}_n(t) - \dot{x}_{n+1}(t)\right] \tag{6.39}$$

Two parameter values must be selected: the reaction time (Δt) and the sensitivity (α). Inspection of the earlier Table 6.1 suggests reaction times of 1.0 to 2.2 seconds and sensitivity parameter values of 0.17 to 0.74. A reaction time of 1.0 seconds is selected for ease of calculations (a multiple of the previously selected update time of 1 second) and a sensitivity parameter value of 0.5, which is just slightly higher than the average value

TABLE 6.3 Car-Following Trajectories

(1)	(2)	(3)	(4)	(5)	(6)	(7)	(8)	(9)
	Lead Vehicle			Following Vehicle			Relative	
Time (sec)	Accel., $\ddot{x}_1$	Speed, $\dot{x}_1$	Distance, x_1	Accel., $\ddot{x}_2$	Speed, $\dot{x}_2$	Distance, x_2	Speed, $\dot{x}_1 - \dot{x}_2$	Distance, $x_1 - x_2$
0		0	0	0.0000	0.0000	−25.00	0.0000	25.00
	3.3			0.0000[a]				
1		3.3	1.6	0.0000	0.0000	−25.00	3.3000	26.60
	3.3			0.8250				
2		6.6	6.6	1.6500	0.8250	−24.59	5.7750	31.19
	3.3			2.2688				
3		9.9	14.8	2.8875	3.0938	−22.63	6.8062	37.43
	3.3			3.1453				
4		13.2	26.4	3.4031	6.2391	−17.96	6.9609	44.36
	3.3			3.4418				
5		16.5	41.2	3.4804	9.6809	−10.00	6.8191	51.20
	3.3			3.4450				
6		19.8	59.4	3.4096	13.1259	1.40	6.6741	58.00
	3.3			3.3733				
7		23.1	80.8	3.3370	16.4992	16.21	6.6008	64.59
	3.3			3.3187				
8		26.4	105.6	3.3004	19.8179	34.37	6.5821	71.23
	3.3			3.2957				
9		29.7	133.6	3.2910	23.1136	55.83	6.5864	77.76
	3.3			3.2921				
10		33.0	165.0	3.2932	26.4057	80.59	6.5943	84.41
	3.3			3.2952				
11		36.3	199.6	3.2971	29.7009	108.64	6.5912	90.96
	3.3			3.2983				
12		39.6	237.6	3.2996	33.0000	139.99	6.6008	97.61
	3.3			3.3000				
13		42.9	278.8	3.3004	36.3000	174.64	6.6000	104.16
	3.3							
13.33		44.0	293.3	—	—	—	—	—
	0.0			3.3002				
14		44.0	322.6	3.3000	39.6002	212.59	4.3998	110.01
	0.0			2.7500				
15		44.0	366.6	2.1999	42.3502	253.56	1.6498	113.03
	0.0			1.5124				
16		44.0	410.6	0.8249	43.8626	296.67	0.1374	113.93
	0.0			0.4468				
17		44.0	454.6	0.0687	44.3094	340.76	−0.3094	113.84
	0.0			−0.0430				
18		44.0	498.6	−0.1547	44.2664	385.05	−0.2664	113.55
	0.0			−0.1440				
19		44.0	542.6	−0.1332	44.1224	429.24	−0.1224	113.36
	0.0			−0.0972				

TABLE 6.3　Car-Following Trajectories (continued)

(1)	(2)	(3)	(4)	(5)	(6)	(7)	(8)	(9)
		Lead Vehicle			Following Vehicle		Relative	
Time (sec)	Accel., $\ddot{x}_1$	Speed, $\dot{x}_1$	Distance, x_1	Accel., $\ddot{x}_2$	Speed, $\dot{x}_2$	Distance, x_2	Speed, $\dot{x}_1 - \dot{x}_2$	Distance, $x_1 - x_2$
20		44.0	586.6	−0.0612	44.0252	473.31	−0.0252	113.29
	0.0			−0.0369				
21		44.0	630.6	−0.0126	43.9883	517.32	0.0117	113.28
	0.0		·	−0.0034				
22		44.0	674.6	+0.0058	43.9849	561.31	0.0151	113.29
	0.0			+0.0067				
23		44.0	718.6	+0.0075	43.9916	605.30	0.0084	113.30
	0.0							
23.33		44.0	733.3	—	—	—	—	
	−4.6			0.0058				
24		40.9	761.6	0.0042	43.9974	649.29	−3.0974	112.31
	−4.6			−0.7722				
25		36.3	800.2	−1.5487	43.2252	692.90	−6.9252	107.30
	−4.6			−2.5056				
26		31.7	834.2	−3.4626	40.7196	734.87	−9.0296	99.33
	−4.6			−3.9862				
27		27.1	863.6	−4.5098	36.7334	773.60	−9.6334	90.00
	−4.6			−4.6632				
28		22.5	888.4	−4.8167	32.0702	808.00	−0.5702	80.40
	−4.6			−4.8009				
29		17.9	908.6	−4.7851	27.2693	837.63	−9.3693	70.93
	−4.6			−4.7349				
30		13.3	924.2	−4.6847	22.5344	862.57	−9.2344	61.63
	−4.6			−4.6510				
31		8.7	935.2	−4.6172	17.8834	882.78	−9.1834	52.42
	−4.6			−4.6045				
32		4.1	941.6	−4.5917	13.2789	893.36	−9.1789	43.24
	−4.6			−4.5906				
32.89		0.0	943.4	−4.5895	8.6883	909.34	−8.6883	34.06
	0.0			−4.4668				
33		0.0	943.4	−4.3442	4.2215	915.79	−4.2215	27.61
	0.0			−3.2275				
34		0.0	943.4	−2.1107	0.9940	918.40	−0.9940	25.00

[a]Average acceleration (deceleration) rate during sampling time interval.

found in the field experiments. The lead vehicle (n) is here identified as vehicle 1 and the following vehicle ($n + 1$) is referred to as vehicle 2. With these selected parameter values, equation (6.39) becomes

$$\ddot{x}_2(t + 1) = 0.5 \left[\dot{x}_1(t) - \dot{x}_2(t) \right] \qquad (6.40)$$

Therefore, the acceleration (or deceleration) of the following vehicle can be predicted for a given point in time if the relative speed of the two vehicles is known for the previous second. The speed of the following vehicle during each second can be calculated using equation (6.35) or (6.36), and the distance traveled by the following vehicle during each second can be calculated using equation (6.37) or (6.38).

The trajectory of the following vehicle in the example problem is calculated according to the above-mentioned equations for each second and the results are tabulated in Table 6.3 in columns 5 through 7. The relative speed and relative distance (distance headway) are tabulated in the last two columns. The acceleration rates, speeds, relative speeds, and relative distances are shown graphically in Figure 6.7. Figure 6.7a pictorially shows how the second vehicle at first lags behind the lead vehicle and then overcompensates exceeding the lead vehicle's constant acceleration rate. After time ($t = 8$), the acceleration rates of the two vehicles are essentially constant and equal. This pattern of lagging, overcompensating, and then constant equal acceleration rates can also be observed in the constant speed and deceleration driving cycle. Figure 6.7b shows how the following vehicle has almost a constant increasing speed (later in the deceleration driving cycle a constant decreasing speed), but its speed lags the leading vehicle's speed by about 2 seconds. Figure 6.7c shows that this relative speed during acceleration is almost a constant 6.6 to 6.9 feet per second and during deceleration is almost a constant 9.1 to 9.6 feet per second. Figure 6.7d is perhaps the most critical, since it indicates whether a minimum safe distance headway is maintained, on the one hand, and that a reasonable capacity is obtained, on the other. The distance headway between vehicles is quite stable and is very adequate considering the safe car-following rule of one car length per 10 miles per hour of speed. During the constant speed cycle of 44 feet per second (30 miles per hour), the distance headway is almost constant between 113 and 114 feet. The corresponding density would be about 46 to 47 vehicles per mile, and the resulting flow would be on the order of 1400 vehicles per hour. This is considerably less than the 1800 vehicles per hour capacity value that might normally be encountered. A distance headway of about 90 feet would be needed for a capacity value of 1800 vehicles per hour. This would indicate that the car-following formulation and parameter values used in this example provide for a safe and stable car following between two vehicles but results in a much lower capacity value than might be encountered on the open roadway. This leads to the next discussion on car-following parameter values and their effect on safe and stable headways and the resulting capacities.

6.4 TRAFFIC STABILITY

Consider a driver of a following vehicle who is slow to respond (high reaction time) and when the driver finally responds, over-responds by exerting significantly large acceleration or deceleration rates. This might be called "unstable" behavior, characterized by higher reaction times and higher sensitivity responses. On the other hand, consider a driver of a following vehicle who is very attentive to the car-following process and does not resort to sudden accelerations or decelerations except in extreme emergencies. This might be called "stable" behavior, characterized by lower reaction times and lower sensitivity responses. Thus, if the product (later referred to as the C value) of

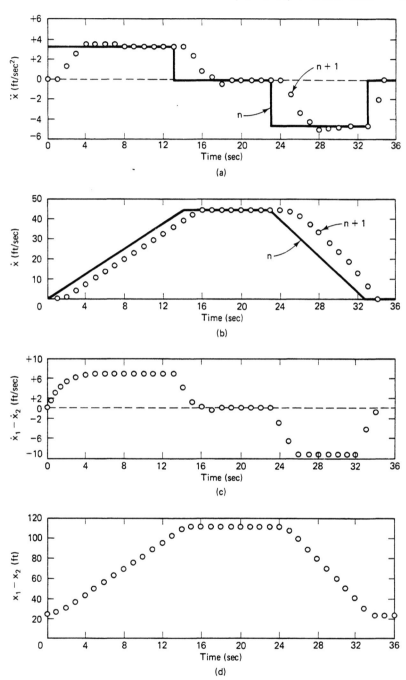

Figure 6.7   Car-Following Trajectories

reaction time and sensitivity is high, unstable traffic conditions are likely to occur, while if the product is low, stable traffic conditions are likely to occur. Another issue is whether only the behavior of one following vehicle is being considered or a long line of following vehicles. The driver of a single following vehicle may behave in a rather erratic or unstable manner, but a collision may not result. However, a line of vehicles following an erratic driver will continuously be accelerating and decelerating, and while the first car-following vehicle may not collide with the lead vehicle, the erratic behavior may have an increasing oscillation and two vehicles somewhere within a line of car-following vehicles may collide. Thus there are two issues in traffic stability, upper limits on car-following parameter values (Δt, α) and the number of vehicles in car-following.

The General Motors researchers studied the question of traffic stability during car-following with particular attention given to the two issues raised in the preceding paragraph [18–21]. The researchers identified two types of traffic stability: local stability and asymptotic stability. Local stability is concerned with the car-following behavior of just two vehicles: the lead vehicle and one following vehicle. This was the case in the example presented in Section 6.3. Asymptotic stability is concerned with the car-following behavior of a line of vehicles consisting of a lead vehicle and theoretically, an infinite number of following vehicles. The influence of parameter values Δt and α in the car-following formulation shown in equation (6.39) on local and asymptotic stability will now be pursued.

The General Motors researchers determined limits for stable and unstable situations involving local and asymptotic stability. The limits were established based on the product (C) of the reaction time (Δt) and sensitivity value (α). The equation is shown below and the limits are shown in Table 6.4.

$$C = \alpha(\Delta t) \tag{6.41}$$

Local stability is divided into three regions: nonoscillatory, damped oscillatory, and increased oscillatory. The example given in Section 6.3 ($C = 0.5$) is a damped oscillatory case in which the following vehicle oscillates (see Figure 6.7a) but over time the oscillation decreases and the car following becomes stable. If the C value had been less than 0.37, no oscillation would have occurred, but if the C value had been greater than 1.57, there would have been increasing oscillation. Graphical examples of local stability under four different C-value situations are shown in Figure 6.8.

Asymptotic stability is divided into two regions: damped oscillatory and increased oscillatory. The example given in Section 6.3 ($C = 0.5$) is exactly on the border between these two states. If the C value had been less, none of the following vehicles would have collided, while if the C value had been greater, two or more of the following vehicles would have eventually collided. A graphical example of asymptotic stability under increased oscillatory conditions ($C = 1.60$) is shown in Figure 6.9. In this example there are nine vehicles in the line and the lead vehicle is moving at a constant speed at an initial distance headway of 40 feet. The lead vehicle decelerates and then accelerates. Since the C value exceeds the limiting value, there is increased oscillation, and each pair of vehicles comes closer to a collision until vehicles 7 and 8 collide.

TABLE 6.4 Limits for Local and Asymptotic Stability

C Value $[C = \alpha(\Delta t)]$	Local Stability	Asymptotic Stability
0.0	Nonoscillatory	Damped oscillatory
(0.37)		
(0.50) 0.5	Damped oscillatory	
1.0		Increased oscillatory
1.5		
1.57		
	Increased oscillatory	
2.0		

Source: Reference 19.

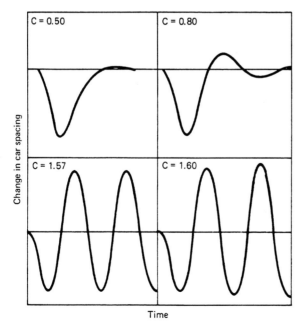

Figure 6.8 Local Stability as a Function of C-Values (From Reference 19)

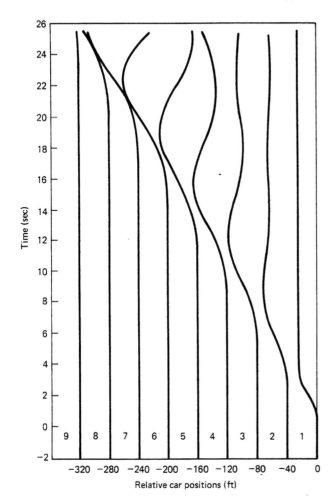

Figure 6.9   Example of Asymptotic Instability (From Reference 19)

In summary, Figure 6.10 shows the influence of parameter values on traffic stability. The shaded area on the figure indicates a combination of parameter values that result in local and asymptotic stability when the first car-following model [equation (6.39)] is employed.

6.5 MEASUREMENTS WITH PRESENCE-TYPE DETECTORS

The most common traffic detector used today is a presence-type detector which detects the presence and passage of vehicles over a short segment of roadway. When a vehicle enters the detection zone, the sensor is activated and remains so until the vehicle leaves the detection zone. The "on" time referred to as the vehicle occupancy time requires the vehicle to travel a distance equivalent to its length plus the length of the detection zone. Note that the length of the detection zone, not necessarily the length of the

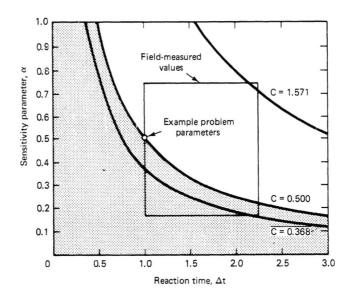

Figure 6.10   Influence of Parameter Values on Traffic Stability (From Reference 19)

physical detector, is used. The absence of a vehicle over a detector may be thought of as giving a "0" signal, while the presence of a vehicle may be thought of as giving a "1" signal. If the detector is scanned at regular intervals, the pictorial representation of the detector output can be illustrated as shown in Figure 6.11. The vehicle occupancy time is a function of vehicle speed, vehicle length, and detection zone length. The off-time between vehicles is the time gap described in Chapter 2. Both microscopic and macroscopic characteristics of traffic flow can be obtained from presence-type detectors.

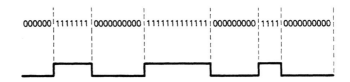

Figure 6.11   Output Signals from a Presence-Type Detector

6.5.1 Measurements with Single Detector

First consider a single presence-type detector and the microscopic traffic flow characteristics that can be obtained. Referring to Figure 6.12, the time headway between vehicles can be easily obtained by

$$h_{n+1} = (t_{on})_{n+1} - (t_{on})_n \qquad (6.42)$$

where h = time headway (seconds)

$\quad t_{on}$ = instant that the vehicle is detected (seconds)

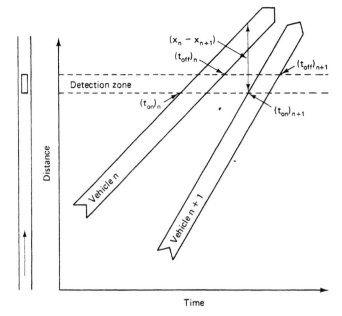

Figure 6.12   Vehicles Passing Over a Single Presence-Type Detector

Referring to Figure 6.12, the vehicle occupancy time can be easily obtained by

$$(t_{occ})_n = (t_{off})_n - (t_{on})_n \tag{6.43}$$

$$(t_{occ})_{n+1} = (t_{off})_{n+1} - (t_{on})_{n+1} \tag{6.44}$$

where t_{occ} is the individual occupancy time (seconds).

As mentioned earlier, vehicle occupancy time is a function of vehicle speed, vehicle length, and detection zone length as shown in the following equations:

$$\dot{x}_n = \frac{L_n + L_D}{(t_{occ})_n} \tag{6.45}$$

$$\dot{x}_{n+1} = \frac{L_{n+1} + L_D}{(t_{occ})_{n+1}} \tag{6.46}$$

where $\dot{x}_n, \dot{x}_{n+1}$ = vehicle speed (feet per second)
L_n, L_{n+1} = vehicle length (feet)
L_D = detection zone length (feet)

With a specific sensitivity adjustment setting, the detection zone length is a constant and can numerically be determined through calibration. If all vehicles passing over the detector have approximately the same length (i.e., passenger vehicles), individual vehicle speed is a function of a constant and the vehicle occupancy time as shown in the following equations.

$$\dot{x}_n = \frac{K}{(t_{occ})_n} \tag{6.47}$$

$$\dot{x}_{n+1} = \frac{K}{(t_{occ})_{n+1}} \tag{6.48}$$

where K is a constant, representing average vehicle length plus length of detection zone in feet. Referring again to Figure 6.12, the distance headway can be estimated as shown in the following equations:

$$\cdot \ \dot{x}_n = \frac{x_n - x_{n+1}}{h_{n+1}} \tag{6.49}$$

and

$$x_n - x_{n+1} = (h_{n+1})(\dot{x}_n) \tag{6.50}$$

The accuracy of this estimation is dependent on the assumed constant vehicle length and the assumption of constant speed of vehicle n during the time period from $(t_{on})_n$ to $(t_{on})_{n+1}$.

Thus keeping in mind the foregoing assumptions, the microscopic traffic flow characteristics of time headways, vehicle occupancy times, vehicle speeds, and distance headways can be estimated from a single presence-type detector. These microscopic flow relationships are shown graphically in a nomograph in Figure 6.13. The right portion of the nomograph is based on equation (6.45) while the left portion of the nomograph is based on equation (6.50). As an example, assume that the occupancy time for a vehicle $[(t_{occ})_n]$ is measured to be 0.34 second and the time headway for the next vehicle (h_{n+1}) is measured to be 3.0 seconds. This particular detector has been calibrated, a fairly constant vehicle length has been found, and the combined length of the average vehicle and the detection zone was found to be 25 feet. Then the graphical solution for estimating the speed of the lead vehicle $(\dot{x}_n)$ and the distance headway of the following vehicle $(x_n - x_{n+1})$ is shown in Figure 6.13 by the arrow lines. The solution gives a speed of 73.5 feet per second (50 miles per hour) and a distance headway of 220 feet. The shaded area indicates unsafe following conditions according to Pipes' theory.

Macroscopic traffic flow characteristics can in turn be estimated from these microscopic traffic flow characteristics. Flow rate can be obtained using either of the following two equations:

$$q = \frac{3600}{[1/(N-1)] \sum\limits_{n=n+1}^{N} h_n} \tag{6.51}$$

where q is the flow rate during passage of N vehicles (vehicles per hour), or

$$q = \frac{\sum\limits_{t=0}^{T} \delta}{T} \tag{6.52}$$

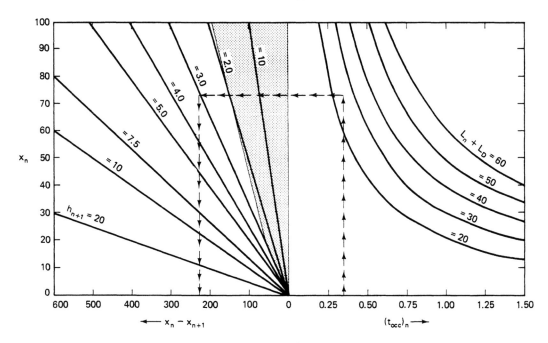

Figure 6.13   Nomograph for Microscopic Flow Relationships Using a Single Presence-Type
Detector

where δ = change in detector signal from state 0 to state 1

T = time period during detector signal change count (hours)

Average traffic speed (time mean speed) at a point can be obtained by averaging the
individual predicted vehicle speeds:

$$\bar{x} = \frac{\displaystyle\sum_{n=1}^{N} \dot{x}_n}{N} = \frac{K}{1/N \displaystyle\sum_{n=1}^{N} (t_{occ})_n} \qquad (6.53)$$

Macroscopic density characteristics will be discussed in the next chapter but at this
point we give simply the equation for density and percent occupancy time that can be
obtained from a single presence-type detector:

$$k = \frac{q}{\mu} \approx \frac{q}{\bar{x}} \qquad (6.54)$$

where k = traffic density (vehicles per mile)

μ = average traffic speed (miles per hour)

(Note that $\bar{x}$ given in equation (6.53) is time-mean-speed, while μ is space-mean-speed.)

Percent occupancy time is a surrogate for density and is obtained by determining the percent of the time that a presence-type detector is occupied. This is easy to measure, transmit, and calculate and is the single most important traffic flow characteristic used as the control variable for freeway control systems:

$$\% \ OCC = \frac{\displaystyle\sum_{n=1}^{N} (t_{occ})_n}{T} \times 100 \qquad (6.55)$$

where $\% \ OCC$ = percent occupancy time
N = number of vehicles detected in time period T
T = selected time period (seconds)

6.5.2 Measurements with Two Detectors

The previous equations of microscopic and macroscopic flow characteristics are based on information from a single presence type detector and the rather strong assumption of a known constant vehicle length. The assumption of a known constant vehicle length can be ignored if two closely-spaced presence-type detectors are employed as shown in Figure 6.14. The nomenclature used in the following equations is identical with that for the single-detector-derived equations except that the subscripts A and B are added to identify detection zones A and B as shown in Figure 6.14.

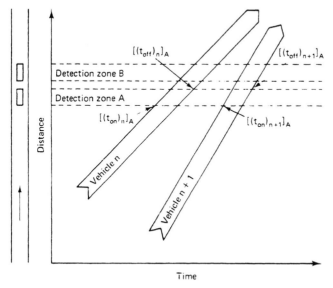

Figure 6.14   Vehicles Passing over Two Closely Spaced Presence-Type Detectors

Time headway and vehicle occupancy times are obtained in the identical way as for the single-detector scheme except that two estimates are obtained for each vehicle (from detection zones A and B). The next step, however, is different from the single-detector scheme because the speed of each vehicle can now be calculated since signals from two detectors are available:

$$\dot{x}_n = \frac{D}{[(t_{on})_n]_B - [(t_{on})_n]_A} \tag{6.56}$$

$$\dot{x}_{n+1} = \frac{D}{[(t_{on})_{n+1}]_B - [(t_{on})_{n+1}]_A} \tag{6.57}$$

where D is the distance from the upstream edge of detection zone A to the upstream edge of detection zone B (feet).

The length of the vehicle can now be calculated using one of the following equations:

$$[L_n]_A = \dot{x}_n[(t_{occ})_n]_A - [L_D]_A \tag{6.58}$$

or

$$[L_n]_B = \dot{x}_n[(t_{occ})_n]_B - [L_D]_B \tag{6.59}$$

$$[L_{n+1}]_A = \dot{x}_{n+1}[(t_{occ})_{n+1}]_A - [L_D]_A \tag{6.60}$$

or

$$[L_{n+1}]_B = \dot{x}_{n+1}[(t_{occ})_{n+1}]_B - [L_D]_B \tag{6.61}$$

The distance headway can be estimated as before using equation (6.50) with the detection zone subscripts added.

$$x_n - x_{n+1} = [h_{n+1}]_A(\dot{x}_n) \tag{6.62}$$

or

$$x_n - x_{n+1} = [h_{n+1}]_B\dot{x}_n \tag{6.63}$$

The macroscopic traffic flow characteristics can be estimated as before with the single-detector scheme.

In summary, the scheme utilizing two closely spaced presence-type detectors provides means for calculating the speed and length of individual vehicles, and in turn the distance headway can be calculated more accurately. This scheme does require that the length of at least one of the detector zones (A or B) and the distance from the upstream edge of the detection zone A to the upstream edge of the detection zone B be determined. In addition, the speed of the individual vehicle is assumed to be constant through the two detector zones.

6.6 SELECTED PROBLEMS

1. Prepare a bibliography and identify techniques for obtaining distance headways.
2. Estimate, through a library search, the average length of vehicles traveling on urban freeways today on a lane basis. Consider lengths by vehicle types and the composition of vehicle types on a lane basis.
3. Study and summarize the early contributions of Reuschel, Kometani, and Sasaki to car-following theory.

4. Pipes based his theory on the suggested rule in the California Motor Vehicle Code. In your driving or observing car-following behavior, do you feel that this rule is applied? If it is not always applied, discuss reasons for variations.

5. Calibrate Pipes' model against the field results shown in Figure 6.3 by varying the assumed vehicle length.

6. According to Pipes' model, minimum safe time headway decreases continuously with increasing speed. Speculate on time headway–speed conditions on futuristic intercity automatic highway vehicle control systems. What capacity level might be obtained?

7. Calibrate Forbes' model against the field results shown in Figure 6.4 by varying the assumed reaction time.

8. Derive the Greenberg macroscopic model from the third microscopic car-following model without the use of the text.

9. A line of vehicles are in car-following mode and all vehicles are traveling at 44 feet per second with distance headways of 114 feet. The lead vehicle suddenly decelerates at a rate of 4.6 feet per second^2 until it stops completely. Calculate and plot the projectory of the lead and only the first following vehicle for every second of time until both vehicles are stopped. Use the GM first car-following model. Consider three different drivers in the following vehicle with the following characteristics.

Driver	Reaction Time, Δt	Sensitivity Parameter, α
1	1.0	0.17
2	1.55	0.37
3	2.2	0.74

10. Consider (but do not perform) solving Problem 9 using higher-level GM models (i.e., second, third, fourth, and fifth models). Compare relative levels of effort for different model levels.

11. Solve Problem 9 but assume that all drivers have characteristics as shown for driver 3. Calculate and plot the projectory of the lead and all following vehicles for every second until the distance headway between two vehicles is less than 25 feet or until all vehicles are stopped. Use the GM first car-following model.

12. Conduct a library study and briefly summarize contributions to car-following theory since 1970.

13. Prepare a list of references about applications with presence-type detectors.

14. Conduct a library search for commercially available presence-type detectors and include detector specifications, including typical detection zone lengths.

15. A new presence-type detector has been installed. Design a field experiment to determine the detection zone length.

16. A presence-type detector with a detection zone length of 6 feet is scanned 30 times per second. A stream of vehicles whose characteristics are shown in the following table pass over the detector. Plot every tenth output signal from the detector (see Figure 6.11).

Vehicle	Length (ft)	Speed (miles/hr)	Distance Headway (ft)
1	18	56	
			300
2	16	54	
			300
3	50	45	
			150
4	30	44 -	
			160
5	15	49	
			200
6	17	51	
			250
7	16	55	

17. Two vehicles travel at fairly constant speeds over a presence-type detector that has a detection zone length of 6 feet. The time on (t_{on}) and time off (t_{off}) for each vehicle recorded from the detector in 1/30-second units are shown in the following table. Calculate all possible microscopic and macroscopic characteristics pertaining to these two vehicles. (Assume vehicle lengths of 16.5 feet.) Check with nomograph (Figure 6.13) and plot projectories (Figure 6.12).

Vehicle	Time On, t_{on}	Time Off, t_{off}
1	16	27
2	141	154

18. Two vehicles pass over two closely spaced presence-type detectors. Each detector has a detection zone of 6 feet and the distance between the upstream edges of the two detection zones is 20 feet. The time on (t_{on}) and time off (t_{off}) for each vehicle and for each detector in 1/30-second units are shown in the following table. Calculate all possible microscopic and macroscopic characteristics pertaining to these two vehicles. Are the vehicles accelerating or decelerating between the two detectors? Plot trajectories (Figure 6.14).

Vehicle	Detector	Time On, t_{on}	Time Off, t_{off}
1	A	16	27
	B	25	36
2	A	141	154
	B	149	163

19. Perform a library search to determine how knowledge of microscopic density characteristics are used in other modes of transport.

6.7 SELECTED REFERENCES

1. Daniel L. Gerlough and Matthew J. Huber, *Traffic Flow Theory—A Monograph,* Transportation Research Board, Special Report 165, TRB, Washington D. C., 1975, Chapter 6.

2. Wolfgang S. Homburger and James H. Kell, *Fundamentals of Traffic Engineering,* University of California, Berkeley, Calif., 1988, pages 4–6 to 4–7.

3. Louis J. Pignataro, *Traffic Engineering—Theory and Practice,* Prentice-Hall, Inc., Englewood Cliffs, N. J., 1973, pages 30–31 and 177–179.

4. Institute of Transportation Engineers, *Transportation and Traffic Engineering Handbook,* 2nd Edition, Prentice-Hall, Inc., Englewood Cliffs, N. J., 1982, pages 454–460 and 590–593.

5. Martin Wohl and Brian V. Martin, *Traffic System Analysis for Engineers and Planners,* McGraw-Hill Book Company, New York, 1967, pages 345–352.

6. Donald R. Drew, *Traffic Flow Theory and Control,* McGraw-Hill Book Company, New York, 1968, pages 337–342.

7. P. S. Parsonson, R. A. Day, J. A. Gaulas, and G. W. Black, Jr., *Use of EC-DC Detectors for Signalization of High-Speed Intersections,* Transportation Research Board, Record 737, TRB, Washington D. C., 1979, pages 17–23.

8. A. Reuschel, Vehicle Movements in a Platoon, *Oesterreichisches Ingenieur-Archir,* Vol. 4, 1950, pages 193–215.

9. A. Reuschel, Vehicle Movements in a Platoon with Uniform Acceleration or Deceleration of the Lead Vehicle, *Zeitschrift des Oesterreichischen Ingenieur-und Architekten-Vereines,* No. 95, 1950, pages 59–62 and 73–77.

10. L. A. Pipes, An Operational Analysis of Traffic Dynamics, *Journal of Applied Physics,* Vol. 24, No. 3, 1953, pages 274–287.

11. E. Kometani and T. Sasaki, On the Stability of Traffic Flow, *Operations Research Society of Japan,* Vol. 2, No. 1, 1958, pages 11–26.

12. E. Kometani and T. Sasaki, Dynamic Behavior of Traffic with a Non-Linear Spacing-Speed Relationship, *Theory of Traffic Flow Symposium Proceedings,* 1961, pages 105–119.

13. E. Kometani and T. Sasaki, A Safety Index for Traffic with Linear Spacing, *Operations Research,* Vol. 7, No. 6, 1959, pages 704–720.

14. E. Kometani and T. Sasaki, Car-Following Theory and Stability Limit of Traffic Volume, *Operations Research Society of Japan,* Vol. 3, No. 4, 1961, pages 176–190.

15. T. W. Forbes, H. J. Zagorski, E. L. Holshouser, and W. A. Deterline, Measurement of Driver Reactions to Tunnel Conditions, *Highway Research Board, Proceedings,* Vol. 37, 1958, pages 345–357.

16. T. W. Forbes, *Human Factor Considerations in Traffic Flow Theory,* Highway Research Board, Record 15, HRB, Washington D. C., 1963, pages 60–66.

17. T. W. Forbes, and M. E. Simpson, Driver and Vehicle Response in Freeway Deceleration Waves, *Transportation Science,* Vol. 2, No. 1, 1968, pages 77–104.

18. R. E. Chandler, R. Herman, and E. W. Montroll, Traffic Dynamics: Studies in Car-Following, *Operations Research,* Vol. 6, No. 2, 1958, pages 165–184.

19. R. Herman, E. W. Montroll, R. Potts, and R. W. Rothery, Traffic Dynamics: Analysis of Stability in Car-Following, *Operations Research,* Vol. 1, No. 7, 1959, pages 86–106.

20. D. C. Gazis, R. Herman, and R. B. Potts, Car-Following Theory of Steady State Flow, *Operations Research,* Vol. 7, No. 4, 1959, pages 499–505.

21. R. Herman and R. B. Potts, Single-Lane Traffic Theory and Experiment, *Theory of Traffic Flow Symposium Proceedings,* 1961, pages 120–146.

22. D. C. Gazis, R. Herman, ahd R. W. Rothery, Non-linear Follow-the-Leader Models of Traffic Flow, *Operations Research,* Vol. 9, No. 4, 1961, pages 545–567.

23. L. C. Edie, Car-Following and Steady-State Theory for Non-congested Traffic, *Operations Research,* Vol. 9, 1961, pages 66–76.

24. J. E. Tolle, Composite Car-Following Models, *Transportation Research,* Vol. 8, 1974, pages 91–96.

25. Adolf D. May, Jr. and Hartmut E. M. Keller, *Non-integer Car-Following Models,* Highway Research Board, Record 199, HRB, Washington D. C., 1967, pages 19–32.

26. R. Herman and R. W. Rothery, Car-Following and Steady State Flow, *Theory of Traffic Flow Symposium Proceedings,* 1963, pages 1–11.

27. R. Rothery, R. Silver, R. Herman, and C. Torner, Analysis of Experiments on Single-Lane Bus Flow, *Operations Research,* Vol. 12, No. 6, 1964, pages 913–933.

28. R. Herman and R. W. Rothery, Microscopic and Macroscopic Aspects of Single-Lane Traffic Flow, *Journal of Operations Research Society of Japan,* Vol. 5, 1962, pages 74–93.

7

Macroscopic Density Characteristics*

Traffic density is a fundamental macroscopic characteristic of traffic flow. It is an important characteristic that can be used in assessing traffic performance from the point of view of users and system operators. It is also employed as the primary control variable in freeway control and surveillance systems. The difficulty in measuring density inhibited its general use until the early 1960s, when presence-type detectors were introduced. The 1985 *Highway Capacity Manual* [31] uses traffic density as the primary measure of level of service for uninterrupted flow situations. Traffic density is expected to play even a more important role in the future in system-wide traffic performance evaluation and in on-line traffic-responsive freeway control systems.

Traffic density is defined as the number of vehicles occupying a length of roadway. The length is usually specified as 1 mile and normally a single lane or file of vehicles is considered. The easiest way to visualize traffic density is to consider an aerial photograph of a section of highway and to count the number of vehicles in a single lane having a length of 1 mile. Traffic densities vary from 0 (the absence of vehicles in a single lane having a length of 1 mile) to values representing vehicles bumper to bumper which are completely stopped. This upper limit, called jam density, is normally on the order of 185 to 250 vehicles per lane-mile, depending on the length of vehicles and the distance gaps between vehicles. For example, if the average distance headway under these conditions (the distance between front bumpers of two consecutive vehicles) is 25 feet, the jam density would be 211.2 vehicles per lane-mile. The earlier-mentioned jam density range of 185 to 250 vehicles per mile implies an average distance headway of 21.1 to 28.5 feet per vehicle. This relationship between traffic

---

*If limited time is available, the more complex example included in Section 7.6.2 can be scanned or omitted.

density and average distance headway can easily be obtained from the following equation:

$$k = \frac{5280}{\overline{d}}$$
(7.1)

where k = density (vehicles per lane-mile)

$\overline{d}$ = average distance headway (feet per vehicle)

Another important parameter on the density scale of 0 to 185–250 is called optimum density and is defined as the density level that exists when the lane of traffic is flowing at capacity. Optimum density is normally on the order of 42 to 67 vehicles per lane-mile, which corresponds to an average distance headway on the order of 79 to 126 feet.

Thus the density scale qualitatively can be subdivided into three regions. Density values from 0 to 42 vehicles per lane-mile denote traffic flow conditions in which traffic demands are less than roadway capacities, and thus users are experiencing a reasonable level of service and the system is serving all the demand that wishes to use the facility. The *Highway Capacity Manual* [31] proposes that this region be further divided into four levels of service designated as A, B, C, and D.

Density values from 42 to 67 vehicles per lane-mile denote flow conditions in which traffic demands are approaching or equal to roadway capacities. While the users are encountering a lower level of service, the system is becoming most productive in terms of vehicle-miles of travel. The *Highway Capacity Manual* [31] designates this condition as level of service E.

The third region occurs at density values greater than 67 vehicles per lane-mile. This region denotes flow conditions in which traffic demands exceed roadway capacities. Not only do the users encounter a very poor level of service, but the system is less productive in terms of vehicle-miles of travel. The *Highway Capacity Manual* [31] designates this condition as level of service F.

Percent occupancy rather than density is used as an indicator of macroscopic density characteristics in freeway control systems because of the ease of measuring percent occupancy as compared with measuring density. Percent occupancy is defined as the percent of time a point or short section of roadway is occupied. Normally, a single lane is considered and the occupancy can vary from 0 percent (the absence of vehicles passing over a point or short section of roadway) to 100 percent (a vehicle completely stopped over a point or short section of roadway). It will be shown later in this chapter that density can be estimated from percent occupancy by the following equation.

$$k = \frac{52.8}{\overline{L_V} + L_D} \% \ OCC$$
(7.2)

where k = density (vehicles per lane-mile)

$\overline{L_V}$ = average vehicle length (feet)

L_D = detection zone length (feet)

$\% \ OCC$ = percent occupancy

The normal combined vehicle length and detection zone encountered is on the order of 20 to 30 feet, which causes the coefficient in equation (7.2) to vary from 1.8 to 2.6.

Table 7.1 summarizes the discussions above by identifying the three regions of traffic flow conditions in terms of density and percent occupancy, and indicates further subregions based on level of service criteria. The importance of density (or percent occupancy) can be seen clearly from this table. A measured density or percent occupancy value provides a clear and specific indication of both the level of service being provided to the users and the productive level of facility use. The measured density or percent occupancy can be used as the control variable in on-line traffic-responsive freeway ramp metering systems. Entry control is not required in the density range of 0 to 30 vehicles per lane-mile (0 to 12 percent occupancy). As density continues to rise, control is initiated and gradually becomes more restrictive. Once densities on the order of 42 to 67 vehicles per lane-mile (17 to 23 percent occupancy) are reached, the metering rate is most restrictive. In addition, density analysis permits the study of shock waves, the estimation of travel times, and the estimation of traffic demands that will be covered in later sections of this chapter.

There are six major sections in this chapter, followed by selected problems and references. First, a historical perspective is given of the study measurement and application of density analysis. Then, attention is given to methods of measuring and calculating density. Next, density contour maps are introduced with examples of applications.

TABLE 7.1   Traffic Flow Conditions Based on Density and Percent Occupancy

Density (vehicles/lane-mile)	Percent Occupancy[a] (%)	Level of Service	Flow Conditions	
0–12	0–5	A	Free-flow operations	
12–20	5–8	B	Reasonable free-flow operations	Uncongested flow conditions
20–30	8–12	C	Stable operations	
30–42	12–17	D	Borders on unstable operations	
42–67	17–28	E	Extremely unstable flow operations	Near-capacity flow conditions
67–100	28–42	F	Forced or breakdown operations	Congested flow conditions
>100	>42		Incident situation operations	

[a]Assuming that $(\bar{L}_V + L_D) = 22$ feet.

Source: Reference 31.

The fourth section will introduce shock wave analysis based on density characteristics. The final two sections will provide techniques for estimating travel time and traffic demand from density analysis.

7.1 HISTORICAL DEVELOPMENT

The measurement and analysis of density characteristics are of particular interest from a historical perspective. The stages of development of density analysis were controlled primarily by measurement techniques. Before the 1950s only photographic techniques were employed. By the early 1960s three approaches were being undertaken in parallel: calculation of density from measured speed and flow, calculation of density from input–output counts, and measurement of percent occupancy. These developments are reviewed in the following paragraphs.

At first, photographic techniques were employed which revealed the importance and significance of density but required considerable planning and time-consuming analysis and obviously could not be analyzed in real time. One of the first studies reported in the literature was published in 1928 and was an aerial photographic study by Johnson of traffic density along the Baltimore–Washington highway [23]. He observed that densities were higher at intersections and on sections of highway approaching the two major cities. In 1934, Greenshields reported on an aerial photographic study that was conducted on a rural two-lane Ohio highway on a Sunday afternoon in the vicinity of a state fair [37]. Speeds and densities were obtained for different time periods which represented different flow intensity levels. The results of this study are presented in Figure 7.1 because of their historical significance and because it is the basis for the first macroscopic traffic flow theory, which will be discussed in Chapter 10. Extensive aerial photographic studies were undertaken in the early 1960s in Chicago [34], Los Angeles [24], Pittsburgh [38], San Francisco [26, 27], and Texas [9, 33] and resulted in the development of density contour maps.

Interest in field studies of speed–flow–density relationships in the 1950s and 1960s focused increased attention on density. As mentioned earlier, density was calculated based on field measurements of speed and flow using the following equation.

$$k = \frac{q}{u} \tag{7.3}$$

where k = density (vehicles per lane-mile)
$\quad q$ = flow rate (vehicles per hour)
$\quad u$ = space-mean-speed (miles per hour)

In 1960, Automatic Signal Division developed a density computer that automatically computed density based on the number and speeds of vehicles measured at a point [5]. Discussion of specific speed–flow–density relationship studies will be presented in Chapter 10.

A unique input–output count algorithm for determining density was developed by the Port of New York Authority in the 1960s in the Lincoln and Holland Tunnels [14,

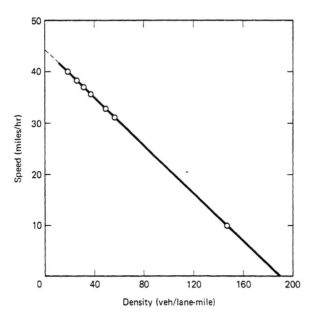

Figure 7.1   Earliest Measured
Speed–Density Relationship (From
Reference 37)

15, 21, 25, 35, 36]. The key to this approach was the unique features of the tunnels and a well-designed detector and software system. The tunnels each had a single entrance and a single exit with no lane changing within the tunnels. Thus the sequence of vehicles in each lane was not changed as the vehicles passed through the tunnel. Pairs of photocell-type detectors spaced 13 feet apart were installed at a number of stations throughout the tunnel. At each station the length of every vehicle was calculated rather precisely. Section density between stations was calculated by determining the number of vehicles within the section at one point in time and then continuously adding arriving vehicles and subtracting departing vehicles. When a unique pair of vehicles (i.e., a long vehicle followed by a very short vehicle) entered the section, the upstream counter began to count vehicles and the downstream counter was reset to zero. When the unique pair of vehicles left the section, the downstream counter began to count vehicles. Since even small counting errors would result in large errors in section densities after relatively short periods of time, the counting procedures were reinitialized at frequent intervals. Section densities were used for control and evaluation purposes.

The most significant advance in the measurement and analysis of density, however, was due to the development of presence-type detectors and the processing of signal pulses to compute percent occupancy in the early 1960s [11, 34]. The cost of equipment, installation, and communications were significantly lower, and the measurements were more reliable. By the mid-1960s analog computers were being replaced by digital computers which were ideally suited for the signal pulses of the presence-type detectors and the on-line calculation of occupancy. By the end of the 1960s, almost all of the major freeway surveillance and control systems utilized presence-type detectors, directly measured percent occupancy, and employed digital computers.

The widespread use of presence-type detectors and percent occupancy calculations has led to new applications. For example, freeway projects now use percent occupancy as a basis for predicting the occurrence of incidents and accidents. Percent occupancy measurements on a freeway system are used to calculate mile-minutes of congestion on a daily basis as a primary measure of performance. Presence-type detectors are commonly employed in traffic signal systems to detect the presence or absence of vehicles and thereby extend, terminate, or skip green phases. The concept of density (the number of stopped vehicles on an approach) is used in field measurements of stopped delay at signalized intersections.

Presence-type detectors and macroscopic density characteristics will continue to be used in the future. Anticipated future directions include: new presence-type detectors which directly measure the number of vehicles occupying a section of roadway; automatic calculation of density and/or occupancy from video images; and extended algorithms using density and/or occupancy measurements for predicting travel times and excess demands, and improving system control and assessment.

7.2 DENSITY MEASUREMENT TECHNIQUES

The purpose of this section is to describe existing techniques for the measurement of macroscopic density characteristics. Macroscopic density characteristics include density and percent occupancy. Existing techniques include photography, input–output counts, speed–flow calculations, and occupancy measurements. These four techniques will be described in the following four paragraphs.

The photographic technique was the earliest technique employed and was the principal technique prior to the 1960s [9, 23, 24, 26, 27, 33, 34, 37, 38]. Gradually, the photographic technique has been replaced by speed–flow calculation and occupancy measurement techniques. Recent research has been directed to automatic extraction of macroscopic density characteristics from video images such as television monitors. Although both ground-mounted and aircraft-mounted cameras have been employed, only the aircraft-mounted photographic technique will be described. A fixed-wing aircraft with a flexible-mounted camera is employed which flies approximately along the centerline of the route to be studied. The speed of the aircraft and the length of roadway covered in each photograph determine the frequency of photographs. The camera can be a time-lapse camera or manually controlled camera if known points along the route are identified as firing points for the camera. An overlap of 10 to 20 percent between photographs is desired. The length of the route to be studied depends on the speed of aircraft and the time interval between sets of photographs. A 10-mile length is a typical length for such studies. The films are developed, and if small negatives are used, either they are enlarged or a projector is employed. The length of roadway is divided into subsections of known length, and in urban areas there might be approximately four subsections per mile. The boundary between subsections are at locations where demands change (on-ramps, off-ramps, intersections, etc.) and at locations where capacities change (lane additions, lane drops, severe grade changes, etc.). The tedious and time-consuming effort is then undertaken to count manually the number of vehicles occupying a subsection of roadway at the time the photograph was taken. Some additional information can be obtained by counting vehicles on a lane basis, distinguishing between cars and trucks, and counting queue lengths at on-ramps and off-ramps. The results can be displayed in a variety of ways, but density contour maps, discussed in Section 7.3, are most frequently used.

The input–output count technique is a rather straightforward approach in concept in which an initial count is made of the number of vehicles along the roadway between two count stations, and over time the number of vehicles entering the section is continuously added and the number of vehicles leaving the section is continuously subtracted from the initial count. If each detector at the two count stations were 100 percent accurate on a continuous basis, the initialization count would be required at only one point in time and the technique would be feasible. The problem is that section density is calculated on the basis of the difference between two large numbers (input and output counts), and detector errors even at the low level of 1 percent can not be tolerated without frequent reinitialization. For example, consider a 1-hour period in which the true number of vehicles entering and leaving the section are exactly equal to 4000 vehicles. The initial number of vehicles in the 1-mile three-lane section is 60, which gives an initial lane density of 20 vehicles per lane-mile. If the set of detectors at the input station undercount by 1 percent and the set of detectors at the output station overcount by 1 percent at the end of 1 hour, the estimated density would be −6.7 vehicles per lane-mile, while the actual density would still be 20 vehicles per lane-mile. The input–output count technique therefore is practical only if the detectors are very reliable and accurate and an automatic initialization process can be employed frequently (on the order of every few minutes). The Port of New York Authority was able to develop such a system in the Holland and Lincoln Tunnels [14, 15, 21, 25, 35, 36]. The tunnel environment permitted the use of photoelectric cells carefully installed in pairs which provided accurate counts and the calculation of individual vehicle lengths. Further, because of the single-lane operation in each of the two lanes (lane changing prohibited), the sequence of vehicles at each detector in the system was unchanged. This permitted the reinitialization of the density count at frequent intervals based on pattern recognization of unusual vehicle length sequences. However, the input–output technique is extremely limited to very special situations.

A third technique is to calculate density from speed and flow measurements using equation (7.3). This calculation technique requires two detectors, count and speed, or two closely spaced detectors with software to convert elapsed travel time to speed. Then, density can be calculated in the field using a microprocessor or transmitted through communication links with density calculated on the central computer. One problem with this approach is that normally time-mean-speed is calculated for this point measurement station while equation (7.3) theoretically requires space-mean-speed.

The fourth technique for determining density is based on the measurement of occupancy and employs equation (7.2). It requires known average vehicle length and detection zone length. In special situations where vehicle lengths do not vary over time and can be assumed, the single presence-type detector for measuring occupancy is adequate. When conditions are encountered in which average vehicle lengths are unknown and vary over time, a second adjacent detector is required for calculating average vehicle length. The calculation of average vehicle lengths from detector pairs was derived in Chapter 6. Equation (7.2) will now be derived for calculating density from measured occupancy.

The speed of an individual vehicle is a function of distance traveled divided by the travel time. A single vehicle passing over a presence-type detector travels a distance equivalent to the length of the vehicle (L_V) plus the length of the detection zone (L_D) during the detector's occupancy time (t_0) and is shown in the equation

$$\dot{x}_i = \frac{L_V + L_D}{t_o} \tag{7.4}$$

where $\dot{x}_i$ = speed of individual vehicle (feet per second)
  L_V = length of individual vehicle (feet)
  L_D = detection zone length (feet)
  t_o = individual vehicle occupancy time (seconds)

The relationship in equation (7.4) can be converted to a macroscopic level by using average values of vehicle lengths and vehicle occupancy times for the N vehicles passing over the detector in some time period T. Note that the resulting average speed is a space-mean-speed since it is based on the average of vehicle occupancy times, not on an average of individual speeds. Converting the lengths to miles and the time to hours, equation (7.4) becomes

$$\bar{\mu}_{SMS} = \frac{3600}{5280} \left(\frac{\overline{L_V} + L_D}{\bar{t}_o} \right) \tag{7.5}$$

The total time the detector is occupied (T_o) in time period T, when N vehicles pass over the detector, is

$$T_o = N \left(\frac{\bar{t}_o}{3600} \right) \tag{7.6}$$

where T_o = total occupancy time in time period T (hours)
  N = number of vehicles passing over the detector in time period T
  T = time period of observations (hours)

Solving for $\bar{t}_o$ in equation (7.5) and substituting into equation (7.6) yields

$$T_o = \frac{N}{\bar{\mu}_{SMS}} \left(\frac{\overline{L_V} + L_D}{5280} \right) \tag{7.7}$$

The percent occupancy (% OCC) during time period T is

$$\% \text{ OCC} = 100 \left(\frac{T_o}{T} \right) \tag{7.8}$$

Substituting T_o from equation (7.7) into equation (7.8) gives

$$\% \text{OCC} = \frac{N}{T}\left(\frac{100}{\overline{\mu}_{\text{SMS}}}\right)\frac{\overline{L}_V + L_D}{5280} \tag{7.9}$$

Since flow rate (q) in vehicle per hour is equal to N/T and $k = q/\overline{\mu}_{\text{SMS}}$, then

$$\% \text{OCC} = \frac{\overline{L}_V + L_D}{52.8}\left(\frac{q}{\overline{\mu}_{\text{SMS}}}\right) = \frac{\overline{L}_V + L_D}{52.8}k \tag{7.10}$$

$$k = \frac{52.8}{\overline{L}_V + L_D}(\% \text{OCC}) \tag{7.11}$$

Thus density can be calculated from measured percent occupancy for a known average vehicle length and detection zone length. A nomograph of this relationship is shown in Figure 7.2. As an illustration, if the detection zone length is 6 feet, the average vehicle length is 18 feet, and the percent occupancy is 20 percent the density would be 44 vehicles per lane-mile. The vertical scale is the conversion factor, which in this case is 2.2.

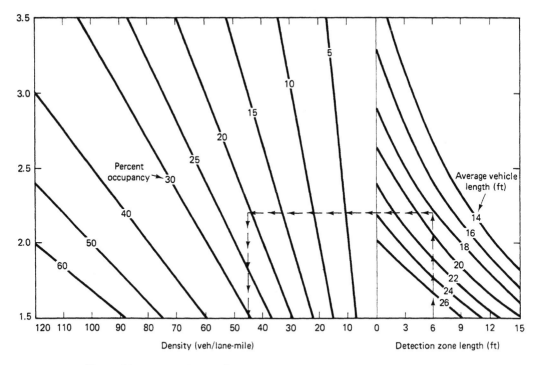

Figure 7.2 Density–Percent Occupancy Relationship as Function of Average Vehicle Length and Detection Zone Length

It should be noted that most traffic control systems that use a macroscopic density characteristic as the control variable, use the measured percent occupancy directly without converting to density. However, density is important in estimating travel time and traffic demands, as will be presented in Sections 7.5 and 7.6.

7.3 DENSITY CONTOUR MAPS

Density contour maps are extremely helpful in understanding traffic flow phenomena on congested uninterrupted traffic facilities such as freeways. These maps also provide an introduction to shock waves and the data base used in constructing these maps are essential in estimating travel times and traffic demands. The first density contour map was developed in 1960–1961 in connection with operational studies of the Hollywood and Pasadena freeways in Los Angeles [24]. Soon thereafter density contour maps were constructed for other freeways and the contour map design was extended to speed and flow as well [9, 26, 27, 33, 34, 38].

Section 7.2 described briefly the procedures for collecting and tabulating data for the density contour map from aerial photographs. These procedures will be reviewed briefly and extended through to the final results by means of the example shown in Figure 7.3. This example was the first density contour map and was based on data collected in February 1961 on the Hollywood Freeway.

The horizontal scale is distance, with traffic moving from left to right. The major cross streets are noted over the 7-mile outbound section of the freeway. Note that the famous four-level interchange is located at mile post 0, and mile post 7 is over the pass leading to the San Fernando Valley. The study section was carefully selected so that it encompassed the upstream congestion, the bottleneck area, and the free-flow conditions downstream. The study section was divided into subsections with boundaries between subsections occurring wherever demand and/or capacity changed. Thus a subsection boundary was designated at each off- and on-ramp, at locations where the number of lanes changed, and so on. There were a total of 22 subsections in this 6.4-mile study section. The length of subsections varied from 0.12 to 0.64 miles long with an average of 0.29 miles (1500 feet). The vertical scale is time beginning at 3:40 P.M. at the bottom and extending to 6:30 P.M. at the top for a total of almost 3 hours. The study period was carefully selected so that it should include before-peak free-flow conditions, the peak period, and after-peak free-flow conditions.

The arrays of numbers within the figure are density values expressed as the number of vehicles per lane-mile in each subsection during each flight. The flight path of the aircraft can easily be seen. For example, at 3:40 P.M. the aircraft was beginning a southbound flight from the north end of the study section. The aircraft reached the south end of the study section at 3:43 P.M.. A total of 29 flights were made, with the last photographs being taken at the south end of the study section at 6:23 P.M.. Note that one additional flight should have been made to photograph the entire study section during after-peak free-flow conditions.

A military-type 9×9 time-lapse camera was employed with a flexible mount so that the photographer could aim the camera in the $x-y$ plane for each photograph. The aircraft flew along the side of the freeway, for it was not necessary to take exactly vertical photographs. Photographs were taken automatically with a time-lapse mechanism, and the timing was set to provide 10 to 20 percent overlap. Films were developed and subsection boundaries were marked in each photograph. Counts of vehicles in each photograph for each subsection were taken and lane densities computed based on

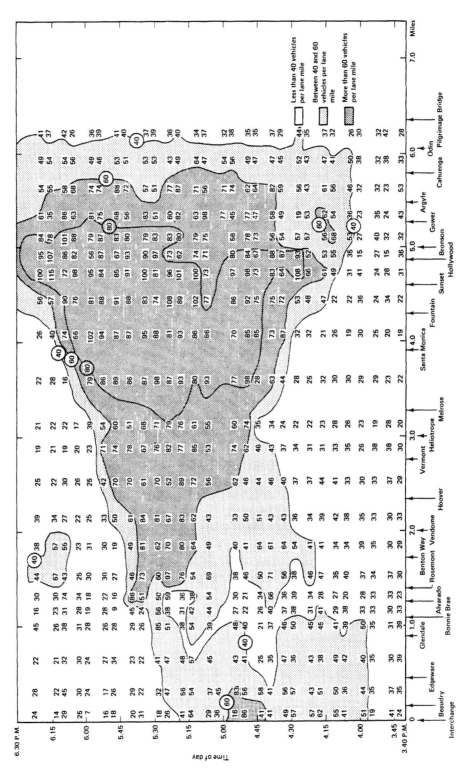

Figure 7.3 Hollywood Freeway Density Contour Map (From Reference 24)

subsection length and number of lanes. Flight trajectories were constructed on the map, and then lane densities were recorded along the flight trajectory in the appropriate subsection and time location. The arrays of density values along the flight trajectories can be seen in Figure 7.3.

Three traffic flow conditions were identified based on density as shown earlier in Table 7.1. Note that slightly different density values were used to distinguish between traffic flow conditions. Uncongested or free-flow traffic conditions were identified in the time–space domain of the density contour map when densities of less than 40 were encountered. Near-capacity flow conditions were identified when densities varied between 40 and 60, while congested flow conditions were identified when densities over 60 were encountered. The 40 and 60 density contour lines were constructed based on the arrays of calculated subsection densities. The unshaded area on the map denotes free-flow conditions, the lightly shaded area denotes near-capacity flow conditions, and the darkly shaded denotes congested flow conditions. No incidents nor accidents that affected traffic flow were observed in the photographic records, nor were there any police reports. Note that the highest densities were only slightly above 100.

Although major attention will be given to the downstream portion of the study section in the following discussion, brief mention will be made of traffic flow conditions in the first 1-mile upstream portion of the study section. Free-flow conditions occurred until almost 4:00 P.M. and after 5:30 P.M.. Near-capacity conditions existed from 4:00 to 5:30 PM almost over the complete 1-mile portion. Slight congestion occurred for 15 minutes in the vicinity of the interchange where three freeways merge together. Two essential points should be made. First, the major downstream congestion is not affecting traffic flow conditions in the vicinity of the interchange. Second, the full effect of the brief bottleneck situation at the interchange cannot be fully evaluated because of the limiting upstream boundary of the study section.

Now consider the middle and downstream portions of the study section which were the main intent of the project. Free-flow conditions occurred until shortly after 4:00 P.M. when the portion between Bronson and Odin began operating near-capacity. Shortly after 4:15 congestion occurred between Sunset and Hollywood, indicating a bottleneck between Hollywood and Bronson. The bottleneck was due to the Hollywood on-ramp demand and a decrease in capacity due to an upgrade. By 4:35 P.M. the congestion had extended 1-mile upstream. Shortly thereafter a rather unique congestion pattern evolved due to the presence of trucks in the traffic stream and the upgrade situation from Hollywood to Cahuenga. While congestion continued to extend farther and farther upstream (to Melrose by 4:45 P.M.), the downstream edge of the congestion moved downstream from Bronson to Cahuenga about 4:35 to 4:40 P.M.. This was due to three factors. First, the portion between Bronson and Cahuenga was already operating near-capacity, and the flow conditions were unstable. The other two factors, trucks and grade, worked together in the following manner. Trucks traveled through the congested region upstream of Hollywood at slow speeds (10 to 20 miles per hour) and once they arrived on the upgrade portion did not have the acceleration capabilities to increase speeds high enough so that the grade portion could maintain its capacity. Consequently, dense platoons of vehicles formed behind the trucks in the traffic stream, and this higher-density lower-capacity condition moved downstream to Cahuenga, where roadway capacities were increased significantly due to lane additions and grade changes.

Except for the situation described above, the shape of the congestion zone (darkly-shaded area) is triangular and is typical of a single isolated bottleneck. The boundaries of the triangular area denote significant changes in traffic flow conditions, and individuals traveling across these boundaries would readily sense these changes. These boundaries can be considered as shock waves since they represent discontinuities in flow conditions and there are three clearly defined shock waves.

The first shock wave, designated as a backward forming shock wave, slopes up and to the left, extending farther and farther upstream over time. The designation "backward forming" is given because the shock wave moves upstream over time in the oppose direction of traffic and is a forming type because over time the zone of congestion is increasing. The backward forming shock wave terminates near Rosemont at about 5:35 to 5:40 P.M. after having traveled about 3.5 miles in 1.3 hours. This shock wave has a speed of -2.7 miles per hour and is given a negative sign since it is moving in the oppose direction of traffic.

The second shock wave, designated as a forward recovery shock wave, slopes up and to the right, extending farther and farther downstream over time. The designation "forward recovery" is given because the shock wave moves downstream over time in the same direction of traffic and is a recovery type because over time the zone of congestion is decreasing. The forward recovery shock wave terminates near Gower at approximately 6:45 P.M. after having traveled about 4.0 miles in approximately 1.1 hours. This shock wave has a speed of $+3.6$ miles per hour and is given a positive sign since it is moving in the direction of traffic.

The third shock wave, designated as a frontal stationary shock wave, is essentially a vertical boundary at the downstream edge of the congested zone which is stationary over time. The shock wave begins at Cahuenga at about 4:35 P.M. and continues at that location until about 6:15 P.M.. There appears to be a shift in the bottleneck from Cahuenga back to the Bronson–Gower subsection from 6:15 until 6:45 P.M.. A stationary-type shock wave has a velocity of zero.

Overall the congestion zone extends over a distance of 4.5 miles and lasts for 2.5 hours. Using the Chicago's freeway systems measure of congestion, which is the product of the length of congestion and the time of congestion, the measure of congestion for this particular bottleneck on this particular day was 330 minute-miles of congestion. The strength of the density contour map is its clear relationship of densities over time and space. For example, a person standing on the Vermont overpass at 5:15 P.M. would observe congestion but would be unaware that the bottleneck problem is 3 miles downstream and started over an hour before. Or consider a vehicle that enters the Hollywood freeway at the upstream four-level interchange at 5:30. At that time and place free-flow conditions are encountered but the driver is unaware that over 4 miles of congested travel lies ahead.

Centralized freeway surveillance systems and computer simulation models have been designed to provide density (or occupancy) contour maps as outputs. Freeway system-wide density contour maps have been developed from aerial photographic studies. An example of each is shown in Figures 7.4, 7.5, and 7.6. The application of density contour maps as an introduction to shock waves, and for estimating travel time and traffic demands are addressed in the following three sections of this chapter.

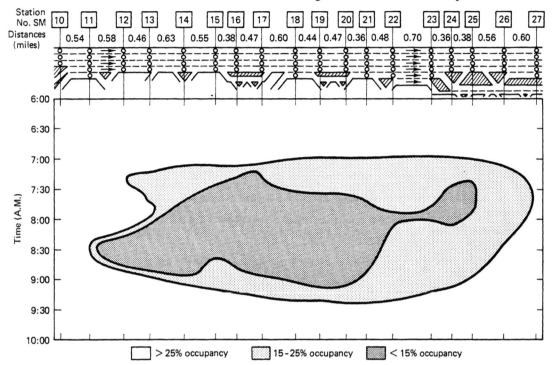

Figure 7.4   Percent Occupancy Contour Map from Santa Monica Freeway Surveillance System

7.4 INTRODUCTION TO SHOCK WAVES

The purpose of this section is to provide a qualitative introduction to shock wave analysis. Quantitative analysis of shock waves will be covered in later sections of this chapter and in Chapter 11. Shock waves are defined as boundary conditions in the time–space domain that demark a discontinuity in flow-density conditions. For purposes of this discussion, the distinct discontinuity between noncongested and congested flow will be considered, and a density contour of 60 vehicles per lane-mile will be used to identify this discontinuity. This single type of shock wave is considered here because of its importance and in order to present the concept of shock waves as clearly and concisely as possible. In Chapter 11 the approach to shock wave analysis is generalized. In the following paragraphs two simple traffic situations are presented and then a matrix of more complex traffic situations is presented and discussed.

```
INSTITUTE OF TRANSPORTATION STUDIES
UNIVERSITY OF CALIFORNIA, BERKELEY
(415)-642-7390

CONTOUR DIAGRAM OF
DENSITY
BEFORE OPTIMIZATION

TIME                                                                                                      BEGIN
SLICE ...................................................................................................  TIME
      .                                                                                                 .
   12 .2222222222222222222222221111111122222222222221112222222222111111111111112222222222222111112222222211111. - 8:45
   11 .3333333333222222222222222222222222222222222222222222222222211122222222222222222222222222333322222222. - 8:30
   10 .3333333333222222223333322222222222222222222222222222222222211122222222222222222222222444443333222222222. - 8:15
    9 .333333333322222222333332222222222222222222222222222222222222222222222222222444444444448888866663332222. - 8:00
    8 .333333333344477777744444222222222222222224444999333333333222222222233333666888888888888*****55553332222. - 7:45
    7 .44444444449999999995555522222224446666688888886666444443333332222666669998888888888899999555533322222. - 7:30
    6 .44444444448888888886666633333333777888887777888666644444333222222233333666777777777779999966663332222. - 7:15
    5 .44444444448888888886666633333333666888887777888666644444333222222233333444666666666667777766663333333. - 7:00
    4 .44444444443338888866666333333335558888877778886666444442222222222222223334444444444455555666633322222. - 6:45
    3 .44444444442225555566666333333334445555566667776666444442222222222222223334444444444444444666633322222. - 6:30
    2 .44444444442224444466666333333334443333344444446666555552222222222222233333333333333333333333333322222. - 6:15
    1 .33333333332223333333333222222233322222333333333333333332221112222222222222222222222222222222222222222. - 6:00
      .                                                                                                 .
      ...................................................................................................
      :         : :   :    :      : :   :  :  :  :    :  :  :  :   :  :          :    :  :  :
      01        02 03 04   05     06 07 08 09 10  11   12 13 14 15  16 17        18   19 20 21

NOTE--THE FIRST FREEWAY SECTION REPRESENTS 29 PERCENT OF THE ENTIRE SYSTEM LENGTH.
ONLY THAT PORTION CONSTITUTING 10 PERCENT OF THE DIAGRAM BASELINE IS SHOWN ABOVE.

THE DIGIT LEVELS DENOTE THE FIRST DIGIT OF DENSITY (EX. 4 MEANS AT LEAST 40.0 BUT LESS THAN 50.
ALL VALUES OVER 100.0 ARE REPRESENTED BY AN ASTERISK.)
```

Figure 7.5 Density Contour Map Output from FREQ Computer Simulation Model

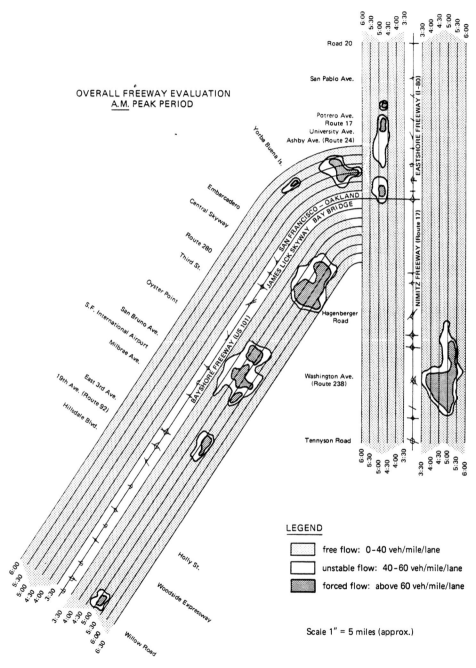

Figure 7.6   Density Contour Map for the San Francisco Bay Area Freeway System (From Reference 26)

7.4.1 Some Simple Examples

Consider a single-lane approach to a pretimed signal-controlled intersection as shown in Figure 7.7. The traffic demand is assumed to be relatively light and arriving at a constant flow rate. The capacity of the traffic signal exceeds the arriving traffic demand but traffic can only be discharged when the signal is green. Consequently, at some distance upstream of the signal and immediately downstream of the signal, free-flow conditions exist with densities less than 60 vehicles per lane-mile. However, just upstream of the signal during the red phase, vehicles will be stopped and densities will exceed 60 vehicles per lane-mile. Therefore, there will be a discontinuity as vehicles join the rear of the standing queue and as vehicles are discharged from the front of the standing queue when the signal is green. The first discontinuity is a backward forming shock wave while the second discontinuity is a backward recovery shock wave. Both shock waves are backward moving because over time the discontinuity is propagating upstream in the opposite direction of the moving traffic. The first shock wave is a forming shock wave because over time the propagation of the shock wave is resulting in the increase of the congested portion while the second shock wave is a recovery shock wave because over time the propagation of the shock wave is resulting in the decrease of the congested portion. There is a frontal stationary shock wave at the stop line during the red phase. The term "frontal" is used to indicate that the shock wave is at the downstream edge of the congested region, and the term "stationary" is used to indicate that the shock wave remains at the same position in space. Consider this example as the case where demand is constant over time, capacity varies over time, and there is an isolated single restriction (bottleneck) with no entrances or exits in the congested region.

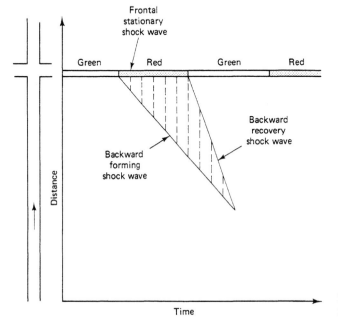

Figure 7.7   Shock Wave Phenomena at a Signalized Intersection

The second simple traffic situation is at a lane-drop location on a long bridge during the afternoon peak period and is shown in Figure 7.8. The capacity of the lane-drop
location is constant over time, but like the typical peak period the traffic demands
increase, causing demands to exceed capacities and then decrease until the peak period
is over. For illustrative purposes the demand flow over time pattern will be assumed to
be equivalent to 1.5, 2.5, 2.0, and 1.5 lanes of capacity (note that the bottleneck capacity is 2 lanes). During the first period of time when demand is equivalent to 1.5 lanes
of capacity, no shock wave (contour of 60 vehicles per lane-mile) will develop. However, as soon as the demand increases to 2.5 lanes of capacity, a backward-forming
shock wave will develop having a constant shock wave velocity. When the demand is
reduced to 2 lanes of capacity, the input equals the output and a rear stationary shock
wave results. As the demand is reduced further to 1.5 lanes of capacity, the length of
the congestion region decreases, as shown by the forward recovery shock wave. A
frontal stationary shock wave occurs at the bottleneck as long as the bottleneck operates
at capacity. The intersection of the frontal stationary shock wave and the forward
recovery shock wave signifies the termination of the congested period. Consider this
example as the case where demand varies over time, the capacity is fixed, and there is
an isolated single restriction with no entrances or exits in the congested region.

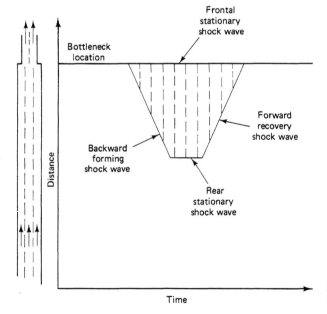

Figure 7.8   Shock Wave Phenomena at a
Freeway Bottleneck during a Peak Period.

Traffic situations in the real world are usually much more complex than the two
examples described in the previous paragraphs. Consider the major assumptions that
were implicitly made or implied in order to develop a matrix of more typical real-world
situations. These assumptions include:

- An isolated bottleneck

- A constant demand (or capacity) over time
- A varying demand (or capacity) over time of a simplistic nature
- "Normal" conditions free of incidents, accidents, and so on
- No forward-moving forming shock wave (as shown in Figure 7.3 between Bronson and Cahuenga at 4:35 to 4:40 P.M.)
- No entrances or exits along the route

7.4.2 Further Examples and Classification

An attempt will now be made to classify the different types of shock waves and to describe situations where and when they might be encountered. Before doing so, a hypothetical and complex density contour map will be presented (Figure 7.9) and the resulting shock waves described. The vertical scale of Figure 7.9 is distance, with traffic moving up the diagram, and six locations are identified in the figure and described below:

A Farthest downstream recurring bottleneck
B Nonrecurring incident site
C Second recurring bottleneck
D Location where demand dropped at time 9
E Third recurring bottleneck
F Upstream end of congestion

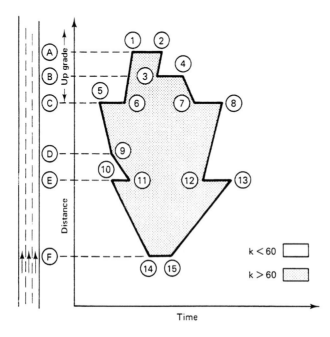

Figure 7.9   Hypothetical Density Contour Map

The horizontal scale is time, with time increasing to the right. A total of 15 points in time are identified in the figure and will be described as the example is presented. The clear area in the time–space domain indicates areas where densities are less than 60 vehicles per lane-mile, while the shaded area indicates where densities are greater than 60 vehicles per lane-mile. The boundaries between these two areas illustrate almost all of the various types of shock waves that can exist.

At time 5, the first congestion is encountered at location C, causing a frontal stationary shock wave (5–6) and a backward forming shock wave (5–9). At time 10 congestion occurs at location E, causing a frontal stationary shock wave (10–11), a backward forming shock wave (10–14), and reduces the velocity of the backward forming shock wave (9–11). At time 6, trucks in the congestion upstream of location C are slowed down, and as the trucks proceed along the upgrade, their speeds are inhibited and a forward forming shock wave resulted (6–1). Location A becomes the bottleneck and a new frontal stationary shock wave is created (1–2). At time 3, an incident occurs at location B that causes a forward recovery shock wave (2–3) and establishes a new frontal stationary shock wave (3–4). At time 4, the incident at location B is removed and a backward recovery shock wave is formed (4–7). At time 7, the bottleneck at location C is reestablished and a frontal stationary shock wave is formed (7–8). The congested length "splits" at time 12, forming a new frontal stationary shock wave (12–13) and lowers the demands between locations E and C, creating a forward recovery shock wave (12–8). Returning to point 14, the demand decreases until the input demand is equal to the flow in the congested region which causes a rear stationary shock wave (14–15). Further reduction in input demand causes a forward recovery shock wave (15–13). The congestion ends at time 13.

In summarizing this introduction to shock waves, Figure 7.10 attempts to classify the various types of shock waves and relate this classification to the earlier examples given in Figures 7.7 through 7.9. The six types of shock waves are identified in Figure 7.10. The high-density area (densities > 60 vehicles per lane-mile) is shown in the center of the figure, while low densities (densities < 60 vehicles per mile) are located on the outside of the individual shock waves. A frontal stationary shock wave must always be present at a bottleneck location and indicates the location where traffic demand exceeds capacity. It may be due to recurrent situations where each workday the normal demands exceed normal capacities during the peak period at specific locations or be due to nonrecurrent situations where the normal demand exceeds reduced capacity (caused by accident or incident) which may occur at any location at any time. The term "frontal" implies that it is at the front (or downstream edge) of the congested region with lower densities farther downstream and higher densities upstream. The term "stationary" means that the shock wave is fixed by location; that is, it does not change location over time. There are examples of frontal stationary shock waves in Figures 7.7 through 7.9 (1–2, 3–4, 5–6, 7–8, 10–11, and 12–13).

Backward forming shock waves must always be present if congestion occurs and indicates the area in the time–space domain where excess demands are being stored. The term "backward" means that over time the shock wave is moving backward or upstream in the opposite direction of traffic. The term "forming" implies that over time

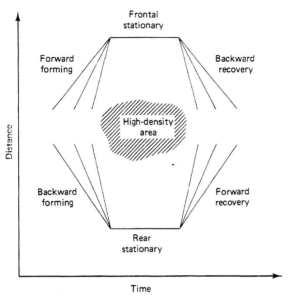

Figure 7.10   Classification of Shock Waves

the congestion is gradually extending to sections farther and farther upstream. The time–space domain to the left of this shock wave has lower densities, and to the right the density levels are higher. There are examples of backward forming shock waves in Figures 7.7 through 7.9 (5–9, 9–11, and 10–14). Note that the slopes of these shock waves represent velocities, with the flatter slopes representing lower velocities and the steeper slopes representing higher velocities. (Note that a horizontal line represents a zero speed, while a vertical line represents an infinite speed.)

The forward recovery shock wave is the next most commonly encountered type of shock wave and occurs when there has been congestion but demands are decreasing below the bottleneck capacity and the length of congestion is being reduced. The term "forward" means that over time the shock wave is moving forward or downstream in the same direction of traffic. The term "recovery" implies that over time free-flow conditions are gradually occurring on sections farther and farther downstream. The time–space domain to the left of this shock wave has higher densities, and to the right the density levels are lower. There are examples of forward recovery shock waves in Figures 7.8 and 7.9 (2–3, 8–12, and 13–15). Note that a forward recovery shock wave does not exist in Figure 7.7 (the situation where the bottleneck capacity changes).

The rear stationary shock wave may be encountered when the arriving traffic demand is equal to the flow in the congested region for some period of time. The term "rear" implies that it is at the rear (or upstream edge) of the congested region with higher densities downstream and lower densities farther upstream. The term "stationary" means that the shock wave does not change location over some period of time. A rear stationary shock wave is shown in Figures 7.8 and 7.9 (14–15). Note that a rear stationary shock wave is not encountered in the situation shown in Figure 7.7.

The backward recovery shock wave is encountered when congestion has occurred but then due to increased bottleneck capacity the discharge rate exceeds the flow rate within the congested region. The term "backward" means that over time the shock wave is moving backward or upstream in the opposite direction of traffic. The term "recovery" implies that over time free-flow conditions are extending farther and farther upstream from the previous bottleneck location. The congested region is to the left of the shock wave and free-flow conditions are to the right. There are examples of backward recovery waves in Figures 7.7 and 7.9 (4–7). Note that a backward recovery shock wave is not encountered in the situation shown in Figure 7.8.

The last type of shock wave is the forward forming shock wave. This type of shock wave is not too common. The term "forward" implies that the shock wave moves in the same direction as traffic, and "forming" means that over time the congestion is gradually extending to sections farther and farther downstream. The time–space domain to the left of this shock wave has lower densities, and to the right, the density levels are higher. A forward forming shock wave is shown in Figure 7.9 (6–1). Note that a forward forming shock wave is not encountered in Figure 7.7 or 7.8.

The purpose of this section was to introduce briefly the subject of shock waves. The reader should be aware that this is only an introduction, for only the most important elements of shock waves are presented in a qualitative manner with strong assumptions. There are other types of waves to be considered in addition to shock waves. Shock waves demark all discontinuities in flow–density conditions not just at a density discontinuity of a specific density level such as 60 vehicles per lane-mile. Finally, many assumptions were made in order to enhance the simplicity and clarity of the presentation. The reader is referred to Chapter 11 for further discussion and analysis of shock waves.

7.5 ESTIMATING TOTAL TRAVEL TIME

The total travel time (TTT) expended in a system during a selected time period T can be estimated by determining the average number of vehicles in the system and multiplying by the length of the time period.

$$\text{TTT} = \bar{V} \cdot T \tag{7.12}$$

where TTT = total travel time expended in the system during the selected time period (vehicle-hours)

$\bar{V}$ = average number of vehicles in the system during the selected time period

T = time period of observations (hours)

The average number of vehicles in the system can be estimated by determining the number of vehicles in the system at frequent intervals during the time period selected.

$$\bar{V} = \frac{\sum_{t=1}^{m} v_t}{m} \tag{7.13}$$

where m = number of observations during the time period selected

v_t = number of vehicles in the system at time t

The number of vehicles in the system at time interval t can be estimated by knowing the densities, lengths, and number of lanes in each subsection of the system, and then summing over all subsections in the system:

$$v_t = \sum_{i=1}^{n} k_{it} \left(L_i N_i \right)$$ (7.14)

where n = number of subsections in the system
$\quad k_{it}$ = density in subsection i at time t (vehicles per lane-mile)
$\quad L_i$ = length of subsection i (miles)
$\quad N_i$ = number of lanes in subsection i

Substituting equation (7.14) into equation (7.13) and in turn, into equation (7.12), the total travel time expended in a system during a time period T can be estimated from the following equation.

$$\text{TTT} = \frac{\sum_{t=1}^{m} \sum_{i=1}^{n} k_{it} \left(L_i N_i \right)}{m} T$$ (7.15)

Consider a situation on a 3.8 mile section of directional freeway during the afternoon peak period. The freeway section is divided into seven subsections and the afternoon peak period from 3:45 until 6:30 P.M. is divided into eleven 15-minute time periods. A density contour map is available and the density (k_{it} in vehicles per lane-mile) is determined for each cell (subsection i during time period t), as shown in Table 7.2. Given the subsection length and number of lanes for each subsection i, the number of vehicles in each subsection and time period (v_{it}) can be calculated as shown in Table 7.2. The number of vehicles in the system at observation period t (v_t) is obtained as the sum of the vertical v_{it} values and are shown in Table 7.2 in the bottom row. The average number of vehicles in the system $\overline{V}$ during the selection time period is shown in Table 7.2 in the bottom right cell and found to be 813.2 vehicles. The total travel time expended in the system during the selected time period (3:45 to 6:30 P.M.) is estimated to be 2236.3 vehicle-hours using equation (7.12).

7.6 ESTIMATING TRAFFIC DEMAND

Traffic demand rate is defined as the number of vehicles that currently wish to pass a point or short section in a given period of time and normally is expressed as a rate of flow in vehicles per hour. Traffic demand does not include latent demand, but only the demand that currently uses the transportation system. If the traffic demand rates at various points in the system do not exceed their corresponding capacities, all vehicles that currently wish to pass can be accommodated, and the measured flow rate in the field is the traffic demand rate. However, if traffic demand rates exceed their corresponding capacities, the measured flow rates in the congested sections and downstream sections represent the number of vehicles (per unit of time) that *can* be handled, not the number

TABLE 7.2 Estimating Total Travel Time

Subsection Number, i		Length of Lanes, L_i	Number of Lanes, N_i	15-Minute Time Interval Ending at:[a]											v_{ti}
				4:00	4:15	4:30	4:45	5:00	5:15	5:30	5:45	6:00	6:15	6:30	
1	Alvarado to Rampart	0.39	4	35 54.6	34 53.0	42 65.5	38 59.3	48 74.9	51 79.6	61 95.2	44 68.6	27 42.1	30 46.8	28 43.7	683.3
2	Rampart to Silver Lake	0.50	4	39 78.0	38 76.0	59 118.0	52 104.0	44 88.0	72 144.0	71 142.0	34 68.0	27 54.0	28 56.0	32 64.0	992.0
3	Silver Lake to Vermont	0.58	4	32 74.2	40 92.8	39 90.5	47 109.0	49 113.7	66 153.1	75 174.0	63 146.2	50 116.0	24 55.7	32 74.2	1199.4
4	Vermont to Normandie	0.64	4	24 61.4	29 74.2	29 74.2	40 102.4	60 153.6	67 171.5	73 186.9	66 169.0	68 174.1	26 66.6	21 53.8	1287.7
5	Normandie to Western	0.83	4	29 96.3	26 86.3	48 159.4	82 272.2	82 272.2	88 292.2	93 308.8	95 315.4	85 282.2	64 212.5	22 73.0	2370.5
6	Western to Sunset	0.47	4	36 67.7	47 88.4	62 116.6	79 148.5	81 152.3	95 178.6	79 148.5	91 171.1	78 146.6	79 148.5	29 54.5	1421.3
7	Sunset to Hollywood	0.39	3	40 46.8	61 71.4	86 100.6	80 93.6	85 99.4	100 117.0	90 105.3	85 99.4	96 112.3	75 87.8	49 57.3	990.9
All v_t	Alvadaro to Hollywood	3.8	3 and 4	479.0	542.1	724.8	889.0	954.1	1136.0	1160.7	1037.7	927.3	673.9	420.5	813.2 $\bar{v}$

[a]The first value given is the average density in vehicles per lane-mile; the second value is the average total number of vehicles in the subsection.

Source: Reference 24.

of vehicles (per unit of time) that *wish* to be handled. Hence traffic demand rates are greater than measured flow rates and are unknown quantities.

It is important that the analyst be able to estimate traffic demand rates in such situations in order to quantify the need for capacity improvements and/or demand controls. For example, consider a situation in which the traffic demand rate for some period of time exceeds the capacity at a bottleneck. High densities will be observed upstream of the bottleneck and measured flow rates in the high-density sections, at the bottleneck, and downstream of the bottleneck do not represent traffic demand rates. In such situations the analyst will be interested in estimating either how much increase in bottleneck capacity is needed or how much traffic demand must be controlled in order to eliminate the adverse traffic situation.

In simple systems where all traffic demand inputs and outputs are upstream of the congested region and there are no other congested regions farther upstream, input and output traffic flow rates can be measured, and with a time adjustment for the travel time from measuring points to the bottleneck, the traffic demand rate at the bottleneck can be estimated. However, in real life more than one bottleneck often exists in transportation systems, and the congestion often extends upstream and encompasses input and output locations. Estimating traffic demands under these more typically encountered situations are more difficult, yet are more important to analyze. The purpose of this section of the chapter is to provide a technique for estimating traffic demands in such situations. Two examples will be presented in the following paragraphs. The first example will deal with a transportation system consisting of a single input and a single bottleneck, and while simplistic will permit a clear and concise presentation of the technique. The second example will be much more complex but will provide a more typical and generalized application of the technique.

7.6.1 A Simple Example

Consider a 4-mile section of freeway with a single input at the upstream end of the section and a single bottleneck at the downstream end of the section. Lane densities have been calculated from aerial photographs or from a traffic detector system for each 1-mile subsection of the 4-mile section at 15-minute time intervals. The 15-minute flow rates in vehicles per hour at the bottleneck were measured by traffic detectors. These lane densities and measured flow rates $(q)_{t,t+1}$ are shown in Table 7.3. The problem is to estimate the traffic demand rates $(\text{DEM})_{t,t+1}$ at the bottleneck.

Earlier in this chapter optimum density (k_0) was introduced and defined as the density level when the flow rate is maximum or operating at capacity. Optimum densities are usually on the order of 40 to 70 vehicles per lane-mile and for purposes of these examples an optimum density of 60 vehicles per lane-mile will be assumed. Therefore, if the lane density is less than 60, the traffic demand is less than the capacity and all of the demand is being served (none is being "stored"). However, if the lane density is over 60, the *excess* lane density is indicative of the quantity of traffic that is *not* being served but is being *stored*. The excess lane densities of the example problem are shown in Table 7.3. The excess lane densities are now converted to the number of excess

TABLE 7.3 Estimating Traffic Demands (First Example)

Time Interval	Subsection[a]				Bottleneck Conditions			
	1	2	3	4	$(V_{EXC})_t$	$(q_{EXC})_{t,t+1}$	$(q)_{t,t+1}$	$(DEM)_{t,t+1}$
	30	35	32	38	0			
1						0	5100	5100
	32	38	40	60	0			
2						360	5400	5760
	30	40	70	80	90			
			(10)	(20)				
3						−240	5400	5160
	35	38	50	70	30			
				(10)				
4						−120	5400	5280
	30	32	35	35	0			
5						0	4800	4800
	15	25	30	34	0			
6						0	3900	3900
	14	25	32	0				

[a]Values given are for lane density; the values in parentheses are excess lane density.

Source: Reference 24.

vehicles for each 15-minute time period by the following equation:

$$\left(V_{EXC}\right)_t = \sum_{i=1}^{n} \left(k_{it} - k_o\right) L_i N_i \tag{7.16}$$

where $(V_{EXC})_t$ = number of excess vehicles in the system at time t

n = number of subsections in the system

k_{it} = density in subsection i at time t (vehicles per lane-mile)

k_o = optimum density in vehicles per lane-mile (for this example k_o is assumed to be 60)

L_i = length of subsection i in miles (for this example all subsections are 1 mile long)

N_i = number of lanes in subsection i (for this example all subsections are three lanes wide)

The number of excess vehicles are shown for the example problem in Table 7.3. Note that equation (7.16) is identical to the earlier equation (7.14) except that k_o is introduced in equation (7.16) to obtain *excess* number of vehicles.

The next step is to convert excess number of vehicles at specific points in time to excess or unserved flow rates at the bottleneck for the various time intervals as shown in the following equation.

$$(q_{\text{EXC}})_{t,t+1} = \frac{(V_{\text{EXC}})_{t+1} - (V_{\text{EXC}})_t}{\text{TI}} \qquad (7.17)$$

where $(q_{\text{EXC}})_{t,t+1}$ = excess or unserved flow rate at the bottleneck between time t and $t+1$

TI = time interval between time t and $t+1$ in hours (for this example TI = 0.25)

The excess flow rates at the bottleneck are shown for the example problem in Table 7.3. Note that the excess flow is positive during time interval 2, which means that the demand rate exceeds the bottleneck capacity. However, in time intervals 3 and 4 the excess flow rate is negative, which implies that fewer vehicles are being stored and therefore the traffic demand rate is less than the bottleneck capacity. Estimates of the traffic demand rates at the bottleneck can be obtained by adding the excess flow rate to the measured flow rate. These results are also shown in Table 7.3 and are depicted graphically in Figure 7.11.

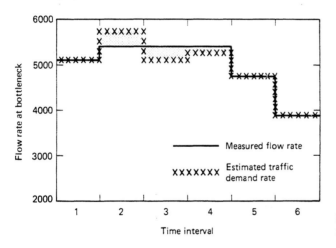

Figure 7.11   Flow Rates at Bottleneck (First Example)

Figure 7.11 provides a comparison between the flow rates that can pass through the bottleneck and the traffic demand that wishes to pass through the bottleneck. In time interval 2 the demand exceeds the capacity by a flow rate of 360 vehicles per hour, while in time intervals 3 and 4 the demands are less than the bottleneck capacity (240 and 120 vehicles per hour, respectively). Converting flow rates to numbers of vehicles, 90 too many vehicles arrive in time interval 2, and this excess demand is transferred to time intervals 3 and 4 when it is discharged (60 and 30 vehicles, respectively). In terms of capacity improvements, a 6 to 7 percent increase in bottleneck capacity would eliminate the adverse flow conditions. Another approach for improving flow conditions would be to control the demand (i.e., upstream ramp control). Assessment of improvements can be a very complex process because of the interactions of system changes and traveler responses and will be discussed more fully in Chapter 8.

7.6.2 A More Typical Example

Consider now the second example, which will be much more complex but will provide a more typical and generalized application. The same data set used for estimating total travel time in Section 7.5 will be used in this example for estimating traffic demand. The initial data of the previous Table 7.2 is rearranged for purposes of this example and are shown in Table 7.4. There are a total of eight subsections, with traffic traveling along the freeway from subsection 1 to 8 and with the bottleneck at the downstream end of subsection 8. The measured flow rate (vehicles per hour) and the accumulative flow (vehicles) is shown for each time interval at the bottom of the table. There are a total of eleven 15-minute time intervals, and lane densities are specified for each subsection at the end of each time interval. The length and number of lanes in each subsection are also given. An optimum density of 60 vehicles per lane-mile is again selected, and the excess lane densities for each subsection and time interval are calculated and shown in Table 7.4. The excess lane densities are now converted to the number of vehicles stored by using the earlier equation (7.16). The numbers of vehicles stored in each subsection and for each time interval are calculated and shown in Table 7.4.

Now there are a few additional complexities. When the freeway is congested, vehicles may also be stored on the on-ramps. Ramp stored vehicles were identified in the aerial photographs for each on-ramp at the end of each time interval, and the number of ramp vehicles stored is recorded in Table 7.4. The total storage is the sum of the storage on the freeway and the on-ramps and is shown in Table 7.4. Another complexity is that all the stored vehicles are not destined to pass through the bottleneck because they may exit on off-ramps along the study section before reaching the bottleneck. The destination factors may vary over time and generally increase as the distance between on-ramps and the bottleneck decrease. The destination factor is an estimate of the proportion of stored vehicles that will travel through the bottleneck. The estimated destination factors are shown in Table 7.4 and the total stored vehicles revised.

The next step is to total the number of vehicles stored for the entire study section for each time interval. These results are shown at the bottom of Table 7.4. Then the unserved flow rate at the bottleneck between observation times t and $t + 1$ is calculated using equation (7.17). The calculated unserved flow rate for each time interval was calculated and results shown at the bottom of Table 7.4. Note again that some unserved flow rates in the later time intervals are negative, which means than the arriving traffic demands are less than the bottleneck capacity and the storage along the study section is decreasing. Finally, by the end of the last time interval no vehicles are stored, and thereafter the measured flow rate at the bottleneck is the same as the traffic demand rate.

The traffic demand rate at the bottleneck can now be estimated by adding the unserved flow rate to the measured flow rate. These results are shown at the bottom of Table 7.4. The accumulative demand flows (vehicles) are also shown for each time interval. The importance of the relationship between accumulative demand flows and accumulative measured flows will be discussed in Chapter 12.

The results of this second example of estimating traffic demands are shown graphically in Figure 7.12. The total number of vehicles stored is shown at the bottom of the

TABLE 7.4 Estimating Traffic Demands (Second Example)

Subsection Number, i	Length of Lanes, L_i	Number of Lanes, N_i	Traffic Characteristic	4:00	4:15	4:30	4:45	5:00	5:15	5:30	5:45	6:00	6:15	6:30
1	0.39	4	Lane density	35	34	42	38	48	51	61	44	27	30	28
Alvarado to Rampart			Excess density	0	0	0	0	0	0	1	0	0	0	0
			Vehicles stored	0	0	0	0	0	0	2	0	0	0	0
			Ramp storage	0	0	0	0	0	0	2	0	0	2	0
			Total stored	0	0	0	0	0	0	2	0	0	2	0
			Destination factor	0.52	0.52	0.52	0.52	0.52	0.52	0.52	0.52	0.52	0.52	0.52
			Revised storage	0	0	0	0	0	0	1	0	0	1	0
2	0.50	4	Lane density	39	38	59	52	44	72	71	34	27	28	32
Rampart to Silver Lake			Excess density	0	0	0	0	0	12	11	0	0	0	0
			Vehicles stored	0	0	0	0	0	24	22	0	0	0	0
			Ramp storage	0	0	0	0	0	0	0	0	0	2	0
			Total stored	0	0	0	0	0	24	22	0	0	2	0
			Destination factor	0.54	0.54	0.54	0.54	0.54	0.54	0.54	0.54	0.54	0.54	0.54
			Revised storage	0	0	0	0	0	13	12	0	0	1	0
3	0.58	4	Lane density	32	40	39	47	49	66	75	63	50	24	32
Silver Lake to Vermont			Excess density	0	0	0	0	0	6	15	3	0	0	0
			Vehicles stored	0	0	0	0	0	14	35	7	0	0	0
			Ramp storage	0	0	0	0	0	0	0	0	0	0	0
			Total stored	0	0	0	0	0	14	35	7	0	0	0
			Destination factor	0.57	0.57	0.57	0.57	0.57	0.57	0.57	0.57	0.57	0.57	0.57
			Revised storage	0	0	0	0	0	8	20	4	0	0	0
4	0.64	4	Lane density	24	29	29	40	60	67	73	66	68	26	21
Vermont to Normandie			Excess density	0	0	0	0	0	7	13	6	8	0	0
			Vehicles stored	0	0	0	0	0	18	33	15	20	0	0
			Ramp storage	0	0	0	0	0	0	0	0	0	0	0
			Total stored	0	0	0	0	0	18	33	15	20	0	0
			Destination factor	0.72	0.72	0.72	0.72	0.72	0.72	0.72	0.72	0.72	0.72	0.72
			Revised storage	0	0	0	0	0	13	24	11	14	0	0
5	0.83	4	Lane density	29	26	48	82	82	88	93	95	85	64	22
Normandie to Western			Excess density	0	0	0	22	22	28	33	35	25	4	0
			Vehicles stored	0	0	0	73	73	93	110	116	83	13	0
			Ramp storage	0	0	0	0	0	0	0	0	0	0	0
			Total stored	0	0	0	73	73	93	110	116	83	13	0
			Destination factor	0.80	0.80	0.80	0.80	0.80	0.80	0.80	0.80	0.80	0.80	0.80
			Revised storage	0	0	0	58	58	74	88	93	66	10	0

TABLE 7.4 (cont'd.) Estimating Traffic Demands (Second Example)

Sub-section Number, i	Length of Lanes, L_i	Number of Lanes, N_i	Traffic Characteristic	15-Minute Time Interval Ending at:										
				4:00	4:15	4:30	4:45	5:00	5:15	5:30	5:45	6:00	6:15	6:30
6	0.47	4	Lane density	36	47	62	79	81	95	79	91	78	79	29
Western to Sunset			Excess density	0	0	2	19	21	35	19	31	18	19	0
			Vehicles stored	0	0	4	36	39	66	36	58	34	36	0
			Ramp storage	0	0	0	2	0	0	0	0	0	0	0
			Total stored	0	0	4	38	39	66	36	58	34	36	0
			Destination factor	0.80	0.80	0.80	0.80	0.80	0.80	0.80	0.80	0.80	0.80	0.80
			Revised storage	0	0	3	30	31	53	29	46	27	29	0
7	0.39	3.5	Lane density	40	61	86	80	85	100	90	85	96	75	49
Sunset to Hollywood			Excess density	0	1	26	20	25	40	30	25	56	15	0
			Vehicles stored	0	1	35	27	34	55	41	34	49	20	0
			Ramp storage	0	0	0	0	0	0	0	0	4	2	0
			Total stored	0	1	35	27	34	55	41	34	53	22	0
			Destination factor	1.00	1.00	1.00	1.00	1.00	1.00	1.00	1.00	1.00	1.00	1.00
			Revised storage	0	1	35	27	34	55	41	34	53	22	0
8	0.60	3	Lane density	21	54	88	70	80	74	85	81	68	74	60
Hollywood to Franklin			Excess density	0	0	26	10	20	14	25	21	3	14	0
			Vehicles stored	0	0	47	18	36	25	45	38	14	25	0
			Ramp storage	0	0	0	2	0	0	0	0	0	0	0
			Total stored	0	0	47	18	36	25	45	38	14	25	0
			Destination factor	1.00	1.00	1.00	1.00	1.00	1.00	1.00	1.00	1.00	1.00	1.00
			Revised storage	0	0	47	18	36	25	45	38	14	25	0
SUMMARY			Total vehicles stored	0	1	25	133	159	241	260	226	174	88	0
			Unserved flow rate	0	4	336	192	104	328	76	-136	-208	-344	-352
			Measured flow rate	4,816	5,028	5,292	5,224	5,396	5,380	5,264	5,384	5,100	5,340	4,120
			Accumulative flow	1,204	2,461	3,784	5,090	6,439	7,784	9,100	10,446	11,721	13,056	14,086
			Traffic demand rate	4,816	5,032	5,628	5,416	5,500	5,708	5,340	5,248	4,392	4,996	3,768
			Accumulative demand	1,204	2,462	3,869	5,223	6,598	8,025	9,360	10,672	11,395	13,144	14,086

Source: Reference 24.

illustration and increases from 0 to a maximum of 260 vehicles at the end of the time interval 7 (5:30 P.M.) and returns to no stored vehicles at the end of time interval 11 (6:30 P.M.). The unserved flow rate is shown in the middle of the illustration and is positive (demand rate exceeds the capacity) from time interval 2 through time interval 7

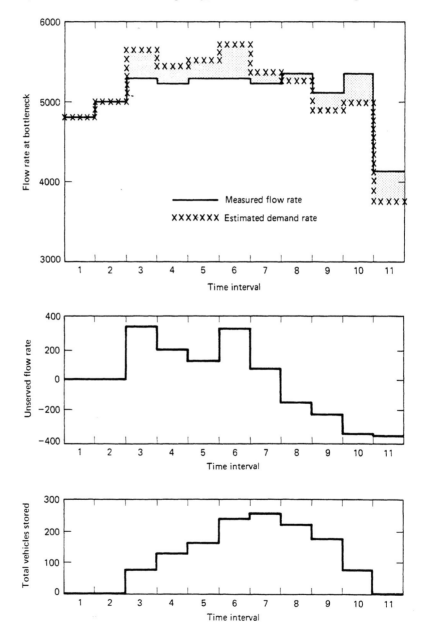

Figure 7.12   Estimating Traffic Demands (Second Example)

(4:00 to 5:30 P.M.) and is negative (demand rates less than capacity) from time interval 8 through time interval 11 (5:30 to 6:30 P.M.). The measured flow rate and estimated demand rate are shown in the upper portion of the illustration, and the shaded area denotes differences between flows and demands.

It should be noted that while congestion begins in time interval 2 and extends into time interval 11, the demand rate exceeds the capacity only from time interval 2 through time interval 7. The congestion from time interval 8 through time interval 11 is *not* due to demand rates exceeding capacities during this period of time but to unsatisfied demands during earlier time intervals being transferred to this period of time. Capacity increases at the bottleneck on the order of 400 vehicles per hour (7 to 8 percent) would eliminate the congestion under the current estimated demand pattern. However, such an improvement would cause some temporal and spatial driver responses which would modify the demand pattern and thereby affect the full benefits of the improvement. Controlling the input demand to the bottleneck would be another possible approach to reducing congestion. The unserved flow rate gives an indication of the time period needed for demand control and the magnitude of the diversion.

7.7 SELECTED PROBLEMS

1. The use of density as a macroscopic traffic flow characteristic has increased significantly during the past 20 to 30 years. Review the 1950, 1965, and 1985 *Highway Capacity Manuals* to observe this increased use of density.

2. The metering rate at local traffic-responsive ramp signals are determined on-line as a function of percent occupancy measured on the freeway. Plot metering rate as a function of percent occupancy graphically based on a review of freeway entry control literature. Consider maximum and minimum metering rate limits.

3. Develop an equation for the speed–density relationship shown in Figure 7.1. What would be the resulting numerical values for optimum density, jam density, and capacity? Also determine the resulting speeds when flow is near zero and when the flow is at capacity.

4. Derive the equation for density as a function of percent occupancy without the use of the text.

5. Assuming a reasonable relationship between speed and density, convert the density contour map shown in Figure 7.3 to a speed contour map. Give particular attention to the regions around the 40 and 60 density contour lines.

6. Interpret the percent occupancy contour map shown in Figure 7.4 with particular attention to the identification of bottlenecks and shock waves.

7. Interpret the density contour map shown in Figure 7.5 with particular attention to the identification of bottlenecks and shock waves.

8. Identify the bottlenecks and assess the severity of congested locations as shown in Figure 7.6 of the San Francisco Bay Area Freeway System.

9. Design and conduct a field experiment that will demonstrate the typical shock wave phenomena of a single lane at a signalized intersection (see Figure 7.7).

10. Assume that the field experiment described in Problem 9 is undertaken at a signalized intersection having a 90-second cycle with 67 percent effective green time. The backward

forming and backward recovery shock waves are found to be 4 and 10 miles per hour, respectively. How far upstream of the stop line will the stopped queue of vehicles extend? Is this approach oversaturated? What effect would that have on the maximum stopped queue distance?

11. Congestion is observed at a freeway bottleneck location from 7:00 until 9:00 A.M.. Maximum queueing extends 2 miles upstream and is observed at 8:15 A.M.. Assuming no rear stationary shock wave, identify and calculate the velocities of all other shock waves.

12. Draw a hypothetical density contour map that includes all types of shock waves and describe the traffic flow situations.

13. Estimate the total travel time for the situation presented in Table 7.3.

14. Estimate the total delay time per lane during one cycle of the traffic signal described in Problem 10. Assume a density of 200 vehicles per lane-mile in the stopped queue and a constant arrival rate of vehicles. What proportion of the vehicles will have to stop?

15. Estimate the total travel time in the congestion upstream of the freeway bottleneck described in Problem 11. Assume an average density of 80 vehicles per lane-mile in the congested region.

16. Estimate the traffic demand at the bottleneck for the problem described in Table 7.3, except use an optimum density value of 50 rather than 60 vehicles per lane-mile. Plot results similar to Figure 7.11.

17. Estimate the traffic demand at the bottleneck for the problem described in Table 7.4, except use an optimum density value of 70 rather than 60 vehicles per lane-mile. Plot results similar to Figure 7.12.

18. In the closing paragraph of Section 7.6.2 a capacity increase at the bottleneck on the order of 400 vehicles per hour was recognized as needed to eliminate the congestion under the current estimated demand pattern. Then an alert was given that this improvement might modify the temporal and spatial patterns of traffic and affect the full benefits of the improvement. Explain. Would controlling the input demand have the same effect?

7.8 SELECTED REFERENCES

1. Wolfgang S. Homburger and James H. Kell, *Fundamentals of Traffic Engineering*, University of California, Berkeley, Calif. 1988, pages 4–6 to 4–7.

2. Institute of Transportation Engineers, *Transportation and Traffic Engineering Handbook*, 2nd Edition, Prentice-Hall, Inc., Englewood Cliffs, N. J., 1982, pages 472–473 and 533–534.

3. Martin Wohl and Brian V. Martin, *Traffic System Analysis for Engineers and Planners*, McGraw-Hill Book Company, New York, 1967, pages 322–329 and 391–395.

4. Daniel L. Gerlough and Matthew J. Huber, *Traffic Flow Theory—A Monograph*, Transportation Research Board, Special Report 165, TRB, Washington D. C., 1975, pages 10–12.

5. Bruce D. Greenshields, The Density Factor in Traffic Flow, *Traffic Engineering*, Vol. 30, No. 6, March 1960, pages 26–28, and 30.

6. Frank A. Haight, The Volume Density Relation in Theory of Road Traffic, *Operations Research*, Vol. 8, No. 4, July–August 1960, pages 512–573.

7. John J. Haynes, *The Development of the Use of Vehicular Density as a Control Element for Freeway Operations*, Ph.D. Dissertation, Texas A&M University, August 1964, 162 pages.

8. Great Britain Road Research Laboratory, *Research on Road Traffic,* Her Majesty's Stationary Office, London, 1965, pages 181–182.

9. John Haynes, *Some Considerations of Vehicular Density on Urban Freeways,* Highway Research Board, Record 99, HRB, Washington D. C., 1965, pages 59–80.

10. W. Leutzbach, *Testing the Applicability of the Theory of Continuity on Traffic Flow after Bottlenecks,* Karlsruhe Technische Hockschule, International Symposium–Traffic Flow, 1968, 5 pages.

11. Gordan F. Paesani, *The Relationship between the Density and Occupancy Concepts,* Research Report 26, Detroit National Proving Groups for Freeway Surveillance, Control and Electronic Traffic Aids, Detroit, Mich., 1966, 12 pages.

12. George Weiss, Statistical Properties of Low Density Traffic, *Quarterly of Applied Mathematics,* Vol. XX, No. 2, 1968, pages 121–130.

13. Frank C. Andrews, Statistical Theory of Traffic Flow in Highways, Steady State Flow in Low Density Limit, *Transportation Research,* Vol. 4, No. 4, December 1970, pages 359–366.

14. D. C. Gazis and Charles H. Knapp, On-Line Estimation of Traffic Densities from Time-Series of Flow and Speed Data, *Transportation Science,* Vol. 5, No. 3, August 1971, pages 283–301.

15. D. C. Gazis and M. W. Szeto, Estimation of Traffic Densities at the Lincoln Tunnel for Time Series of Flow and Speed Data, in P. A. W. Lewis, ed., *Stochastic Point Processes,* 1971, pages 151–165.

16. B. B. Lientz, Estimation of Mass Concentration for Problems in Traffic Flow, *Transportation Research,* Vol. 5, No. 2, June 1971, pages 75–81.

17. P. K. Munjal and L. A. Pipes, Propagation of On-Ramp Density Perturbation on Unidirectional Two- and Three-Lane Freeways, *Transportation Research,* Vol. 5, No. 4, December 1971, pages 241–255.

18. P. K. Munjal and L. A. Pipes, Propagation of On-Ramp Density Perturbation on Unidirectional Two- and Three-Lane Freeways, *Transportation Research,* Vol. 5, No. 4, November 1971, pages 390–402.

19. Charles H. Knapp, Traffic Density Estimation for Single- and Multi-lane Traffic, *Transportation Science,* Vol. 7, No. 1, February 1973, pages 75–84.

20. N. E. Nahi and A. N. Trivedi, Recursive Estimation of Traffic Variables: Section Density and Average Speed, *Transportation Science,* Vol. 7, No. 3, August 1973, pages 269–286.

21. Man-Feng Chang and Denos C. Gazis, Traffic Density Estimation with Consideration of Lane-Changing, *Transportation Science,* Vol. 9, No. 4, November 1975, pages 308–320.

22. Adolf D. May, Traffic Characteristics and Phenomena on High Density Control Access Facilities, *Traffic Engineering,* Vol. 31, No. 6, March 1961, pages 11–19.

23. A. N. Johnson, Maryland Aerial Survey of Highway Traffic Between Baltimore and Washington, *Highway Research Board, Proceedings,* 1928, Vol. 8, pages 106–115.

24. Adolf D. May, *California Freeway Operations Study,* Thompson Ramo Wooldridge, Chatsworth, Calif., 1962, pages 71–81 and 93–100.

25. Robert S. Foote and Kenneth W. Crowley, *Developing Density Controls for Improved Traffic Operations,* Highway Research Board, Record 154, HRB, Washington D. C., 1966, pages 24–37.

26. Brian L Allen and Adolf D. May, *System Evaluation of Freeway Design and Operations,* Highway Research Board, Special Report 107, HRB, Washington D. C., 1969, pages 45–59.

27. R. R. Jacobs, *Bay Area Freeway Operations Study*, 7th Interim Report, University of California, Berkeley, Calif., April 1969, 66 pages.

28. Roger P. Roess, William R. McShane, and Louis J. Pignataro, *Freeway Level of Service: A Revised Approach*, Transportation Research Board, Record 699, TRB, Washington D. C., 1979, pages 7–16.

29. J. Uren and J. B. Garner, *Integrated System for Urban Traffic Data Collection*, Transportation Research Board, Record 699, TRB, Washington D. C., 1979, pages 41–49.

30. A. V. Gafarian, J. Pahl, and T. L. Ward, *Some Properties of Freeway Density as a Continuous-Time Stochastic Process*, Transportation Research Board, Record 667, TRB, Washington D. C., 1978, pages 79–83.

31. Transportation Research Board, *Highway Capacity Manual*, Special Report 209, TRB, Washington D. C., 1985.

32. American Association of State Highway and Transportation Officials, *A Policy on Geometric Design of Highways and Streets*, AASHTO, Washington D. C., 1984, 1087 pages.

33. William T. McCasland, *Comparison of Two Techniques of Aerial Photography for Applications in Freeway Traffic Operations Studies*, Highway Research Board, Record 65, HRB, Washington D. C., 1965, pages 95–115.

34. A. D. May, P. Athol, W. Parker and J. B. Rudden, *Development and Evaluation of Congress Street Expressway Pilot Detection System*, Highway Research Board, Record 21, HRB, Washingyon D. C., 1963, pages 48–68.

35. Lucien Duckstein, *Control of Traffic in Tunnels to Maximize Flow*, Highway Research Board, Record 154, HRB, Washington D. C., 1967, pages 1–23.

36. L. C. Edie and R. S. Foote, Effect of Shock Waves on Tunnel Traffic Flow, *Highway Research Board, Proceedings*, Vol. 39, 1960.

37. Bruce Greenshields, A Study of Traffic Capacity, *Highway Research Board, Proceedings*, Vol. 14, 1933, pages 468–477.

38. Adolf D. May and David G. Fielder, *Squirrel Hill Tunnel Operations Study*, Highway Research Board, Bulletin 324, HRB, Washington D. C.,1962, pages 12–37.

8

Demand–Supply Analysis

Essentially all analytical techniques of traffic systems are structured in a demand–supply framework. The demand–supply framework can be at the microscopic level of analysis, in which individual traffic units are studied, or at the macroscopic level of analysis, in which attention is given to groups of traffic units in aggregate form.

These demand–supply analytical techniques vary, and the analyst must not only learn these various techniques but must develop skills in selecting the most appropriate technique for the problem at hand. These analytical techniques are described in the following chapters and include capacity analysis, traffic stream models, shock wave analysis, time–space diagrams, queueing analysis, and simulation.

This chapter is devoted to presenting an analytical framework for demand–supply analysis. The major elements of the analytical framework are described and their interactions identified. The analytical framework is presented as having two processes: an initial process and one or more feedback processes. At the end of the chapter, a simple example is used to demonstrate this approach.

8.1 AN ANALYTICAL FRAMEWORK

An analytical framework for demand–supply analysis is shown Figure 8.1. The input includes demand, supply, control, and environment. The analytical technique processes the input data in order to predict the performance of the traffic system. Once the initial performance is predicted, the input can be modified as a feedback process. These modifications can be user selected or be part of an optimization and/or demand-related process.

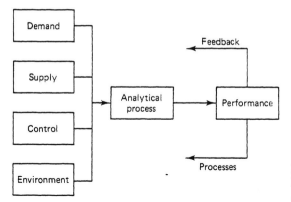

Figure 8.1   An Analytical Framework for Demand–Supply Analysis of Traffic Systems

8.1.1 Initial Analysis

The demand input represents the number of units that *would like to be served* and is always equal to or greater than the number of units that *can be served*. In microscopic analysis the demand input is expressed as the arrival time of individual units, while in macroscopic analysis the demand input is expressed as an arrival rate (units per time interval). In undersaturated situations, that is, when demands are less than capacities, single time periods can be analyzed! However, in oversaturated situations, where demands exceed capacities, multitime periods must be analyzed. In this way, excess demands in earlier time periods can be transferred and served in later undersaturated time periods. The time frame for the demand can be for a current point in time, for some future year, or simply a demand modification due to supply or control changes.

The supply input represents the maximum number of units that can be served. In microscopic analysis the supply input is expressed as the minimum time headway between units to be served, while in macroscopic analysis the supply input is expressed as the maximum flow rate in units per time interval. This maximum flow rate is often referred to as capacity and generally remains constant over time periods or traffic states. However, capacity can change between time periods or traffic states due to changes in demand patterns, vehicle mix, driver characteristics, control states, and/or environment. Often the analyst begins with the actual physical dimensions of the traffic facility and must translate these features into capacity values. This can be an involved process, and the next chapter is devoted to capacity analysis.

The control input consists of a set of rules as to how the units interact with each other and with the system. The control can be constant over time, such as car-following rules, rules of service [i.e., first in, first out (FIFO)], traffic restrictions, vehicle performance capabilities, and so on. The control can also be time dependent, such as at traffic signals and during peak period restrictions. The control can be traffic responsive, such as at railroad crossings, ramp control, traffic signals, and when drivers modify their behavior due to impatience and frustration. Finally, it should be noted that "no control" is a control state that can be encountered.

The environment input generally serves to modify the other inputs; namely, demand, supply, and control. Situations encountered include visibility, weather conditions, pavement conditions, unusual distractions, and others. In practice the performance may be overestimated because the traffic system is often assumed to be under "normal" environmental conditions, that is, under near-perfect conditions.

The analytical technique processes the above-mentioned input data to predict the performance of the traffic system. In the microscopic level of analysis, analytical techniques may include car-following theories, time–space diagrams, stochastic queueing analysis, and microscopic simulation. At the macroscopic level of analyses, analytical techniques may include capacity analyses, speed–flow–density relationships, shock wave analysis, deterministic queueing analysis, and macroscopic simulation. These analytical techniques can be accomplished manually or by computer, can utilize mathematical expressions or simulation, and can lead to solutions of a deterministic or a stochastic form. Later chapters are devoted to these analytical techniques.

The objective of the demand–supply analysis is to predict the performance of the traffic system from the viewpoint of users and from the perspective of the total system. The users would be interested in their travel times, delays, queueing, comfort, risks, and energy consumption. The system manager would be concerned with system productivity, level of service, air pollution, noise generation, safety record, and total transport costs. The formulation often takes the form of a multiobjective function with a set of constraints. For example, at an isolated signalized intersection the formulation might be to minimize a combination of delays and stops with the constraint of a maximum individual delay. Generally, as the size and complexity of the traffic system increases, the formulation becomes more comprehensive.

8.1.2 Feedback Processes

Once the initial analysis is completed and the performance is predicted, a feedback process (or an iterative process) may be required or desired. Feedback processes take many forms and include determining demand–performance equilibrium, answering "what if" statements, conducting sensitivity investigations, optimizing an objective function, and/or investigating traveler responses due to system changes.

Often the inputs to the analysis, such as demand and supply, are developed independently without knowledge as to how the traffic system will perform. When the initial analysis is completed, the anticipated demand level may not be compatible or appropriate for the resulting predicted performance. Consider a very simple single-link traffic system with demand–supply–cost functions as shown in Figure 8.2. An initial demand level of D_1 was assumed with an anticipated user cost of C_1. However, once the initial process was completed, a user cost of C_2 (rather than C_1) was predicted. The feedback process required modifying the demand input until the demand and specified supply were in equilibrium as denoted as point 3 in Figure 8.2, where the demand level would be reduced to D_3 with an associated average cost of C_3.

A second type of feedback process would be answering "what if" statements. For example, in a rural highway investigation, the analyst might want to estimate the change

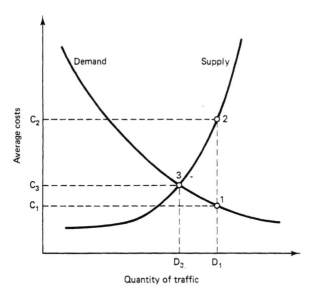

Figure 8.2   Demand–Supply Equilibrium

in predicted performance *if* a 1500-foot passing lane is added at the midpoint of the upgrade. The analyst would perform the initial process and then in the feedback process, the supply input would be modified to include the 1500-foot passing lane.

Another type of feedback process would be for sensitivity analysis. The analyst would generate supply input modifications either manually or by mathematical formulation. For example, the previous two-lane rural highway problem could be addressed to find the effect of the location of the 1500-foot passing lane from the bottom of the upgrade to the top of the upgrade. A practical example of this approach is shown in Figure 8.3.

The optimization of an objective function is another form of the feedback process. The analyst normally employs some type of mathematical programming technique. The technique may vary from a rigid linear programming approach to a brute-force branch-and-bound approach. One example is the use of the TRANSYT model for optimizing a signal timing plan in an arterial network. After the initial process of evaluating the arterial network using the existing signal timing plan, an iterative optimizing process is followed to improve the splits, offsets, and cycle length. The rather-simplistic but effective search process proceeds in incremental steps for each control parameter separately. In each step of the feedback process a control parameter is incrementally changed and the analytical technique (in this case a macroscopic simulation model) is applied to predict the new system performance. Comparisons are made between the new state and the previous state to determine if an optimum setting has been reached. The feedback process continues until all control parameters for all intersections have reached their "optimum" value.

A final feedback process is the investigation of traveler responses to system changes. Initially, there is equilibrium between supply and demand. However, over time the input to the demand–supply analysis is modified due to demand growth, supply changes, and/or control implementation. These input modifications change the

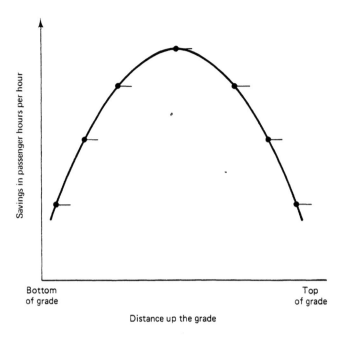

Figure 8.3   A Sensitivity Analysis Feedback Process for Passing Lanes on Two-Lane Rural Highways (From Reference 3)

predicted performance of the traffic system. The users then may respond by spatial, temporal, modal, or total travel changes. Users may change their paths of travel through the network, resulting in a spatial response. Users may change their time of travel, resulting in temporal responses. Users may change their mode of travel (car pool, bus, etc.), resulting in modal response. Finally, users may eliminate, combine, or create trips, resulting in total travel responses. As users respond in these various ways, the input to the demand–supply analysis changes, and correspondingly, the predicted performance changes. This process of user responses and system performance changes becomes an iterative process and requires special procedures in order to reach equilibrium. An added complexity is that changes in travel patterns may modify the previously optimized control strategy. An example of this type of feedback process would be a natural extension of the optimization feedback process described previously. Once the signal timing plan has been optimized, the system users may modify their paths of travel through the arterial network. Users will select cheaper cost routes, and flows change from some links to other links. If a sufficient number of users modify their paths through the arterial network, the previously "optimized" signal timing plan may no longer be optimum. A number of interesting research studies have been directed to this feedback process [4].

An analytical framework for demand–supply analysis has been presented in this section and included an initial process and a series of possible feedback processes. The major elements of the analytical framework were described and their interactions identified. Space does not permit an extensive treatment of the subject, but perhaps this introduction of an analytical framework and the following example will encourage the reader to study further this complex but extremely important and interesting subject.

8.2 AN EXAMPLE OF DEMAND–SUPPLY ANALYSIS

This example attempts to demonstrate the application of the analytical framework for demand–supply analysis to a directional freeway corridor. Three points in time will be analyzed: day −1, day +1, and after traveler response. Day −1 may represent either existing conditions or the day before some change is made to the input to the demand–supply analysis. Day +1 represents the day after some change was made to the input to the demand–supply analysis, but users have not changed their travel pattern. The last point in time is after the users have considered changing their travel pattern because of the change in the traffic system.

8.2.1 Base Conditions

A directional freeway corridor is to be analyzed to predict its performance for a specified demand–supply situation. The following paragraphs first describe the demand, supply, control, and environment elements. Then the analytical framework and numerical solution are presented.

The demand pattern for the directional freeway corridor is given in Figure 8.4a. There are three origins along the directional freeway (O_1, O_2, and O_3) and all the freeway traffic is destined to the downstream end of the freeway segment. During the afternoon peak period, the traffic demand hourly rates are given for each origin in each 15-minute period. To simplify this example problem, the nonfreeway demands for the parallel and perpendicular arterials are assumed to be negligible.

The supply side of the freeway corridor is depicted in Figure 8.4b. The directional freeway segment is divided into three subsections. The first subsection is three lanes wide extending from the upstream end of the freeway segment to the first on-ramp. The second subsection is also three lanes wide and 1 mile long and extends from the first on-ramp to the second on-ramp. The third and last subsection, which extends from the second on-ramp to the downstream end of the freeway segment, is 1 mile long and four lanes wide. Other supply-side information is needed, such as design speed, grade profile, vehicle mix, and geometric design features, to convert the supply characteristics to capacity values. However, to simplify the procedure, capacity values of 1800 vehicles per hour per lane are assumed for the freeway, and capacity values of 900 and 1600 vehicles per hour are assumed for the two ramps, respectively. The speed on all arterials is assumed to be 25 miles per hour regardless of flow rate, while the speeds on the freeway are a function of the volume–capacity ratios, as depicted in Figure 8.5.

The control situation for the base condition is essentially a "no control" state. That is, there is no control at the intersections nor on the ramps. At freeway ramp merges, the freeway and ramp vehicles alternate their use of the merge area, while at all other locations it is assumed that the first vehicle to arrive is the first vehicle to be served. The environmental situation is "normal," that is, there are no unusual incidents, and the weather and pavement conditions are ideal.

A demand–supply matrix is constructed as shown in Figure 8.6. The horizontal scale is distance with freeway vehicles traveling from left to right through the three

| Time | Time | Input Origin | | |
Slice	Period	O_1	O_2	O_3
1	4:45–5:00	4800	400	800
2	5:00–5:15	5000	800	800
3	5:15–5:30	5000	800	800
4	5:30–5:45	3800	800	800
5	5:45–6:00	3200	200	600

(a) Input demand pattern

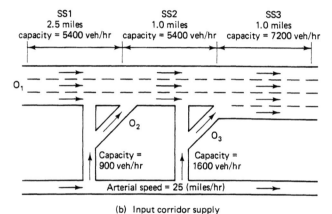

(b) Input corridor supply

Figure 8.4   Demand and Supply Inputs
for Example Problem

freeway subsections. Each subsection has unique capacity values, and these values can be thought of as penetrating down through the matrix and the various time slices. The vertical scale is time, with the five time slices extending down the matrix. Each time slice has unique demand values, and these demand values can be thought of as moving horizontally across the matrix and the various subsections.

There are a total of 15 cells in the matrix (three subsections by five time slices), and demand and capacity values are entered in each cell assuming no congestion. The analysis begins in the upper left-hand corner of the matrix (subsection 1 in time slice 1) and continues horizontally to the right through all subsections during the first time slice. Then, as with a carriage return, the analysis continues in the second time slice of the first subsection and continues horizontally to the right through all subsections during the second time slice. This procedure is followed until all subsections in all time slices are analyzed. However, this straightforward procedure is followed only if the demands in each cell are equal to or less than their corresponding capacities. If demand exceeds capacity in a cell, the procedure is interrupted and a four-step subprocedure is introduced, consisting of bottleneck analysis, downstream demand modification, upstream vehicle storage, and excess demand transfer to the next time slice.

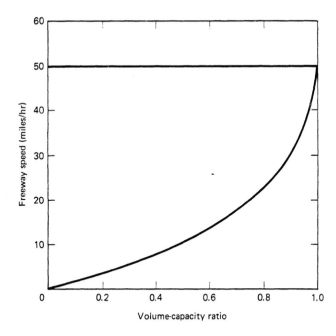

Figure 8.5 Freeway Speed as a Function Volume–Capacity Ratio

Bottleneck analysis is performed when demand exceeds capacity. The flow is no longer equal to demand but is limited by the capacity. The volume is equal to the capacity, and the v/c ratio is 1.00. When bottlenecks are encountered, the original downstream demands are incorrectly calculated. These demands are recalculated since excess

Time Slice	SS1	SS2	SS3
1	$D_{11} = 4800$ $C_{11} = 5400$	$D_{12} = 5200$ $C_{12} = 5400$	$D_{13} = 6000$ $C_{13} = 7200$
2	$D_{21} = 5000$ $C_{21} = 5400$	$D_{22} = 5800$ $C_{22} = 5400$	$D_{23} = 6600$ $C_{23} = 7200$
3	$D_{31} = 5000$ $C_{31} = 5400$	$D_{32} = 5800$ $C_{32} = 5400$	$D_{33} = 6600$ $C_{33} = 7200$
4	$D_{41} = 3800$ $C_{41} = 5400$	$D_{42} = 4600$ $C_{42} = 5400$	$D_{43} = 5400$ $C_{43} = 7200$
5	$D_{51} = 3200$ $C_{51} = 5400$	$D_{52} = 3400$ $C_{52} = 5400$	$D_{53} = 4000$ $C_{53} = 7200$

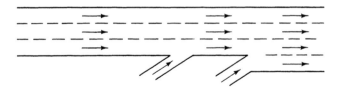

Figure 8.6 Demand–Supply Matrix for Example Problem

demands at the bottleneck are being stored upstream and cannot reach downstream sub-sections. The excess demand at a bottleneck is stored in upstream subsections, and the previous calculated performances of upstream subsections are recalculated because of the upstream storage. Finally, the excess bottleneck demand cannot be served in this time period and is transferred to the demand schedule for the next time slice. When this subprocedure is completed, the original procedure is continued. Bottlenecks can be encountered again in downstream subsections and later time slices and each time they are encountered, the subprocedures are employed. The process becomes very complex when the effects of one bottleneck affects upstream bottlenecks, but this discussion will be postponed to Chapter 11. .

A flowchart of this procedure and associated subprocedure is shown in Figure 8.7. The numerical solution of the example problem is given in Figure 8.8. In each cell the demand, flow, capacity, v/c ratio, and speed are indicated. Note that in the second time slice, the demand for subsection 2 exceeds its capacity (5800 > 5400 vehicles per hour rate). Therefore, the flow in this subsection is 5400 and the v/c ratio is 1.00. The demand in subsection 3 is reduced from 6600 to 6200 because of the bottleneck in sub-section 2. The first on-ramp has a demand rate of 800 vehicles per hour and a capacity of 900 vehicles per hour so that all ramp vehicles can enter the freeway. Therefore, the maximum flow that can cross from subsection 1 to subsection 2 is 4600 vehicles per hour rate (5400 − 800). Therefore, while the upstream portion of subsection 1 is under free-flow conditions, the downstream portion of subsection 1 is congested, with a v/c

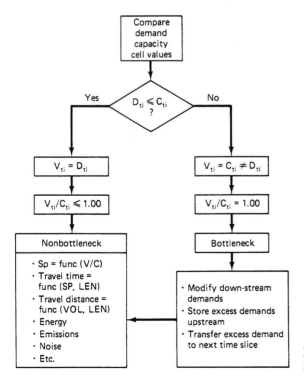

Figure 8.7   Flowchart of Demand–Supply Matrix Procedures

ratio of –0.85 (the minus indicates congested flow) and a speed of 25 miles per hour (see Figure 8.5). There is a discontinuity in flow conditions in subsection 1 from free-flow conditions to congested conditions but this is discussed in Chapter 11. The shaded area of Figure 8.8 denotes the congested region.

Subsection 2 continues to be the bottleneck in time slice 3 and the congestion extends the farthest upstream at the end of time slice 3. During the fourth time slice the demands decrease and the congestion is dissipated at the end of time slice 4. Hence the congestion lasts for 45 minutes and extends upstream a distance of approximately 2 miles.

Once the flows and speeds are determined for each cell of the matrix, a wide selection of measures of performance can be predicted. Total travel time, total travel

Time Slice		SS1		SS2	SS3
1	D	4800	4800	5200	6000
	V	4800	4800	5200	6000
	C	5400	5400	5400	7200
	v/c	0.89	0.89	0.96	0.83
	s	50	50	50	50
2	D	5000	5000	5800	~~6600~~ 6200
	V	5000	4600	5400	6200
	C	5400	5400	5400	7200
	v/c	0.93	–0.85	1.00	0.86
	s	50	25	50	50
3	D	5000	5400 ~~5000~~	5800	~~6600~~ 6200
	V	5000	4600	5400	6200
	C	5400	5400	5400	7200
	v/c	0.93	–0.85	1.00	0.86
	s	50	25	50	50
4	D	3800	4600 ~~3800~~	4600	~~5400~~ 6200
	V	3800	4600	5400	6200
	C	5400	5400	5400	7200
	v/c	0.70	–0.85	1.00	0.86
	s	50	25	50	50
5	D	3200	3200	3400	4000
	V	3200	3200	3400	4000
	C	5400	5400	5400	7200
	v/c	0.59	0.59	0.63	0.56
	s	50	50	50	50

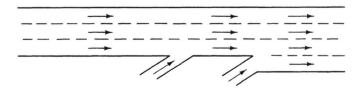

Figure 8.8   Numerical Solutions of Supply–Demand Matrix for Base Conditions

distance, average system speed, total nondelay travel time, mainline delay, ramp delay, and percent delay are a few examples of these system measures of performance. These measures of performance were calculated for the base conditions and are summarized in Table 8.1.

TABLE 8.1   System Performance under Base Conditions

System Performance Measure	Result
Total travel time (vehicle-hours)	614.5
Total travel distance (vehicle-miles)	26,975
Average system speed (miles/hr)	43.9
Total nondelay travel time (vehicle-hours)	539.5
Mainline delay (vehicle-hours)	75.0
Ramp delay (vehicle-hours)	0.0
Percent delay	12.2

Subsection 2 is identified as the bottleneck; that is, it is the cause of the congestion. Note, however, that the effect of the congestion is displayed upstream of the bottleneck. It is important to distinguish between cause and effect. Imagine traveling along a freeway section under free-flow conditions and up ahead brake lights can be seen, speeds are reduced, and congestion is encountered. This section of the freeway is *not* the bottleneck, therefore *not* the cause of the problem. Traveling farther along the freeway, speeds begin to increase. When speeds on the order of 30 to 40 miles per hour are encountered, this is the clue that the bottleneck is here. Applying this graphic description of freeway travel to the results shown in Figure 8.8, subsection 2 is confirmed as the bottleneck. Demand exceeds capacity in subsection 2 for two time slices (time slices 2 and 3) by 400 vehicles per hour rates (5800 minus 5400 vehicles per hour). Note that demands exceeds capacities for two time slices but congestion lasts for three time slices. Two solutions are possible. Either the capacity of subsection 2 should be increased by at least 400 vehicles per hour, or the demand of subsection 2 should be reduced by at least 400 vehicles per hour in time slices 2 and 3. The selection of the best alternative can be a very complex process and is not germane to this example problem. Since controlling the demand better illustrates the feedback process, ramp control rather than increasing the size of the facility will be selected as the improvement plan. Ramp control has some special merits in this case since demands exceed capacity only for two time slices, and underutilized parallel arterial routes are available.

8.2.2 Ramp Control without User Response

A freeway ramp control strategy is to be imposed on this traffic system, but at this point in time (day +1), the users will not have modified their travel plans. The objective of the control could be to serve as many users as possible, but with the restriction that no freeway congestion will exist. Because of the simplistic nature of the traffic system, the numerical solution is intuitively straightforward. However, to provide a more generalized approach, a linear programming formulation will be presented first and then a numerical solution obtained.

The solution to an optimal freeway ramp control strategy can be formulated as a linear programming problem with an objective and several sets of constraints. The objective of the freeway ramp control strategy is to maximize the vehicle input to the directional freeway. The objective can be expressed as

$$\text{MAX} \sum_{i=1}^{N} X_i \tag{8.1}$$

where i = freeway input origin

X_i = allowable input flow rate at origin i

N = number of input origins

Three sets of constraints are often incorporated into the linear programming formulation. The first set of constraints ensures that no freeway congestion will exist. This set of constraints can be expressed as

$$\sum_{i=1}^{N} a_{ij}X_i \leq c_j \tag{8.2}$$

where j = freeway subsection

a_{ij} = proportion of traffic that enters at origin i and is destined to pass through subsection j

c_j = capacity of subsection j

A set of these equations are employed: one equation for each freeway subsection j.

The second set of constraints ensures that if ramp control is selected, the metering rate will fall within certain practical limits. This set of constraints can be expressed as

$$\text{MAX } R_i \geq X_i \geq \text{MIN } R_i \tag{8.3}$$

where MAX R_i = maximum allowable metering rate for freeway input origin i (usually, 720 to 900 vehicles per hour per lane)

MIN R_i = minimum allowable metering rate for freeway input origin i (usually, 180 to 240 vehicles per hour per lane)

The third set of constraints ensures that freeway input origins which should not be controlled from a practical point of view are not selected for control by the linear programming formulation. This set of constraints can be expressed as

$$X_{it} = D_{it} \tag{8.4}$$

where X_{it} = allowable input flow rate at origin i in time slice t
    D_{it} = input demand rate at origin i in time slice t

This constraint is often applied for freeway input origins that are on the freeway itself and for freeway connectors.

More sophisticated linear programming formulations have been developed [6, 7]. Objectives have included maximizing vehicle-miles, people input, and person-miles. Constraints have included queue constraints, special ramp treatment, and capacity buffers. These are not required in the present example, but readers can refer to the literature for greater in depth coverage.

Returning to the example problem, the linear program is applied to the first time slice. Since demands do not exceed capacities in any subsection, as shown in Figure 8.8, the objective is best met by exerting no control in the first time slice. In the second time slice, demand exceeds capacity in subsection 2 by a rate of 400 vehicles per hour. To eliminate congestion through ramp control, either origin 1 and/or origin 2 must be metered. Since origin 1 is a mainline freeway input and is normally not controlled from a practical point of view, the demand at origin 2 must be metered at a rate of 400 vehicles per hour (one vehicle release every 9 seconds). A metering rate of 400 vehicles per hour is obtained by subtracting the mainline input (5000 vehicles per hour) from the capacity of subsection 2 (5400 vehicles per hour). This will just fill up the freeway so that the volume–capacity ratio of subsection 2 will just reach 1.00. There will be no freeway congestion, but at the end of time slice 2 there will be a queue of 100 vehicles at the first on-ramp (a storage rate of 400 vehicles per hour for 15 minutes).

In the third time slice results are identical to those obtained in the second time slice except since 100 vehicles were in the queue at the first on-ramp at the end of time slice 2, the queue at the end of time slice 3 would be 200. In the fourth time slice (because of the previous metering rates) no freeway congestion is encountered, and the metering rate at the first on-ramp can be increased. Three points need to be considered to determine the metering at the first on-ramp in time slice 4. From the point of view of eliminating freeway congestion, the maximum metering rate is 1600 vehicles per hour (the mainline freeway demand of 3800 vehicles per hour subtracted from the capacity of subsection 2 of 5400 vehicles per hour). Considering the ramp users, a metering rate of at least 1600 vehicles per hour is needed to discharge the ramp demand of 800 vehicles per hour in time slice 4 in addition to discharging the transferred queue of 200 vehicles. However, the maximum metering rate and the ramp capacity limits the metering rate to 900 vehicles per hour. Thus, in time slice 4 the bottleneck subsection will not be fully utilized, and the queue at the end of time slice 4 at the first on-ramp will be reduced to 175 vehicles.

In the fifth and final time slice (because of the previous metering rates) no freeway congestion is encountered, and the metering rate at the first on-ramp can be increased. Three points again need to be considered to determine the metering rate at the first on-ramp in time slice 5. From the point of view of eliminating freeway congestion, the maximum metering rate is 2200 vehicles per hour (the mainline freeway demand of 3200 vehicles per hour subtracted from the capacity of subsection 2 of 5400 vehicles per hour). Considering the ramp users, a metering of at least 900 vehicles per hour is needed to discharge the ramp demand of 200 vehicles per hour in time slice 5 in addition to discharging the transferred queue of 175 vehicles. Fortunately, the maximum metering rate and the ramp capacity will permit a metering rate of 900 vehicles per hour, so that with this metering rate the ramp queue will just dissipate at the end of time slice 5 and no further control is needed. Figure 8.9 summarizes the numerical

Time Slice		SS1		SS2	SS3
1	D	4800	4800	5200	6000
	V	4800 Ramp	4800	5200	6000
	C	5400 queue	5400	5400	7200
	v/c	0.89	0.89	0.96	0.83
	s	50	50	50	50
			0		
2	D	5000	5000	~~5800~~ 5400	~~6600~~ 6200
	V	5000	5000	5400	6200
	C	5400	5400	5400	7200
	v/c	0.93	0.93	1.00	0.86
	s	50	50	50	50
			100		
3	D	5000	5000	~~5800~~ 5400	~~6600~~ 6200
	V	5000	5000	5400	6200
	C	5400	5400	5400	7200
	v/c	0.93	0.93	1.00	0.86
	s	50	50	50	50
			200		
4	D	3800	3800	~~4600~~ 4700	~~5400~~ 5500
	V	3800	3800	4700	5500
	C	5400	5400	5400	7200
	v/c	0.70	0.70	0.87	0.76
	s	50	50	50	50
			175		
5	D	3200	3200	~~3400~~ 4100	~~4000~~ 4700
	V	3200	3200	4100	4700
	C	5400	5400	5400	7200
	v/c	0.59	0.59	0.76	0.65
	s	50	50	50	50
			0		

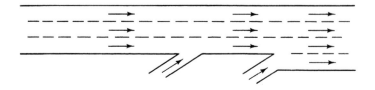

Figure 8.9   Numerical Solution of Supply–Demand Matrix with Ramp Control

solution of the supply–demand matrix under ramp control without user response. The structure of Figure 8.9 is similar to Figure 8.8 to compare the results between day −1, the base conditions, and day +1, under ramp control without user response. The number of vehicles queued at the end of each time slice on the first on-ramp is also indicated. Note that ramp queues begin at the beginning of time slice 2 and continues until the end of time slice 5. This is 15 minutes longer than the previous freeway congestion.

Table 8.2 is constructed like Table 8.1 in order to compare system performance between day −1 and day +1. The total travel distance and the total nondelay travel time remain unchanged since all vehicles ultimately use the freeway and use it at a speed of 50 miles per hour. However, the mainline delay of 75 vehicle-hours, is replaced by a ramp delay of 118.75 vehicle-hours, which increased the total travel time, reduced the average system speed, and increased the percent delay. The net result is an increase of 43.75 vehicle-hours in total travel time, which represents a 7 percent increase in total travel time. Based on these analyses, it would hardly be worth while adding ramp control in order to exchange a 75-vehicle-hour delay on the freeway for a 118.75-vehicle-delay on the ramp. But before making a decision, has equilibrium been reached? Will vehicles currently using the first on-ramp simply wait their turn and encounter the ramp delay? Returning to Figure 8.4, the freeway corridor includes a parallel arterial and current users of the first on-ramp might consider diverting to the parallel arterial and entering the freeway at the second on-ramp. This is a type of feedback process mentioned earlier in the chapter, one that will now be addressed.

TABLE 8.2 System Performance with Ramp Control

System Performance Measure	Result
Total travel time (vehicle-hours)	658.25
Total travel distance (vehicle-miles)	26,975
Average system speed (miles/hr)	41.0
Total nondelay travel time (vehicle-hours)	539.5
Mainline delay (vehicle-hours)	0.0
Ramp delay (vehicle hours)	118.75
Percent delay	18.0

8.2.3 Ramp Control with User Response

A freeway ramp control strategy was imposed on this traffic system at day +1 and the effect of this control strategy without user response was evaluated in Section 8.2.2. After a period of experiencing delays at the first on-ramp, ramp users may begin to experiment and search for alternative routes for possible diversion. In this simple example the leading alternative facing the ramp user is to divert to the parallel arterial and enter the freeway at the second on-ramp. The users' criteria is a selfish one, that is, what alternative is best for them. In this example, minimizing travel time will be used as the criterion and travel time expended, whether in motion or stopped, will be considered equal.

The key question is: How much of a delay at the ramp will the user accept before diverting to the parallel arterial? Again, because of the simplistic freeway corridor characteristics and assumptions, this delay value can easily be determined. The potential diversion would be for 1 mile and the speeds on the freeway and arterial route would be 50 and 25 miles per hour. Therefore, as soon as the ramp delay exceeds 1.2 minutes, the ramp users would divert to the parallel arterial.

Another important step in this analysis is to relate ramp queue length with ramp delay and metering rate. For example, using the ramp delay value of 1.2 minutes and selecting a metering rate of say, 400 vehicles per hour, the resulting ramp queue length would be 8 vehicles. Thus there would be equilibrium between the two alternatives when the ramp delay was 1.2 minutes (and a queue length of 8 vehicles if a metering rate of 400 vehicles per hour is encountered).

Returning now to the example problem, consider the first time slice. Since the freeway is not congested, ramp control is not required, and all users of the first on-ramp would use the freeway. The analyses of the second time slice is a little more involved. Ramp control is required and a metering rate of 400 vehicles per hour (100 vehicles per time slice) is determined. The original users of the first on-ramp would continue to use this ramp until a queue of 8 vehicles is formed on the ramp, and after that the arrival rate at the ramp would decrease to 400 vehicles per hour so as to maintain a queue length of 8 vehicles throughout the second time slice. Thus, at the end of time slice 2 the queue length would be 8 vehicles, 100 vehicles would have entered the freeway, and 92 vehicles would have diverted to the parallel arterial.

In the third time slice identical results are obtained as in the second time slice except that since 8 vehicles were in the ramp queue, at the beginning of the third time slice, none can be added to the queue, and the number served on the freeway is 100 vehicles. This means that 100 vehicles would be diverted to the parallel arterial.

In the fourth slice no ramp control is needed, and 208 vehicles, would enter the freeway at the first on-ramp. There would be a slight delay during the first few seconds discharging the 8 queued vehicles, but it would be negligible. No further congestion, ramp control, ramp queueing, nor diversion would be encountered.

Figure 8.10 summarizes the numerical solution of the supply–demand matrix under ramp control and with user response. The structure of Figure 8.10 is similar to Figures 8.8 and 8.9 in order to compare results between the base conditions, ramp

Time Slice		SS1		SS2	SS3
	D	4800	4800	5200	6000
	V	4800 Ramp	4800	5200	6000
1	C	5400 queue	5400	5400	7200
	v/c	0.89	0.89	0.96	0.83
	s	50	50	50	50
			`0 │ 0`		
	D	5000	5000	~~5800~~ 5400	~~6600~~ ~~6200~~
	V	5000	5000	5400	6568 6568
2	C	5400	5400	5400	7200
	v/c	0.93	0.93	1.00	0.91
	s	50	50	50	50
			`8 │ 92`		
	D	5000	5000	~~5800~~ 5400	~~6600~~ ~~6200~~
	V	5000	5000	5400	6600 6600
3	C	5400	5400	5400	7200
	v/c	0.93	0.93	1.00	0.92
	s	50	50	50	50
			`8 │ 100`		
	D	3800	3800	~~4600~~ 4632	~~5400~~ 5432
	V	3800	3800	4632	5432
4	C	5400	5400	5400	7200
	v/c	0.70	0.70	0.86	0.75
	s	50	50	50	50
			`0 │ 0`		
	D	3200	3200	3400	4000
	V	3200	3200	3400	4000
5	C	5400 Diverted	5400	5400	7200
	v/c	0.59 vehicles	0.59	0.63	0.56
	s	50	50	50	50

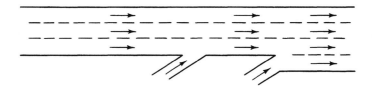

Figure 8.10   Numerical Solution of Supply–Demand Matrix with Ramp Control and User Response

control without user response, and ramp control with user response. The number of vehicles queued at the end of each time slice and the number of vehicles diverted to the parallel arterial in each time slice are also indicated. Note that ramp queues are encountered during only slightly more than two time slices, and ramp users divert during only two time slices. Also note that the diverted vehicles reenter the freeway at the second on-ramp in the second and third time slices. Thus the demands in subsection 3 are first reduced because of the ramp control at the first on-ramp and then are increased because of the diverted traffic.

Table 8.3 is constructed like Tables 8.1 and 8.2 in order to compare system performance between the base conditions, ramp control without user response, and ramp

TABLE 8.3   System Performance with Ramp Control and User
Response

System Performance Measure	Freeway/ Ramps	Arterial	Freeway Corridor
Total travel time (vehicle-hours)	539.7	7.7	547.4
Total travel distance (vehicle-miles)	26,783	192	26,975
Average system speed (miles/hr)	49.6	25	49.3
Total nondelay travel time (vehicle-hours)	535.7	—	539.5
Mainline delay (vehicle-hours)	0.0	4.0	—
Ramp delay (vehicle-hours)	4.0	—	4.0
Percent delay	0.7	—	2.1

control with user response. One change in Table 8.3 is subdividing the calculated results between freeway/ramp, arterials, and freeway corridor. In regard to the arterial, 192 vehicles are diverted to it and enter the freeway 1 mile downstream. The vehicles travel at 25 miles per hour on the arterial, and 7.7 vehicle-hours are expended. The total travel distance in the freeway corridor has remained unchanged, so the portion of total travel distance on the freeway can be calculated. The ramp delay is caused by an average queue length of 8 vehicles over a 30-minute time period. The total travel time in the freeway corridor consists of 4 vehicle-hours of delay on the ramp, 7.7 vehicles-hours of additional travel time on the parallel arterial, and 535.7 vehicle-hours of travel time on the freeway.

The total travel time for the freeway corridor for the base conditions, ramp control without user response, and ramp control with user response was 614.5, 658.25, and 547.4 vehicle-hours, respectively. This represents a reduction of 67.1 vehicle-hours per afternoon peak period compared with the base conditions or a 10.9% reduction in total travel time.

8.3 SUMMARY

This chapter has served as a transition between the first part of the book which deals with traffic flow characteristics, and the second part, which addresses traffic system analysis. The first part dealt with individual microscopic and macroscopic traffic flow characteristics and the ability to analyze each separately in considerable detail. Now a broader view is developed which attempts to give structure to traffic systems in order to analyze their performance and the consequences of system changes and users responses. The remaining chapters are devoted to traffic systems analysis techniques, which are an integral part of the analytical framework for demand–supply analysis.

This chapter has two major sections. First, an analytical framework was presented which provided structure in order to predict system performance through an analytical process based on specified demand, supply, control, and environment input elements. These input elements and their interactions were identified. The analytical framework provided for the initial process as well as a series of possible feedback processes which were described.

The second major section of this chapter was devoted to an example problem to demonstrate the application of the previously described analytical framework for demand–supply analysis to a directional freeway corridor. After specifying the input elements, the freeway corridor was analyzed at three points in time: day −1, representing base conditions; day +1, representing ramp control implementation without user response; and some time later, after users had responded to the ramp control implemented.

The proposed analytical framework is simple and effective. Its application to real-life traffic systems can be very complex and requires knowledge of system analysis techniques and the ability to select the right technique for the problem at hand. While containing many simplifying assumptions, the example problem provides a practical and numerical guide to the initial and feedback processes. The remaining chapters are devoted to traffic system analysis techniques.

8.4 SELECTED PROBLEMS

1. Apply the analytical framework depicted in Figure 8.1 to a macroscopic analysis of a single-lane approach at a fixed-time traffic signal. Identify and describe the possible contents in each element of the analytical framework.

2. Select a simple traffic system and apply the analytical framework depicted in Figure 8.1. Identify and describe the possible contents in each element of the analytical framework.

3. Conceptualize a wide variety of traffic systems. Do any of these traffic systems *not* fit the analytical framework proposed in Figure 8.1? List the traffic systems identified and describe any traffic systems that do not seem to fit.

4. Assume that the demand and supply curves shown in Figure 8.2 are the base condition. Draw three sets of demand–supply curves that would result if (a) the traffic route is improved; (b) the demand for the area increases and (c) the traffic route is improved and the demand for the area increases. Identify equilibrium states for each.

5. Apply the analytical framework depicted in Figure 8.1 to a macroscopic analysis of a single-lane approach at a fixed-time traffic signal. Identify and describe the various types of feedback process that might be encountered.

6. Select a simple traffic system and apply the analytical framework depicted Figure 8.1. Identify and describe the various types of feedback processes that might be encountered.

7. Conceptualize a wide variety of traffic systems. Consider the feedback processes covered in the text. Can you identify any additional types of feedback processes? Are certain feedback processes more prevalent than others?

8. A speed versus volume–capacity ratio relationship is shown in Figure 8.5. The upper portion of the curve is shown as a horizontal line. Is this a typical curve? Does it simplify the example problem?

9. Can a demand–capacity ratio ever exceed 1.0? Can a volume–capacity ratio ever exceed 1.0? Explain your answers.

10. Referring to Figure 8.8, in time slice 2 the demand in subsection 3 is modified. In time slice 3 the demand in the downstream portion of subsection 1 is modified. Explain each.

11. Confirm by calculation the various system performance measures shown in Table 8.1.

12. Freeway congestion is identified under the base conditions. The alternative solution investigated in the example problem was ramp control. Could the congestion be alleviated through design improvements? Generate a minimum design improvement and calculate the total travel time. Describe possible feedback processes.

13. Attempt to modify the linear programming formulation as stated in equations (8.1) through (8.4) to extend the objectives and constraints as mentioned in the text.

14. Confirm by calculations the ramp queue lengths shown in Figure 8.9.

15. Confirm by calculations the ramp delay shown in Table 8.2.

16. A spatial user response is investigated in the example problem after ramp control is implemented. Describe other user responses that might occur and their likelihood of occurring.

17. Minimizing travel time is the criterion used in the example problem for considering diversion. Describe more realistic and comprehensive criteria, and their complexity for use.

18. Confirm by calculations the number of diverted vehicles shown in Figure 8.10.

19. Confirm by calculations the ramp delay and the arterial total travel time shown in Table 8.3.

8.5 SELECTED REFERENCES

1. Adolf D. May, *Demand–Supply Modeling for Transportation System Management*, Transportation Research Board, Record 835, TRB, Washington D. C., January 1980, pages 80–86.

2. Frederick A. Wagner and Keith Gilbert, *Transportation System Management: An Assessment of Impacts*, Office of Policy and Programs Development, Urban Mass Transportation Administration, Washington D. C., 1978, 195 pages.

3. Jan L. Botha, Richard S. Bryant, and Adolf D. May, A Decision-Making Framework for the Evaluation of Climbing Lanes on Two-Lane, Two-Way Rural Roads, *Eighth International Symposium of Transportation and Traffic Theory, Proceedings*, June 1981, pages 91–120.

4. Joyce Holroyd and D. I. Robertson, *Strategies for Area Traffic Control Systems: Present and Future*, Transport and Road Research Laboatory, Crowthorne, Berkshire, England, 1973.

5. K. Moskowitz and L. Newman, *Notes on Freeway Capacity*, Highway Research Board, Record 27, HRB, Washington D. C., 1963, pages 44–68.

6. Tsutomu Imada and Adolf D. May, *FREQ8PE–A Freeway Corridor Simulation and Ramp Metering Optimization Model*, UCB-ITS-RR-85-10, University of California, Berkeley, Calif., June 1985, 263 pages.

7. Khosrow Ovaici, Roger F. Teal, James K. Ray, and Adolf D. May, Developing Freeway Priority Entry Control Strategies, *Sixth International Symposium on Transportation and Traffic Theory Proceedings*, August 1974, pages 125–160.

9

Capacity Analysis

The capacity of facilities has always interested builders and operators of engineering systems. Common questions are: *How much will it hold? How much can it bear?* or *How much can it carry?* Literature on the capacity of highway facilities dates back to the 1920s with studies dealing with such topics as the effect of street cars on automobile capacity and the discharge rate at signalized intersections. In recent years the field of capacity analysis has been extended to include levels of service. Hence, today capacity analysis not only includes the determination of capacity but also the trade-off between the quantity of traffic and the resulting level of service to the users.

This chapter is divided into four major parts. The first two sections provide a historical background of the subject of capacity analysis and describe the scope and limitations of the chapter contents. In the second part, attention is given to the capacity analysis of uninterrupted flow situations and includes multilane facilities, ramps and ramp junctions, and weaving sections. The third part of the chapter deals with interrupted flow situations and focuses particularly on signalized intersections. The chapter is concluded with a summary, selected problems, and selected references.

9.1 HISTORICAL BACKGROUND

As mentioned earlier, capacity analysis dates back at least to the early 1920s with capacity studies and the formation of a capacity analysis committee within the newly formed Highway Research Board. However, the primary attention in those years was given to building the highway system and surfacing the primary highways. Perhaps the mid-1930s can be characterized as the period of reassessment, with federal requirements for taking inventory of the physical and user service of the entire highway system. World

War II temporarily postponed capacity analysis research, but soon thereafter this became a major area of activity. This activity culminated with the publication of the first U.S. *Highway Capacity Manual* [1]. The HRB Committee on Highway Capacity authored the manual and it was published and widely distributed by the Bureau of Public Roads. It became the highway capacity document for the United States and for many countries of the world.

Again attention turned toward expanding the highway system with particular attention given to freeways and the Interstate system. But, traffic continued to grow at even a faster rate than new facilities, and engineers and planners were called on more and more to perform capacity analysis. Realizing that the 1950 HCM was becoming outdated and limited in scope, the HRB Committee on Highway Capacity took on the task of preparing a second *Highway Capacity Manual*. This manual (called the 1965 HCM) was published in 1966 [2] and dedicated to Mr. O. K. Norman, who served as chairman from 1944 until his death in 1964, and he was designated by his committee members as "Mr. Capacity." Probably the most significant advancement made in the second edition was the introduction of the concept of level of service.

While the 1965 HCM was widely distributed and considered as the document for capacity analysis in the United States, researchers in other countries (e.g., England and Australia) were developing capacity analysis procedures. The need for capacity analysis techniques was recognized worldwide and researchers were drawn to the subject in many countries.

By the mid-1970s the need for a third edition of the capacity manual was recognized and serious work began [3]. Two important changes had taken place. First, the work of the Research Board was broadened and the name was changed from Highway Research Board to the Transportation Research Board. Second, the name of the Committee was changed from Highway Capacity to Highway Capacity and Quality of Service. These name changes have affected the scope of the third edition and have given added emphasis to level of service. Also, emphasis was changing from the design of new facilities to the operations of existing facilities.

The next major event was the publication of *Interim Materials on Highway Capacity* in 1980 [4]. This publication attempted to summarize the then-current knowledge in highway capacity and to identify needs for immediate research in order to complete the planned third edition of the *Highway Capacity Manual*. As the title of the document indicated, this publication was intended to serve as a interim document.

The third edition of the *Highway Capacity Manual* was published in 1985 [5]. As indicated in the introduction to the 1985 HCM, "this third edition reflects over two decades of comprehensive research conducted by a variety of research agencies with the sponsorship of a number of agencies, primarily the National Cooperative Highway Research Program and the Federal Highway Administration. Its development has been guided by the Transportation Research Board's Committee on Highway Capacity and Quality of Service." The scope and comprehensiveness of the 1985 HCM can be readily seen by reviewing its table of contents which is shown in Table 9.1.

The foreword of the 1985 HCM concluded with the following statement, which gives some insights as to the future: "The Committee views this publication as a

TABLE 9.1   1985 HCM Table of Contents

milestone in the growing body of knowledge of highway capacity—not the conclusion. Research will continue." Research is continuing and opportunities have been provided for users and researchers to work together. The loose-leaf form of the 1985 HCM gives evidence that new developments are on the horizons and that the frontier of knowledge in highway capacity will continue to expand.

9.2 SCOPE AND LIMITATIONS

The table of contents of the 1985 HCM presented in Table 9.1 provides a scope of the area known as highway capacity analysis. Highways, transit, pedestrians, and bicycles are included, but capacity analysis of other modes, such as air and water transportation, is not. Like the 1985 HCM, this chapter will be limited to land transportation, particularly to highway transportation. Readers with an interest in capacity analysis of other modes of transportation are encouraged to review the literature. Some are identified with the selected references [6–9].

With an attempt to balance between scope and depth, this chapter will be limited to specific areas of capacity analysis that provide an opportunity to cover the fundamental concepts, and at the same time, cover areas that the reader will most likely encounter in practice. Therefore, unfortunately, areas such as two-lane roads, unsignalized intersections, urban and suburban arterials, transit, pedestrians, and bicycles will not be covered. Readers with an interest in these topics are encouraged to review selected chapters of the 1985 HCM and the references listed at the end of each chapter of the Manual [5].

Emphasis in this chapter is given to four areas of capacity analysis. These areas include multilane facilities, ramps and ramp junctions, weaving sections, and signalized intersections. Even with this limitation, space will not permit exhaustive treatment of each area and again the reader is encouraged to refer to the 1985 HCM [5], references listed at the end of each chapter in the 1985 HCM, and to selected references at the end of this chapter for greater depth of coverage.

9.3 MULTILANE FACILITIES

Multilane facilities have two or more lanes available for use for each direction of travel. A multilane facility may vary from undivided rural highways without access control to freeways, which are divided access-controlled facilities. The key is that multilane facilities provide for "uninterrupted" flow conditions away from the influence of ramps or intersections. Multilane facilities are sometimes referred to as basic highway or freeway segments. The 1985 HCM covers these types of facilities in Chapters 3 and 7, but in this text a unified approach is proposed. Two excellent references served as a basis for Chapter 3 in the 1985 HCM [10, 11].

In this proposed approach, capacity analysis of multilane facilities is presented in two parts. First, capacity analysis of multilane facilities under *ideal conditions* is covered, followed by capacity analysis of multilane facilities under *nonideal conditions*. Before proceeding further the term "ideal conditions" must be defined. Ideal conditions for multilane facilities require that *all* of the following conditions exist:

- Essentially level and straight roadway

- Divided highway with opposing flows not influencing each other

- Full access control

- Design speed of 50 miles per hour or higher

- Twelve-foot minimum lane widths

- Six-foot minimum lateral clearance between the edge of the travel lanes and the nearest obstacle or object

- Only passenger cars in the traffic stream

- Drivers are regular users of such facilities

9.3.1 Capacity Analysis Under Ideal Conditions

The speed–flow relationships for multilane facilities under ideal conditions are shown in Figure 9.1. This is the only diagram needed to perform capacity and level of service analysis for this ideal condition situation.

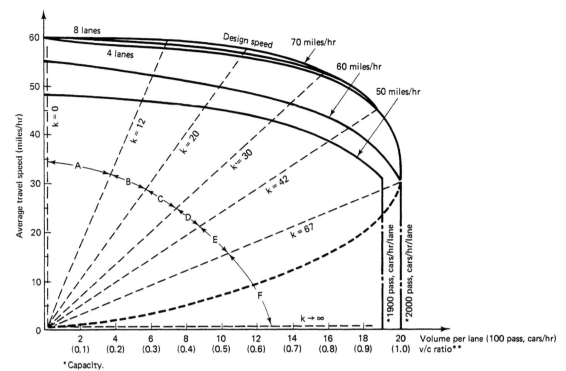

Figure 9.1   Speed–Flow Relationship for Multilane Facilities under Ideal Conditions

There are three scales on this diagram: speed, flow, and density. The average travel speed in miles per hour is shown on the vertical scale. It is an indication of the level of service provided to the users. Traffic flow per lane is shown on the horizontal scale. These lane flows are expressed as the hourly rate during the 15-minute period that is being investigated. It is an indication of the quantity of traffic using the facility. Note that the volume–capacity ratio is also given on the horizontal scale, but it is based on a capacity flow of 2000 vehicles per hour per lane.

The third and last traffic characteristic shown in Figure 9.1 is density. The lane densities are shown as diagonal straight lines radiating from the origin and extending up and to the right. Lane densities are expressed as the numbers of vehicles per mile of roadway on a per lane basis. It is an indication of the level of service provided to users since it is a surrogate for freedom of movement. Since flow is the product of speed and density, density values can be determined for specified values of flow and speed. Seven density values were calculated and are shown on the diagram. The density values selected were 0, 12, 20, 30 42, 67, and infinity because these density values demark levels of service as proposed by the 1985 HCM. The proposed relationships between density ranges and levels of service are given in Table 9.2. The highest level of service, LOS A, occurs with lane densities between 0 and 12 vehicles per mile per lane. Levels

TABLE 9.2 Relationship between
Density Levels
and Proposed Levels
of Service

Lane Density Range (veh/mile/lane)	Level of Service
0–12	A
12–20	B
20–30	C
30–42	D
42--67	E
> 67	F

Source: Reference 5.

of service B, C, D, E, and F are encountered as lane densities increase, as indicated in Table 9.2. Note that the upper boundary of LOS E (67 vehicles per mile per lane) intercepts the 70-mile per hour design speed curve at its highest lane flow value of 2000 vehicles per hour per lane. This maximum flow value is the capacity value; hence the upper boundary of LOS E denotes capacity conditions. Capacity is defined as the maximum sustained (15-minute) rate of flow at which traffic can pass a point or uniform segment of a multilane facility under prevailing roadway and traffic conditions. Capacity is defined for a single direction of flow and is expressed in vehicles per hour. Later, the lane flows and lane capacities will be converted to directional flows and directional capacities by considering the number of directional lanes. Finally, when lane densities of greater than 67 vehicles per mile per lane are encountered, the level of service is F. This occurs when a downstream bottleneck exists and queues extend upstream into the study section. The lane densities exceed 67 vehicles per mile per lane and are determined by the flow within the downstream subsection and the characteristics of the subsection being studied.

The basic equation needed for capacity and level of service analysis under ideal conditions combined with Figure 9.1 is

$$SF_i = \left(\frac{v}{c_j}\right)_i \left(c_j N\right) \qquad (9.1)$$

where i = level of service

SF_i = maximum service flow rate for level of service i (vehicles per hour)

j = design speed (miles per hour)

c_j = lane capacity under ideal conditions for design speed j (vehicles per hour per lane)

N = number of directional lanes

$(v/c_j)_i$ = maximum volume to capacity ratio associated with LOS i

The equation can be used in one of three ways. By solving for SF_i, the maximum service flow can be determined for a given designed multilane facility under a specified level of service requirement. By solving for $(v/c_j)_i$, the level of service can be determined for a given designed multilane facility carrying a specified service flow rate. Finally, by solving for $(c_j N)$, the design of a multilane facility can be determined when the level of service and service flow rate are specified. These three types of analyses might be referred to as quantitative analysis, qualitative analysis, and design analysis.

This completes the capacity and level of service analysis procedures for multilane facilities under ideal conditions. Attention will now turn to multilane facilities that are less than ideal. This conversion from ideal conditions to something less than ideal conditions, although numerically complex, is rather straightforward from a conceptual viewpoint and builds on the concepts just discussed under ideal conditions.

9.3.2 Capacity Analysis under Nonideal Conditions

The beginning point for capacity and level-of-service analysis for multilane facilities under less than ideal conditions is to return to equation (9.1). The $c_j N$ term should be reduced by some factor or a series of factors. Each factor would represent one nonideal condition listed at the beginning of the multilane facilities section and would take on a value of 1.0 if that particular condition was ideal, and values of less than 1.0 if nonideal conditions were encountered. The resulting series of factors would be multiplied together to obtain the composite reduction factor. Then this composite reduction factor is multiplied by the capacity for ideal conditions $c_j N$ and used in equation (9.1). It should be noted that this treatment of combining factors assumes that the factors are independent and their combined independent effects are multiplicable. These are strong assumptions and worthy of further research investigations.

Following the developed logic of a series of factors multiplied together to reduce capacity from ideal conditions to something less than ideal conditions, equation (9.1) can be rewritten as follows:

$$ SF_i = \left(\frac{v}{c_j} \right)_i \left(c_j N \right) \left[f_1 \times f_2 \times f_3 \times \cdots \times f_n \right] \tag{9.2} $$

where f_i, f_2, f_3, ... f_n are reduction factors for nonideal conditions. The 1985 HCM proposes that four reduction factors be considered: width factor (f_W), heavy vehicle factor (F_{HV}), driver population factor (f_P), and environment factor (f_E). These four factors are discussed in the following four paragraphs.

The width factor (f_W) is used to consider the reduction in capacity due to less than ideal lane widths and side clearances. If lane widths equal to or greater than 12 feet combined with side clearances equal to or greater than 6 feet are encountered, the condition is classified as ideal and the width factor is 1.0 $(f_W = 1.0)$. If narrower lanes and/or less side clearances are encountered, the condition is classified as nonideal and the width factor is less than 1.0 ($f_W < 1.0$). Tables are available in Chapters 3 and 7 of the 1985 HCM to determine the numerical value of the width factor. For most circumstances, the width factor lies between 0.9 and 1.0. For example, on a four-lane divided multilane facility, the width factor is 0.89 when lane widths are 10 feet and side clearances are 4 feet.

The heavy vehicle factor (f_{HV}) is used to consider the reduction in capacity due to the presence of heavy vehicles under different vertical alignment situations. If no heavy vehicles are present, the heavy vehicle factor is 1.0 ($f_{HV} = 1.0$). If heavy vehicles are present, the condition is classified as nonideal and the heavy vehicle factor is less than 1.0 ($f_{HV} < 1.0$). Tables are available in Chapters 3 and 7 of the 1985 HCM to determine the numerical value of the heavy vehicle factor. The process of determining the heavy vehicle factor is complicated since there are various classes of heavy vehicles (i.e., recreational vehicles, light trucks, medium trucks, heavy trucks, and buses). Also analyses can be undertaken for general terrain situations and for specific grade situations, and a wide range of values for the heavy vehicle factor can be encountered. As an example, a multilane facility in rolling terrain with 10 percent of the vehicles classified as trucks would have a heavy vehicle factor of 0.77.

The driver population factor (f_P) is used to consider the reduction in capacity due to the presence of nonregular users. If all drivers are regular users of the facility, the driver population factor is 1.0 ($f_P = 1.0$). When nonregular users are encountered, considerable judgment is required on the part of the analyst. Driver population factors as low as 0.75 are suggested, but on most multilane facilities a high proportion of the drivers are regular users and the driver population factor is closer to 1.0.

The last factor to be considered is the environment factor (f_E). The environment factor is used to consider the reduction in capacity due to the lack of a median and/or the lack of access control. On a multilane facility such as a freeway that has a median and access control, the enviroment factor is 1.0 ($f_E = 1.0$). On nonfreeway multilane facilities the environment factor is 0.95 on a undivided rural roadway, 0.90 on a divided suburban roadway, and 0.80 on a undivided suburban roadway.

Equation (9.2) can now be put in its most specific form as follows:

$$\text{SF}_i = \left(\frac{v}{c_j}\right)_i (c_j N) \left[f_W \times f_{HV} \times f_P \times f_E\right] \qquad (9.3)$$

If ideal conditions are encountered, all the f-factors equal 1 and the last term can be eliminated. If less than ideal conditions are encountered, some or all of the f-factors take on values of less than 1, as described in previous paragraphs. As with the three analytical procedures for ideal flow situations, equation (9.3) can be employed to perform quantitative analysis, qualitative analysis, or design analysis.

As mentioned at the beginning of this section, the key to the capacity and level of service analysis of multilane facilities is the assumption of "uninterrupted" flow conditions away from the influence of ramps or intersections. In the next sections we address the situation when ramps and intersections begin to affect these capacity and level-of-service analytical techniques.

9.4 RAMPS AND RAMP JUNCTIONS

Ramps are sections of roadway that provide connections from one highway facility to another. Entering and exiting traffic causes disturbances to traffic on multilane facilities and affect the capacity and level of service of basic highway and freeway segments. The existence of ramps causes complex flow situations at several distinct locations which must be carefully studied. Materials from both the 1965 HCM [2] and the 1985 HCM [5], as well as an original piece of research [12], serve as a basis for this presentation.

A typical section of a directional multilane facility with an on-ramp followed by an off-ramp is shown in Figure 9.2 and provides a basic configuration for understanding and analyzing the effects of ramps on capacity and level of service. Three locations on each ramp must be carefully studied and are depicted as A, B, and C for the on-ramp and D, E, and F for the off-ramp in Figure 9.2.

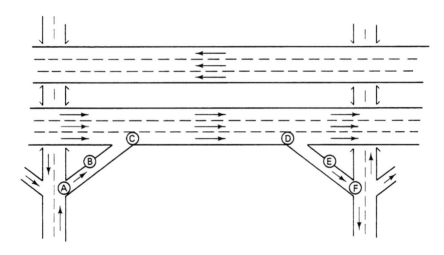

Figure 9.2   Typical Freeway–Ramp Configuration

Location A is the entrance to the on-ramp and is affected by the ramp itself and/or by the at-grade intersection. Since the dimensions and geometrics of on-ramps at location A are normally better or at least as good as that of location B, the effect of the physical on-ramp will be studied further at location B. Normally, the at-grade intersection controls the entrance to the on-ramp, and this potential restriction will be studied in a later section when the capacity and level-of-service analysis of intersections are covered.

Location B is on the on-ramp itself and its capacity is affected by the number and width of lanes as well as by the length and grade of the ramp. Most on-ramps will have capacities on the order of 1200 to 1600 vehicles per hour per lane. As long as the demand is less than the ramp capacity, level of service is not of serious concern. The reason for this is the relative short length of most ramps and the fact that queues will be negligible and will not affect the at-grade intersection.

Locations E and F are mirror images of locations A and B in an analytical sense. Location E is the off-ramp itself, and its capacity is also on the order of 1200 to 1600 vehicles per hour per lane. Its level of service is not of major concern unless the demand exceeds its capacity. Location F is at the exit of the off-ramp where it connects to a crossing arterial at an at-grade intersection. An important difference between locations A and F is the location of queues if they exit. At location A, any queues will extend into the at-grade intersection, whereas at location F, any queues will extend up the off-ramp and if serious enough, into the multilane facility.

Locations C and D are the merge and diverge areas and require special analytical procedures. The concept and basic principles are straightforward, although the numerical solution can be rather tedious. The substance of the analytical procedure is to compare the actual demands in the merge and diverge areas with the allowable service flow rates. This comparison is then used to determine the resulting level of service.

The allowable service flow rates for merging and diverging areas for ideal conditions for the various levels of services are shown in Table 9.3. The upper limit of level of service E corresponds to the capacity of the rightmost lane under ideal conditions, which is shown to be 2000 passenger-cars per hour (15-minute period). As noted in Table 9.3, the levels of service of merge and diverge areas diminish as the traffic demands in the rightmost lane increase. These allowable service flow rates should be reduced when nonideal conditions are encountered. The reduction factors discussed under multilane facilities should be employed. If the capacity and service flow rates of the basic multilane segment between the merge and diverge area have been computed, the multilane service flow rates divided by the number of lanes in the basic segment can be used as the allowable lane service flow rates in the merge and diverge analysis.

TABLE 9.3 Allowable Service Flow Rates
for Merging and Diverging Areas
(Passenger Cars per Hour)

Level of Service	Merge Flow Rate	Diverge Flow Rate
A	≤ 600	≤ 650
B	≤ 1000	≤ 1050
C	≤ 1450	≤ 1500
D	≤ 1750	≤ 1800
E	≤ 2000	≤ 2000
F	—	—

Source: Reference 5.

The major difficulty is in estimating the traffic demands in the rightmost lane. The key to the solution is to consider that traffic in the rightmost lane is made up of subgroups of traffic each having a unique origin and destination along the multilane

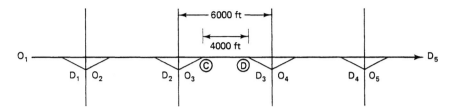

Figure 9.3   Extended Typical Freeway–Ramp Configuration

facility. Figure 9.3, which is an extension of Figure 9.2, can be used to illustrate this approach. Note that the merge area (location C) and the diverge area (location D) in Figure 9.3 are identical to those shown in Figure 9.2. Origins (O's) along the freeway are denoted as O_1, O_2, etc., and destinations (D's) along the freeway are denoted as D_1, D_2, etc.

All possible O–D freeway movements are identified in Figure 9.4. Certain O–D movements do not pass through the merge and diverge areas and can be ignored. Those

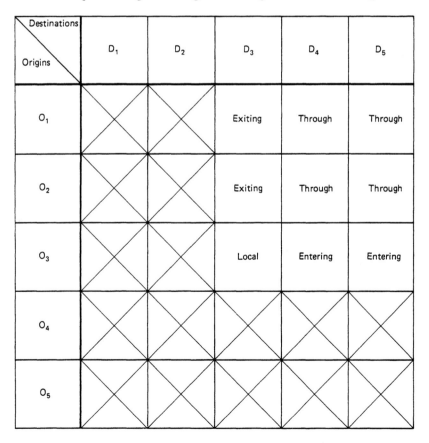

Figure 9.4   Possible Freeway O–D Movements

$O–D$ movements have been crossed out in Figure 9.4. The remaining nine $O–D$ movements can be combined into four groups: through, entering, exiting, and local. Each
will now be addressed in order to determine its share of the traffic demand in the rightmost lane in the vicinity of the merge and diverge areas in question. For demonstration
purposes the distance between the on-ramp nose and the off-ramp gore is assumed to be
4000 feet and its share of traffic in the rightmost lane will be calculated at 1000-foot
intervals. All results are summarized in Figure 9.5.

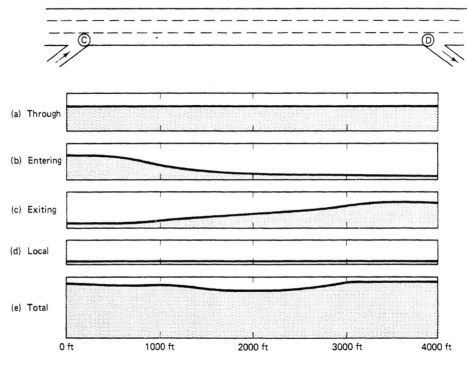

May 9.5

Figure 9.5   Estimated Traffic in Rightmost Lane

Through traffic is that traffic that enters the freeway at least 4000 feet upstream of
the merge area and exits the freeway at least 4000 feet downstream of the diverge area.
Assuming interchange spacing on the order of 1 mile in this example, the following
$O–D$ movements in Figure 9.4 are combined and classified as through traffic:
O_1D_4, O_1D_5, O_2D_4, and O_2D_5. The question now is to determine how much of this
traffic is in the rightmost lane. This is accomplished by using Table 9.4, in which the
percent of through traffic in the rightmost lane is given as a function of the size of the
freeway and total directional through traffic. This percentage is assumed constant
between the on-ramp and off-ramp and is depicted in the first sub-graph in Figure 9.5a.

As noted in Figure 9.4, entering traffic is that traffic that enters the freeway in the
merge area (location C) and has a destination beyond the diverge area (location D). The
following $O–D$ movements are combined and classified as entering traffic: O_3D_4 and

TABLE 9.4   Approximate Percentage of Through Traffic
in Rightmost Lane

Total Directional Through Traffic (veh/hr)	Through Traffic in Rightmost Lane (%)		
	8-Lane Freeway	6-Lane Freeway	4-Lane Freeway
≥6500	10	—	—
6000–6499	10	—	—
5500–5999	10	· —	—
5000–5499	9	—	—
4500–4999	9	18	—
4000–4499	8	14	—
3500–3999	8	10	—
3000–3499	8	6	40
2500–2999	8	6	35
2000–2499	8	6	30
1500–1999	8	6	25
≤1499	8	6	20

O_3D_5. All of this entering traffic is in the rightmost lane in the merge area and as the entering traffic moves farther and farther downstream, a smaller and smaller proportion remain in the rightmost lane. The percentage of on-ramp traffic present in the rightmost lane at various distances downstream of the merge area is shown in Figure 9.6a. The amount of entering traffic in the rightmost lane is shown graphically in Figure 9.5b.

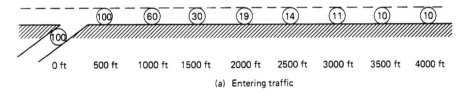

(a) Entering traffic

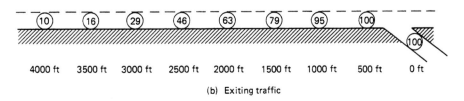

(b) Exiting traffic

Figure 9.6   Percentage of Entering and Exiting Traffic in Rightmost Lane

As noted in Figure 9.4, exiting traffic is that traffic that exits the freeway in the diverge area (location D) and has an origin upstream of the merge area (location C). The following O–D movements are combined and classified as exiting traffic: O_1D_3 and O_2D_3. All of this exiting traffic is in the rightmost lane in the diverge area, but as the exiting traffic is farther and farther upstream of the off-ramp, a smaller and smaller proportion of the exiting traffic is in the rightmost lane. The percentage of the exiting traffic present in the rightmost lane at various distances upstream of the off-ramp is shown in Figure 9.6b. The amount of exiting traffic in the rightmost lane is shown graphically in Figure 9.5c.

Finally, local traffic is shown in Figure 9.4 as traffic that enters in the merge area (location C) and exits in the diverge area (location D), and is identified as O–D movement O_3D_3. It is assumed that all local traffic remains in the rightmost lane. The amount of local traffic in the rightmost lane is shown graphically in Figure 9.5d.

Therefore, the total traffic in the rightmost lane at various points between the on-ramp and the off-ramp is the sum of the through, entering, exiting, and local traffic and is shown graphically in Figure 9.5e. This demand is compared with the allowable service flow rates given in Table 9.3. The highest demand in the vicinity of the merge area is compared with the allowable merge service flow rates, and the highest demand in the vicinity of the diverge area is compared with the allowable diverge service flow rates. The resulting level of service can then be determined.

Although the principles set forth earlier for capacity and level-of-service analysis of merging and diverging areas are straightforward, their applications can be complicated and tedious. The complications can be caused by unusual ramp geometrics and are particularly difficult at near capacity or oversaturated situations. The 1985 HCM contains many nomographs that can be used to estimate the level of service provided in merge and diverge areas under a wide variety of geometric configurations. At near capacity or oversaturated situations, professional judgment is required in applying these principles and an iterative type of analytical procedure will be required.

9.5 WEAVING SECTIONS

Traffic entering and leaving multilane facilities can also interrupt the normal flow of basic highway and freeway segments by creating weaving sections. Weaving is defined as the crossing of two or more traffic streams traveling in the same direction along a significant length of highway without the aid of traffic control devices. Weaving vehicles that are required to change lanes cause turbulence in the traffic flow and by so doing reduce the capacity and level of service of weaving sections. Thus, analytical techniques are needed to evaluate this reduction in capacity and level of service.

A variety of weaving analysis techniques are available and are being used such as the 1965 HCM method [2], PINY method [13], Leisch method [14], JHK method [15], 1985 HCM method [5], and Fazio method [16]. Even so, the TRB Committee on Highway Capacity and Quality of Service has recognized that further research on the capacity and level of service of weaving sections is one of the highest research priorities.

Several weaving capacity research projects have already been initiated. It is anticipated that this chapter of the 1985 HCM will probably be one of the first chapters revised. Because of the variety of methods available and the anticipated 1985 HCM chapter revision, only highlights of the 1965 HCM and 1985 HCM methods will be presented. The emphasis of this section will be given to the characteristics of weaving sections and some fundamental concepts of approaches to analyzing them. Because of the large number of symbols and definitions, Table 9.5 has been prepared to aid the reader. A typical simple weaving section is shown in Figure 9.7. Figure 9.7a depicts a three-lane directional facility with an entrance ramp preceding an exit ramp. Figure 9.7b identifies four types of movements in this simple weaving section. Normally, the heaviest flow is from A to C and is referred to as outer flow 1 (v_{o1}), while the lightest flow is usually from B to D and is referred to as outer flow 2 (v_{o2}). Neither of these movements is a weaving movement and their sum ($v_{o1} + v_{o2}$) is referred to as the total nonweaving flow (v_{nw}). The flow from B to C and A to D cross each other's path over a longitudinal distance and are the two weaving flows. The higher weaving flow is referred to as v_{w1} and the lower weaving flow as v_{w2}. The sum of the two weaving flows is referred to as v_w. The total flow (v) in the section is the sum of the nonweaving flow and the weaving

TABLE 9.5 Symbols and Definitions Used in Weaving Analysis

v_{o1}	=	heavier outer flow in a weaving section (passenger car equivalents per hour)
v_{o2}	=	lighter outer flow in a weaving section (passenger car equivalents per hour)
v_{nw}	=	total nonweaving flow rate in the weaving section [passenger car equivalents per hour (sum of v_{o1} and v_{o2})]
v_{w1}	=	weaving flow rate for the larger of the two weaving flows (passenger car equivalents per hour)
v_{w2}	=	weaving flow rate for the smaller of the two weaving flows (passenger car equivalents per hour)
v_w	=	total weaving flow rate in the weaving area [passenger car equivalents per hour (sum of v_{w1} and v_{w2})]
v	=	total flow rate in the weaving area [passenger car equivalents per hour (sum of v_w and v_{nw})]
L	=	length of weaving area (feet)
k	=	weaving influence factor
N	=	total number of lanes in the weaving area
N_w	=	number of lanes used by weaving vehicles in the weaving area
N_{nw}	=	number of lanes used by nonweaving vehicles in the weaving area
N_w (max)	=	maximum number of lanes that may be used by weaving vehicles for a given configuration
VR	=	volume ratio, v_w/v
R	=	weaving ratio, v_{w2}/v_w
S_w	=	average running speed of weaving vehicles in the weaving area (miles per hour)
S_{NW}	=	average running speed of nonweaving vehicles in the weaving area (miles per hour)

flow. The length of the weaving area is denoted as L and is defined as shown in Figure 9.7.

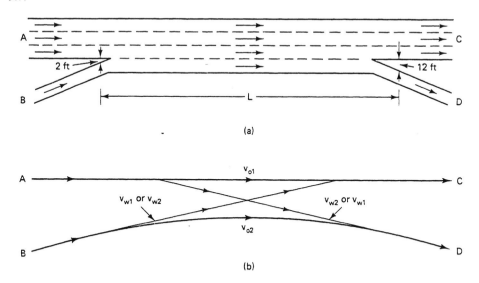

Figure 9.7   Typical Simple Weaving Section

The first step in the weaving analysis process is to determine if weaving causes more than the normal amount of lane changing. For example, referring to Figure 9.7, if the weaving length (L) is large, say several miles, and the sum of the weaving movements (v_w) is small, say 100 or so vehicles per hour, then only the normal amount of lane changing would take place and the section would be "out of the realm of weaving." This fundamental issue is addressed in the 1965 HCM [2] and is illustrated in Figure 9.8. The area of Figure 9.8 that falls below the curve labeled $k = 1.0$ is determined to be "out of the realm of weaving." Thus, for simple weaving sections based on the 1965 HCM, the analyst can determine when weaving and the resulting lane changing is inhibiting basic highway and freeway segments.

The 1965 HCM proposed the following equation, which when combined with Figure 9.8 can be used to calculate the capacity of simple weaving sections.

$$SF = \frac{v_{w1} + kv_{w2} + v_{o1} + v_{o2}}{N} \tag{9.4}$$

where N = number of lanes in the weaving section
       k = weaving influence factor
      SF = service flow rate (vehicles per hour per lane)
and where v_{w1}, v_{w2}, v_{o1} and v_{o2} are as defined earlier.

The 1985 HCM [5] method provided a more comprehensive approach to the analysis of level of service of weaving sections. Three types of weaving configurations were addressed, and further extensions involved the introduction of unconstrained and constrained operations, and prediction of speeds of weaving and nonweaving vehicles.

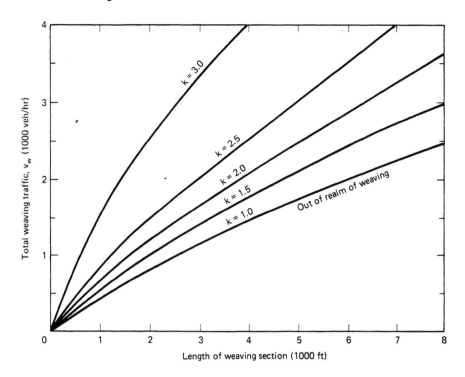

Figure 9.8   Operating Characteristics of Weaving Sections (From Reference 2)

Based on empirical field study results, 12 multiple regression equations were proposed. From these speed predictions, levels of service can be determined.

Many weaving configurations can be encountered on multilane facilities. The 1985 HCM proposed that weaving sections on multilane facilities be classified as one of three types: A, B, and C. This classification is based on lane-changing requirements. These three types of weaving sections are shown in Figures 9.9, 9.10, and 9.11 and discussed in the following paragraphs.

Some of the weaving configurations classified as type A weaving sections are presented in Figure 9.9. Type A weaving sections require that each weaving vehicle make one lane change to execute the desired movement. More than one lane change may be required if weaving vehicles on entry roadway A are not in the rightmost lane or weaving vehicles on entry roadway B in the lower diagram are not in the leftmost lane. The minimum number of total lane changing operations is equal to $v_{w1} + v_{w2}$ and the minimum rate of lane changes is equal to $[(v_{w1} + v_{w2})/L]$.

Some of the weaving configurations classified as type B weaving sections are shown in Figure 9.10. Type B weaving sections require that one weaving movement may be accomplished without making any lane changes, and the other movement requires at most one lane change. This can be a more effective design if the weaving flow from A to D is relatively small. The minimum number of total lane changing operations is equal to (v_{w2}) and the minimum rate of lane changes is equal to $(v_{w2})/L$.

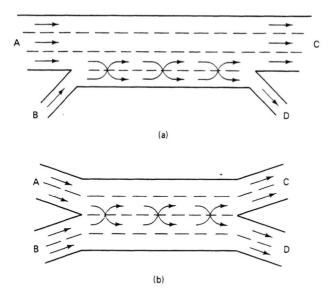

(a)

(b)

Figure 9.9   Type A Weaving Sections

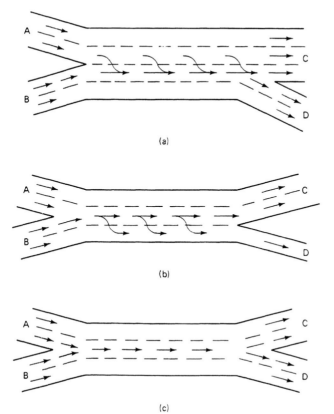

(a)

(b)

(c)

Figure 9.10   Type B Weaving Sections

Type C weaving sections, shown in Figure 9.11, require that one weaving movement may be accomplished without making a lane change, and the other weaving movement requires at least two or more lane changes. This can be an effective design if the second weaving flow is small, but it can have very adverse effects if the second weaving flow is large, the number of lane changes is large, and the weaving length is short. The minimum number of total lane changing operations is equal to v_{w2} times 2 or more, and the minimum rate of lane changes is equal to $[(v_{w2})2+]/L$.

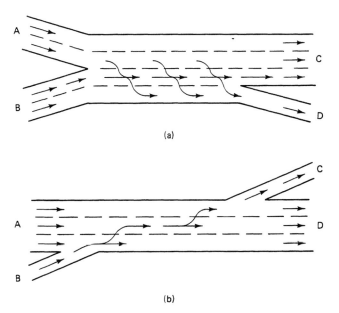

(a)

(b)   **Figure 9.11**   Type C Weaving Sections

The 1985 HCM also considered unconstrained and constrained operations of weaving sections. If the weaving configuration in combination with the traffic demand pattern permits the weaving and nonweaving vehicles to spread out evenly across the lanes in the weaving section, the flows will be somewhat balanced between lanes and the operation is more effective and is classified as unconstrained. On the other hand, if the configuration and demand limits the ability of weaving vehicles to occupy their proportion of available lanes to maintain balanced operations, the operation is less effective and is classified constrained. For example, consider the weaving configuration shown in Figure 9.9a. If the flow from A to C is relatively light and the other flows are relatively heavy, the lanes on the left side of the weaving section will be underutilized and the lanes on the right side will be overutilized. Such imbalanced or constrained operations will result in weaving vehicles traveling at a lower speed (hence lower level of service) and nonweaving vehicles traveling at a higher speed.

Determination of the type of operation is made by comparing two variables:

N_w = number of lanes that must be used by weaving vehicles in order to achieve balanced or unconstrained operation

$N_W(\text{max})$ = maximum number of lanes that may be used by weaving vehicles for a given configuration

If $N_w \leq N_w$ (max), the operation is defined as unconstrained, while if $N_w > N_w$ (max), the operation is defined as constrained. Based on empirical evidence, the procedures for calculating these two variables are summarized in Table 9.6 for types of weaving configurations.

TABLE 9.6   Criteria for Unconstrained and Constrained Operations of Weaving Sections

Type of Configuration	Number of Lanes Required for Unconstrained Operation, N_w	Maximum Number of Weaving Lanes, N_w (max)
Type A	$2.19 N \text{VR}^{0.571} \left(\dfrac{L_H^{0.234}}{S_W^{0.438}} \right)$	1.4
Type B	$N \left[0.085 + 0.703\,\text{VR} + \left(\dfrac{234.8}{L} \right) - 0.018(S_{NW} - S_w) \right]$	3.5
Type C	$N \left[0.761 - 0.011\,L_H - 0.005(S_{NW} - S_W) + 0.047\,\text{VR} \right]$	3.0

Source: Reference 5.

The next step in the analysis is to select appropriate multiregression type equations for predicting weaving and nonweaving speeds based on types of weaving configurations and types of operation. Again, empirically derived equations were developed and the parameter values for the basic equation are summarized in Table 9.7. For specified weaving configuration and type of operation, parameter values are selected from Table 9.7, and substituted into the base equation

$$S_w \text{ or } S_{nw} = 15 + \frac{50}{1 + a(1 + \text{VR})^b \left[(v/N)^c / L^d \right]} \tag{9.5}$$

Note that weaving and nonweaving speeds are required as input to determine unconstrained and constrained operations, yet these speeds have not been determined. The suggested procedure is to assume unconstrained operations, calculate weaving and nonweaving speeds based on parameters shown in Table 9.7, and then use the equations in Table 9.6 to see if the assumption of unconstrained operations is correct. If not, the process is repeated assuming constrained operations. The final step in determining level of service of weaving sections is to enter Table 9.8 with the predicted weaving and nonweaving speeds.

Although rather extensive field studies were undertaken, a limited set of the wide variety of weaving sections were studied, and thus the application of this procedure is

TABLE 9.7 Parameter Values for Prediction of Weaving and Nonweaving Speeds in Weaving Sections

Type of Configuration and Operation	Parameter Values for Weaving Speeds, S_W				Parameter Values for Nonweaving Speeds, S_{NW}			
	a	b	c	d	a	b	c	d
Type A								
Unconstrained	0.226	2.2	1.00	0.90	0.020	4.0	1.30	1.00
Constrained	0.280	2.2	1.00	0.90	0.020	4.0	0.88	0.60
Type B								
Unconstrained	0.100	1.2	0.77	0.50	0.020	2.0	1.42	0.95
Constrained	0.160	1.2	0.77	0.50	0.015	2.0	1.30	0.90
Type C								
Unconstrained	0.100	1.8	0.80	0.50	0.015	1.8	1.10	0.50
Constrained	0.100	2.0	0.85	0.50	0.013	1.6	1.00	0.50

Source: Reference 5.

TABLE 9.8 Level of Service Criteria for Weaving Sections

Level of Service	Minimum Average Weaving Speed, S_W (miles/hr)	Minimum Average Nonweaving Speed, S_{NW} (miles/hr)
A	55	60
B	50	54
C	45	48
D	40	42
E	$\frac{35}{30}$	$\frac{35}{30}$
F	$< \frac{35}{30}$	$< \frac{35}{30}$

Source: Reference 5.

rather restricted. For practical applications, the reader is referred to the 1985 HCM, with particular reference to application limitations. In addition, primary attention has been given to simple weaving sections with two input flows and two output flows. Multiple weaving sections involving more than two input flows and/or two output flows are more complicated and the reader is referred to methods identified in the list of selected references [2, 5, 13–16].

9.6 SIGNALIZED INTERSECTIONS

Signalized intersections are the most critical and complicated elements of the arterial network system and can have significant effects at ramps connected to multilane facilities. It is the most often studied element of the highway system, and many methods have been developed for the purpose of capacity and level of service analysis. Although primary attention is given to methods described in the 1985 HCM [5], other methods have been developed in the United States and in many other countries. Readers with special interests are encouraged to study the vast literature on this subject. Some of the other methods include:

- U.S. planning-type method [4, 17]

- U.S. operations-type method [1, 2, 4, 18, 19]

- Australian method [21, 23], British method [20], and Swedish method [22, 24]

In addition, several other studies have been undertaken which are of specific interest. These include the measurement of delay [25], comparative study of capacity analysis methods [26, 27], and field validation of the NCHRP method in South Africa [28].

The 1985 HCM proposes two entirely different methods for analyzing signalized intersections. One is a planning method that requires a minimum amount of input data and simple and not very time-consuming procedures. It does not result in a traditional capacity or level-of-service estimate but in an assessment as to whether the signalized intersection will likely "work" or not. The other method, called the operations method, requires considerable amount of input data, and the procedures are tedious and time consuming. Fortunately, computer programs are available for the operations method. The operations method does result in capacity and level-of-service predictions for each lane group, approach, and for the total intersection.

9.6.1 Planning Method

The planning method requires turning movements, the geometric layout of the intersection, and the designation as to which lanes can be used by which turning movements. A special worksheet has been prepared for recording data input and for the analysis. The worksheet containing a sample problem is shown as Figure 9.12. The worksheet is divided into four parts. The top part is used for intersection identification, the second part for data input, the next part for determining critical lane movements, and the last part for intersection assessment.

The input data are recorded in the second portion of the worksheet and the next step is to record flows in each individual lane approaching the intersection. When all approaches contain exclusive left-turn lanes, the determination of lane flows is relatively simple. An additional worksheet is required if there are lanes with shared left-turn and

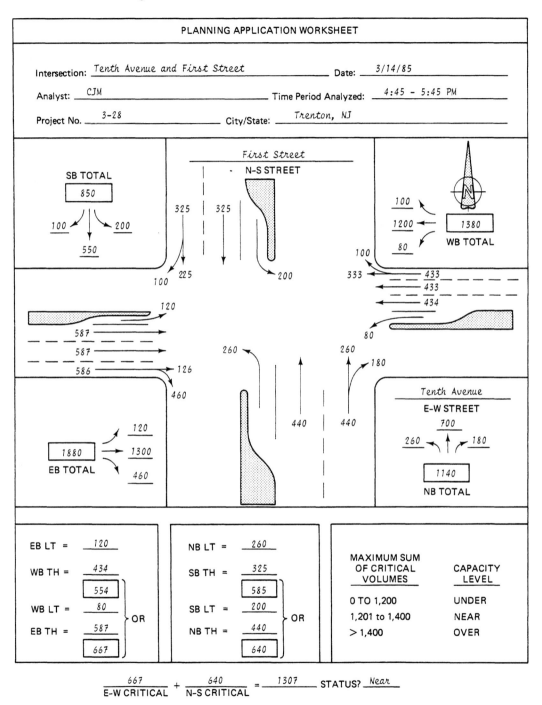

Figure 9.12   Worksheet for Planning Method (From Reference 5)

through movements in order to determine individual lane flows. This situation is discussed a little later.

The analysis is now undertaken in the third portion of the worksheet given in Figure 9.12. Conceptually, assume a two-phase signal operation, and first consider only east and west approaches, and then consider north and south approaches. In the east–west direction, which conflicting critical movements are the largest? The eastbound left-turn-lane movement conflicts with the westbound through lane movement and at the same time the westbound left-turn-lane movement conflicts with the eastbound through lane movement. The sum of these two conflicting movements are 554 and 667 vehicles per hour respectively. The more critical of the two is the larger, 667 vehicles per hour. This process is repeated for the north–south approaches and the critical movement is determined to be 640 vehicles per hour.

The assessment of how well the intersection will work is carried out at the bottom of the worksheet and consists of three steps. First, the east–west critical movement and the north–south critical movement are added together to obtain the sum of the critical movements, which is 1307 vehicles per hour. Next, consider the maximum number of vehicles that could be handled in a single lane in 1 hour. A freeway could handle 2000 vehicles per hour per lane, while a signalized intersection lane under ideal conditions could handle 1800 vehicles per hour per lane. Considering that the amber or intergreen periods probably accounts for 10 percent of the cycle time, and the presence of opposing traffic and other nonideal conditions reduces the figure another 10 percent or so, a maximum of the sums of the critical movements has been selected to be 1400 vehicles per hour (some propose 1500 vehicles per hour). The last step is to compare the calculated sum of the critical lane movements (1307 vehicles per hour in the sample problem) with the maximum allowable sum of the critical lane movements (1400 vehicles per hour). One of three conclusions are reached on the basis of this comparison: undercapacity, nearcapacity, or overcapacity. Stating the conclusions in another way: the intersection *will* work, the intersection *may* work, or the intersection *won't* work.

A complication in the analysis is encountered when the intersection has shared left-turn and through lanes. Then it is difficult to determine lane flows. To do this, a second worksheet, shown in Figure 9.13, can be used to calculate lane flows. Left-turn movements are weighted according to the opposing volume. The lane flows that are then obtained on the second worksheet are transferred to the first worksheet (Figure 9.12) for critical movement analysis. Special procedures are also required for single-lane approaches to signalized intersections. The reader should refer to Chapter 9 of the 1985 HCM for these special procedures.

As mentioned earlier, the planning method requires a minimum of input data and can be applied rapidly. However, the conclusions are rather qualitative, and basically, the planning method tries to answer the question: Will it work? The planning method also requires many assumptions. For example, no signal information is used in the method and a "good" signal timing plan is assumed. Typical demand, design, and environmental situations are assumed to be incorporated by the reduction of the maximum sum of the critical lane movements from 1800 to 1400 vehicles per hour.

LANE DISTRIBUTION FOR SHARED LEFT/THROUGH LANES ON A MULTILANE APPROACH WITH PERMITTED LEFT TURN LANES (OPTIONAL WORKSHEET)										
①	②	③	④	⑤	⑥	⑦	⑧	⑨	⑩	⑪
V_O Opposing Volume (vph)	PCE_{LT}	V_{LT}	LT Equiv. PCE's	Total Volume (TH + RT)	Total	No. of Lanes on Approach	Equiv. Volume Per Lane	Thru Vehicles in LT + TH Lane	Vol. in LT + TH Lane	Vol. in each of the Remaining Lanes
$0 - 199 = 1.1$ $200 - 599 = 2.0$ $600 - 799 = 3.0$ $800 - 999 = 4.0$ $\geqslant 1,000 = 5.0$ APPR.			② × ③		④ + ⑤		⑥ ÷ ⑦	⑧ − ④	③ + ⑨	$\dfrac{(③ + ⑤) - ⑩}{⑦ - 1.0}$
EB LT										
WB LT										
NB LT										
SB LT										

Figure 9.13   Second Worksheet for Planning Method

9.6.2 Operations Method

The operations method, while requiring significant amounts of input data and involving complex multistep analytical procedures, provides for a more accurate and comprehensive assessment of capacity and level of service at signalized intersections. The procedure includes five modules of analysis, as shown in Figure 9.14. Note the similiarity between the structure of the operational analytical procedure and the proposed analytical framework for demand–supply analyses presented in Chapter 8. Space will not permit detailed coverage of the operations method, but emphasis will be given to the structure of the analytical procedures and basic concepts.

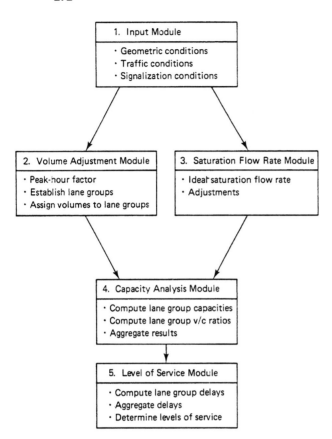

Figure 9.14   Operational Analysis Procedure (From Reference 5)

Module One, referred to as the input module, can best be described by reviewing the input worksheet shown as Figure 9.15. The input worksheet is divided into four parts: intersection identification, volume and geometrics, traffic and roadway conditions, and signal phasing. All input required for the operations method is placed on this worksheet. The pattern of traffic arriving at the intersection is an important consideration in estimating delay, and this is handled by inputting the arrival type for each approach as follows:

- Type 1   Dense platoon arriving at the beginning of the red phase
- Type 2   Dense platoon arriving during the middle of the red phase or a dispersed platoon arriving throughout the red phase
- Type 3   Random arrivals
- Type 4   Dense platoon arriving during the middle of the green phase or a dispersed platoon arriving throughout the green phase
- Type 5   Dense platoon arriving at the beginning of the green phase

The second module, the volume adjustment module, is characterized by the volume adjustment worksheet as shown in Figure 9.16. For each lane group or

---

INPUT WORKSHEET

Intersection: _Third Ave. and Main St._ _____ Date: _8/12/85_

Analyst: _RPR_ _____ Time Period Analyzed: _4 - 5 PM_ Area Type: ☒ CBD ☐ Other

Project No.: _____ City/State: _Central City_ _____

VOLUME AND GEOMETRICS

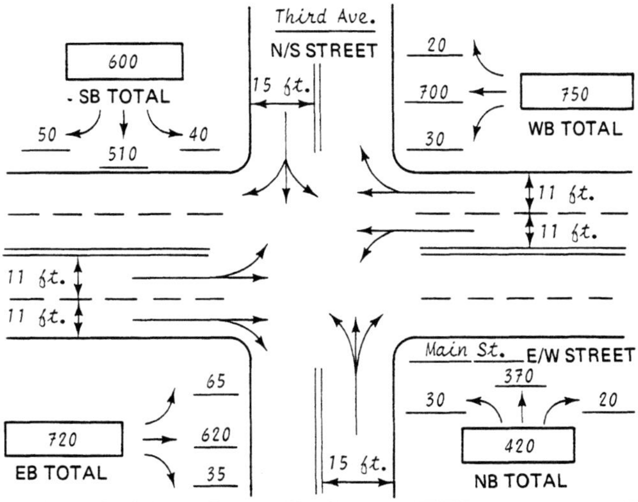

NORTH

Third Ave.
N/S STREET

| 600 |
- SB TOTAL
50 ↙ ↓ ↘ 40
510

20 ↖
15 ft.
700 ← | 750 |
30 ↙   WB TOTAL

‡11 ft.
‡11 ft.

11 ft. ‡
11 ft. ‡

IDENTIFY IN DIAGRAM:

1. Volumes
2. Lanes, lane widths
3. Movements by lane
4. Parking (PKG) locations
5. Bay storage lengths
6. Islands (physical or painted)
7. Bus stops

65

Main St. E/W STREET
370
30 ↖ ↑ ↗ 20
| 420 |
NB TOTAL

15 ft.

| 720 | → 620
EB TOTAL   ↘ 35

TRAFFIC AND ROADWAY CONDITIONS

Approach	Grade (%)	% HV	Adj. Pkg. Lane		Buses (N_B)	PHF	Conf. Peds. (peds./hr.)	Pedestrian Button		Arr. Type
			Y or N	N_m				Y or N	Min. Timing	
EB	0	5	N	-	0	0.90	100	N	9.8	4
WB	0	5	N	-	0	0.90	100	N	9.8	2
NB	0	8	N	-	0	0.90	100	N	13.8	3
SB	0	8	N	-	0	0.90	100	N	13.8	3

Grade: + up, − down
HV: veh. with more than 4 wheels
N_m: pkg. maneuvers/hr

N_B: buses stopping/hr
PHF: peak-hour factor
Conf. Peds: Conflicting peds./hr

Min. Timing: min. green for pedestrian crossing
Arr. Type: Type 1-5

PHASING

D I A G R A M

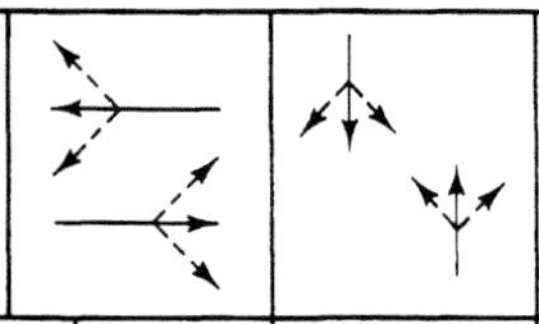

Timing	G = 27 Y + R = 43	G = 37 Y + R = 33	G = Y + R =	G = Y + R =	G = Y + R =	G = Y + R =	G = Y + R =	G = Y + R =
Pretimed or Actuated	P	P						

↗ Protected turns   - - -↗ Permitted turns   - - - - - - Pedestrian   Cycle Length _70_ Sec

---

Figure 9.15   Input Module Worksheet (From Reference 5)

VOLUME ADJUSTMENT WORKSHEET

(1) Appr.	(2) Mvt.	(3) Mvt. Volume (vph)	(4) Peak Hour Factor PHF	(5) Flow Rate v_p (vph) ③ ÷ ④	(6) Lane Group	(7) Flow rate in Lane Group v_g (vph)	(8) Number of Lanes N	(9) Lane Utilization Factor U Table 9-4	(10) Adj. Flow v (vph) ⑦ × ⑨	(11) Prop. of LT or RT P_{LT} or P_{RT}
	LT	65	0.90	72						
EB	TH	620	0.90	689		800	2	1.05	840	0.09 LT 0.05 RT
	RT	35	0.90	39						
	LT	30	0.90	33						
WB	TH	700	0.90	778		833	2	1.05	875	0.04 LT 0.03 RT
	RT	20	0.90	22						
	LT	30	0.90	33						
NB	TH	370	0.90	411		466	1	1.00	466	0.07 LT 0.05 RT
	RT	20	0.90	22						
	LT	40	0.90	44						
SB	TH	510	0.90	567		667	1	1.00	667	0.07 LT 0.08 RT
	RT	50	0.90	56						

Figure 9.16   Volume Adjustment Worksheet (From Reference 5)

intersection approach, the movement volumes are modified by the peak-hour factor and the lane utilization factor. The peak-hour factor converts the 1-hour volumes to 15-minute hourly rates of flow. The lane utilization factor adjusts the average lane volumes to represent that lane which might be utilized more than the average. The results of this module are the adjusted volume and proportion of left turners and right turners in each lane group.

The third module is the saturation flow adjustment module, and the worksheet appearing in Figure 9.17 is used for this analysis. For each lane group or intersection approach, the ideal saturation flow of 1800 passenger car equivalents per hour of green per lane is adjustment by the number of lanes and then by a series of eight potential reduction factors which adjust the saturation flow from ideal conditions to actual conditions. The result of this module is the adjusted saturation flow rate in vehicles per hour of green.

The fourth module is the capacity analysis module, and this analysis is performed on the worksheet shown as Figure 9.18. For each lane group or intersection approach, the adjusted flow rates are transferred from the volume adjustment worksheet, and the adjusted saturation flow rates are transferred from the saturation flow adjustment worksheet. The flow ratios are computed and then with the green time-cycle ratios, lane group capacities and volume–capacity ratios are calculated. Critical lane groups can then be identified and the signal operation evaluated.

The last module is the level-of-service module, and this analysis is undertaken using the worksheet shown as Figure 9.19. For each lane group or intersection approach, the volume–capacity ratio, capacity, and signal timing are transferred from the capacity analysis worksheet. The uniform delay and the incremental delay are calculated, and their sum is adjusted by a progression factor that accounts for arrival pattern type and the quality of signal equipment. The final delay is calculated and from it, the level of service is determined. Composite results for each approach and for the total intersection can be determined.

The delay equation used in the 1985 HCM is as follows:

$$d = PF(d_1 + d_2) \tag{9.6}$$

where PF = progression factor (e.g., equal to 1.00 for random arrivals and pretimed control)
d_1 = uniform stopped delay (seconds per vehicle)
d_2 = incremental stopped delay (seconds per vehicles)

The uniform delay and incremental delay are calculated using the following equations:

$$d_1 = 0.38C \frac{(1 - g/C)^2}{1 - (g/C)(X)} \tag{9.7}$$

$$d_2 = 173X^2 \left[(X - 1) + \sqrt{(X - 1)^2 + \left(16\frac{X}{c} \right)} \right] \tag{9.8}$$

where g = effective green time (seconds)
C = cycle length (seconds)

SATURATION FLOW ADJUSTMENT WORKSHEET												
LANE GROUPS				ADJUSTMENT FACTORS								
①	②	③	④	⑤	⑥	⑦	⑧	⑨	⑩	⑪	⑫	⑬
Appr.	Lane Group Movements	Ideal Sat. Flow (pcphgpl)	No. of Lanes N	Lane Width f_w	Heavy Veh. f_{HV}	Grade f_g	Pkg. f_p	Bus Blockage f_{bb}	Area Type f_a	Right Turn f_{RT}	Left Turn f_{LT}	Adj. Sat. Flow Rate s (vphg)
				Table 9-5	Table 9-6	Table 9-7	Table 9-8	Table 9-9	Table 9-10	Table 9-11	Table 9-12	
EB		1800	2	0.97	0.975	1.00	1.00	1.00	0.90	0.99	0.75	2275
WB		1800	2	0.97	0.975	1.00	1.00	1.00	0.90	0.99	0.85	2579
NB		1800	1	1.10	0.96	1.00	1.00	1.00	0.90	0.89	0.86	1309
SB		1800	1	1.10	0.96	1.00	1.00	1.00	0.90	0.89	0.95	1446

Figure 9.17   Saturation Flow Adjustment Worksheet (From Reference 5)

CAPACITY ANALYSIS WORKSHEET								
LANE GROUP		③	④	⑤	⑥	⑦	⑧	⑨
① Appr.	② Lane Group Movements	Adj. Flow Rate v (vph)	Adj. Sat. Flow Rate s (vphg)	Flow Ratio v/s ③ ÷ ④	Green Ratio g/C	Lane Group Capacity c (vph) ④ × ⑥	v/c Ratio X ③ ÷ ⑦	Critical ? Lane Group
EB		840	.2275	0.369	0.386	878	0.956	✓
WB	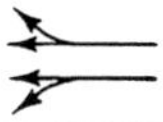	875	2579	0.339	0.386	995	0.879	
NB		466	1309	0.357	0.528	692	0.673	
SB		667	1446	0.461	0.528	765	0.872	✓

Cycle Length, C ___70___ sec

Lost Time Per Cycle, L ___6___ sec

$\sum_i (v/s)_{ci} = $ ___0.835___

$X_c = \dfrac{\sum_i (v/s)_{ci} \times C}{C - L} = $ ___0.84___

Figure 9.18   Capacity Analysis Worksheet (From Reference 5)

LEVEL-OF-SERVICE WORKSHEET												
Lane Group		First Term Delay				Second Term Delay				Total Delay & LOS		
(1)	(2)	(3)	(4)	(5)	(6)	(7)	(8)	(9)	(10)	(11)	(12)	(13)
Appr.	Lane Group Movements	v/c Ratio X	Green Ratio g/C	Cycle Length C (sec)	Delay d_1 (sec/veh)	Lane Group Capacity c (vph)	Delay d_2 (sec/veh)	Progression Factor PF Table 9-13	Lane Group Delay (sec/veh) ((6) + (8)) × (9)	Lane Group LOS Table 9-1	Approach Delay (sec/veh)	Appr. LOS Table 9-1
EB		0.956	0.386	70	15.89	878	15.04	0.90	27.84	D	27.84	D
WB		0.879	0.386	70	15.18	995	6.50	1.18	25.58	D	25.58	D
NB		0.673	0.528	70	9.16	753	1.80	1.00	10.96	B	10.96	B
SB		0.872	0.528	70	10.95	807	7.64	1.00	18.59	C	18.59	C

Intersection Delay ___22.22___ sec/veh   Intersection LOS ___C___ (Table 9-1)

Figure 9.19   Level-of-Service Worksheet (From reference 5)

c = capacity of lane group (vehicles per hour)

X = volume–capacity ratio for the lane group

The level of service is determined based only on the average stopped delay per vehicle in the following manner:

- Delays less than 5 seconds   LOS A
- Delays 5–15 seconds   LOS B
- Delays 15–25 seconds   LOS (
- Delays 25–40 seconds   · LOS D
- Delays 40 –60 seconds   LOS E
- Delays > 60 seconds   LOS F

9.7 SUMMARY

This chapter serves as an introduction to analytical techniques for capacity and level of service determination of critical elements of the highway system. Heavy emphasis has been placed on the 1985 *Highway Capacity Manual* and for more comprehensive and in-depth coverage the reader should refer to the Manual [5]. Other methods for capacity analysis are available in the United States and abroad, and where possible they have been noted in the text and included with the selected references.

The field of capacity is not limited to highway facilities but includes other land transport modes as well as air and water transportation. The reader should refer to the 1985 HCM for discussions of capacity analysis for transit, pedestrians, and bicycles. A few references have been selected that cover capacity analysis of other modes of transportation, and they are contained in the list of selected references at the end of the chapter.

The subject of capacity is a rich and dynamic field of study. Much is known about the subject, but the frontier of knowledge continues to move forward at a rapid rate. It offers promise of significant breakthroughs for the researcher, and it is an essential tool for the practitioner.

9.8 SELECTED PROBLEMS

1. Often the future can be predicted by looking back in history. Compare the 1950, 1965, and 1985 *Highway Capacity Manuals*. What will capacity manuals look like during your professional career?
2. Undertake a literature search of capacity analysis in (a) air transportation, (b) water transportation, and (c) rail transportation.
3. Calculate the volume–capacity ratio for levels of service A, B, C, D, and E of a multilane facility under ideal conditions when the design speed is 50, 60, and 70, miles per hour. Are the volume-capacity ratios quite similar for different design speeds? (*Hint:* Study Figure 9.1.)

4. A bottleneck is caused at a lane drop location along a directional freeway. Draw a sketch and indicate where levels of service C, E, and F might be encountered. Draw profiles of speed, density, and volume–capacity ratio.

5. What service flow rates exist on a directional multilane facility under ideal conditions at levels of service A, B, C, D, and E when the design speeds are 50, 60, and 70 miles per hour? Which service flow rate is higher: (a) when design speed is 50 miles per hour at LOS C or (b) when design speed is 70 miles per hour at LOS B?

6. An urban directional freeway is to be designed for 4000 vehicles per hour with a level of service of D to be provided. The freeway is to be located in rolling terrain, and the truck percentage is expected to be 10 percent. Specify and then quantify design elements that can provide this level of service for the anticipated traffic demand.

7. A typical three-lane directional freeway that is assumed to be under ideal conditions was shown in Figure 9.3. The expected O–D 15-minute hourly demand rates are given in the following table.

Destination Origin	D_3	D_4	D_5
O_1	300	700	2000
O_2	200	600	800
O_3	100	400	500

Considering multilane basic segment procedures only, determine the density, speed, and level of service at a location:

(a) Just upstream of origin 3
(b) Just downstream of origin 3
(c) Just downstream of destination 3

8. With the information given in Problem 7 and assuming single-lane ramps, determine the following:

(a) Adequacy of the on-ramp and off-ramp themselves
(b) Level of service in the merge area
(c) Level of service in the diverge area

9. With the information given in Problems 7 and 8, determine the level of service of the weaving section between origin 3 and destination 3 if an auxiliary lane is added.

10. One method for improving level of service in merge and diverge areas is to increase the distance between on-ramps and adjacent downstream off-ramps. Solve Problem 8 with the distance varying from 2000 feet to 5000 feet in 1000 foot intervals. Plot the flow in the rightmost lane (vertical scale) versus the distance between ramps in the vicinity of the merge area. Denote the resulting level of service on the curve. Repeat for the diverge area.

11. One method for improving level of service in weaving sections is to increase the length of the weaving section. Solve Problem 9 with the length of the weaving section varying from

· 1000 to 6000 feet in 1000 foot intervals. Determine the level of service as a function of the length of the weaving section.

12. The planning application example shown in Figure 9.12 reveals that the intersection is operating very close to capacity. Generate three minimum redesigns of the intersection in order for the intersection to operate "under" capacity. Evaluate all three redesigns with the planning method and recommend your preferred redesign. (*Hint:* Study the four critical movement sums.)

13. Assume the addition of one lane on each approach to the intersection shown in Figure 9.12. All added lanes will be for through traffic only. Evaluate this redesigned intersection using the planning method.

14. The resulting levels of service in the operations application example are not very balanced between the east–west and north–south streets (see Figure 9.19). Without changing the cycle length, adjust the green phases shown in Figure 9.15 until the levels of service on the two crossing streets are approximately the same. How much is the intersection delay of 22.22 seconds per vehicle reduced?

15. How much would the intersection delay of 22.22 seconds per vehicle be reduced in Figure 9.19 if on all approaches the traffic arrived in dense platoons at (**a**) the beginning of the green phase or (**b**) the beginning of the red phase? (*Hint:* Refer to Table 9.13 of 1985 HCM for progression factors.)

16. Compare the operations method of analysis contained in the 1985 HCM with the most recent Australian method of analysis for signalized intersections. Identify differences and similarities.

9.9 SELECTED REFERENCES

1. Bureau of Public Roads, *Highway Capacity Manual*, BPR, Washington D. C., 1950, 147 pages.

2. Highway Research Board, *Highway Capacity Manual 1965*, Special Report 87, HRB, Washington D. C., 1966, 411 pages.

3. A. D. May, *Intersection Capacity 1974: An Overview*, Transportation Research Board Special Report 153, TRB, Washington D. C., 1975, pages 50–59.

4. Transportation Research Board, *Interim Materials on Highway Capacity*, Circular 212, TRB, Washington D. C., January 1980, 272 pages.

5. Transportation Research Board, *Highway Capacity Manual*, Special Report 204, TRB, Washington D. C., 1985, 474 pages.

6. Stephen L. M. Hockaday and Adib K. Kanafani, Developments in Airport Capacity Analysis, Transportation Research, Vol. 8, 1974, pages 171–180.

7. Adib Kanafani, Operational Procedures to Increase Runway Capacity, *Journal of Transportation Engineering*, Vol. 109, No. 3, May 1983, pages 414–424.

8. Robert Horonjeff, *Planning and Design of Airports*, 2nd Edition, McGraw-Hill Book Company, New York, 1975, 460 pages.

9. V. R. Vuchic, *Urban Public Transportation: Systems and Technology*, Prentice Hall, Inc., Englewood Cliffs, N. J., 1981.

10. R. Roess, W. McShane, E. Linzer, and L. Pignataro, Freeway Capacity Analyses Procedures, *ITE Journal*, Vol. 50, No. 12, December 1980, pages 16–20.

11. R. Roess, W. McShane, and L. Pignataro, *Freeway Level of Service: A Revised Approach*, Transportation Research Board Record 699, TRB, Washington D. C., 1980.

12. K. Moskowitz and L. Newman, *Traffic Bulletin No. 4–Notes on Freeway Capacity*, Highway Research Board, Record 27, HRB, Washington D. C., 1963, pages 44–68.

13. L. Pignataro, W. McShane, R. Roess, K. Crowley, and B. Lee, *Weaving Areas—Design and Analysis*, NCHRP Report 159, Transportation Research Board, Washington D. C., 1975, 119 pages.

14. J. Leisch, *Completion of Procedures for Analysis and Design of Traffic Weaving Sections*, Federal Highway Administration, Washington D. C., 1983.

15. W. Reilly, J. Kell, and P.-Johnson, *Weaving Analysis Procedures for the New Highway Capacity Manual*, JHK and Associates, Tucson, Ariz., August 1984.

16. J. Fazio and N. M. Rauphail, *Freeway Weaving Section: Comparison and Refinement of Design and Freeway Analysis Procedures*, Transportation Research Board Presentation, January 1986.

17. C. J. Messer and D. B. Fambro, *Critical Lane Analysis for Intersection Design*, Transportation Research Board, Record 644, TRB, Washington D. C., 1977.

18. *Signalized Intersections Capacity Study*, NCHRP Project 3-28 (2), JHK and Associates, Tucson, Ariz., December 1982.

19. *NCHRP Signalized Intersection Capacity Method*, NCHRP Project 3-28 (2), JHK and Associates, Tucson, Ariz., February 1986.

20. F. V. Webster and B. M. Cobbe, *Traffic Signals*, Her Majesty's Stationery Office, London, 1966.

21. A. J. Miller, *The Capacity of Signalized Intersections in Australia*, Australian Road Research Board, Bulletin 3, ARRB, Victoria, Australia, 1968.

22. B. E. Petersen and E. Imery, *Swedish Capacity Manual*, Statens Vagverk, Stockholm, Sweden, February 1977, 309 pages.

23. R. Akcelik, Traffic Signals: Capacity and Timing Analysis, Australia Road Research Board, Report 123, ARRB, Victoria, Australia, 1981.

24. K. L. Bang, *Capacity of Signalized Intersections*, Transportation Research Board, Record 667, TRB, Washington D. C., 1978, pages 11–21.

25. W. R. Reilly, C. C. Gardner, and J. H. Kell, *A Technique for Measurement of Delay at Intersections*, Federal Highway Administration, Washington D. C., 1976.

26. D. S. Berry, *Other Methods for Computing Capacity of Signalized Intersections*, Transportation Research Board Presentation, January 1977.

27. Adolf D. May, Ergun Gedizlioglu, and Lawrence Tai, *Comparative Analysis of Signalized-Intersection Capacity Methods*, Transportation Research Board, Record 905, TRB, Washington D. C., 1983, pages 118–127.

28. Adolf D. May, Wessel Piernaar, and Cecil A. Rose, *Use of the NCHRP Signalized Intersection Capacity Method—A South African Experience*, Transportation Research Board, Record 971, TRB, Washington D. C., 1984, pages 32–40.

10

Traffic Stream Models

Traffic stream models provide the fundamental relationships of macroscopic traffic stream characteristics for uninterrupted flow situations. The traffic stream characteristics include flow characteristics (Chapter 3), speed characteristics (Chapter 5), and density characteristics (Chapter 7). The relationships are for free-flow and congested-flow conditions away from flow interruptions such as at intersections.

Traffic stream models are used in the planning, design, and operations of transportation facilities. Travel times as a function of traffic load are needed in planning studies of traffic assignment and modal choice. The trade-off between levels of service and service flows are needed in design of new facilities and the redesign of existing ones. Operational control of facilities requires the knowledge of relationships between on-line measurements of performance such as density or occupancy and optimum level of control. Although field-derived empirical relationships are useful, expressing these relationships mathematically on a theoretical basis provides a sound generalized framework for understanding and application.

This chapter is presented in five major parts followed by selected problems and references. Stream flow fundamentals followed by practical considerations are first presented to establish a theoretical base with an understanding of the real world. Discrete single- and multiregime stream models are then presented with emphasis on the mathematical formulation of each model and their suitability to field conditions. Attention is then given to expanding these discrete stream models into a continuum of both single- and multiregime families of models with emphasis on formulation and appropriateness. The final section is devoted to stream models for pedestrian facilities.

10.1 STREAM FLOW FUNDAMENTALS

Basic stream flow diagrams are shown in Figure 10.1. A linear speed–density relationship is assumed to simplify this presentation of stream flow fundamentals. Later sections of this chapter present more sophisticated relationships, some of which better represent actual field conditions.

The three flow characteristics shown in Figure 10.1 are flow, speed, and density. Flow (q) is defined as the number of vehicles passing a specific point or short section in a given period of time in a single lane. Flow is expressed as an hourly rate on a per lane basis (veh/hr/lane). One unique flow parameter is maximum flow or capacity (q_m). Speed (u) is defined as the average rate of motion and is expressed in miles per hour (mi/hr). From a theoretical perspective, space-mean-speed rather than time-mean-speed should be employed [1]. The two unique speed parameters are free-flow speed (u_f) and optimum speed (u_o). Free-flow speed is that speed which exists when flows approach zero under free-flow conditions while optimum speed is that speed which exists under maximum flow conditions. Density (k) is defined as the number of vehicles occupying a section of roadway in a single lane. Density is expressed on a per mile and a per lane basis (veh/mile/lane). The two unique density parameters are jam density (k_j) and optimum density (k_o). Jam density is that density that occurs when both flow and speed approach zero while optimum density occurs under maximum flow conditions.

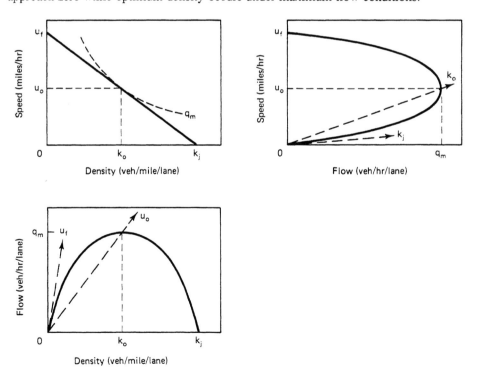

Figure 10.1   Basic Stream Flow Diagrams

Consider the speed–density relationship shown in the upper-left corner of Figure 10.1. As mentioned earlier a linear speed–density relationship is assumed to simplify the presentation. The equation for this relationship is

$$u = u_f - \left(\frac{u_f}{k_j}\right)k \tag{10.1}$$

This relationship indicates that speed approaches free-flow speed (u_f) when density (and flow) approach zero $(k \rightarrow 0$ and $q \rightarrow 0)$. As density (and flow) increases, speeds are reduced until flow is maximum (q_m), and speed and density approach their optimum values $(u \rightarrow u_o$ and $k \rightarrow k_o)$. Further increases in density result in lower speeds (and lower flows) until density reaches its maximum value (k_j) and correspondingly speed (and flow) approach zero $(u \rightarrow 0$ and $q \rightarrow 0)$. Note that flows can be represented on the speed–density diagram as contour lines with the maximum flow contour (q_m) just touching the speed–density line at optimum values of speed and density $(u_o$ and $k_o)$.

Before proceeding with the flow–density relationship, the following equation needs to be presented and described:

$$q = uk \tag{10.2}$$

The flow is shown to be equivalent to the product of speed and density. For example, if a single lane 1-mile long, contains five vehicles $(k = 5)$ and the space-mean-speed of these vehicles is 50 miles per hour $(u = 50)$, after 1 hour of steady flow conditions 50 one-mile blocks of traffic would pass, and 250 vehicles would be counted $(q = 5 \times 50)$. Note that the dimensions of the product of speed (miles per hour) and density (vehicles per mile per lane) are identical to flow (vehicles per hour per lane).

The flow–density relationship is shown directly below the speed–density relationship in Figure 10.1 because of their common horizontal scales. The equation for this relationship can be derived by substituting q/k for u [based on equation (10.2)] into equation (10.1), and solving for flow (q). The resulting flow–density equation is

$$q = u_f k - \left(\frac{u_f}{k_j}\right)k^2 \tag{10.3}$$

Under very low density conditions $(k \rightarrow 0)$, flow approaches zero $(q \rightarrow 0)$ and speed approaches free-flow speed $(u \rightarrow u_f)$. As flow increases, density increases while speed is decreasing. When optimum density is reached, flow becomes maximum. Further increases in density result in decreased flow until finally, as jam density is reached, flow approaches zero. Note that speeds can be represented on the flow–density diagram as radial lines extending up to the right from the origin. Steeper-sloped lines represent higher speeds; that is, a vertical line represents a speed of infinity while a horizontal line represents a speed of zero. The slope of the flow–density curve is zero when maximum flow occurs. Therefore, the relationship between optimum and jam density can be determined in the following manner beginning with equation (10.3). Since $dq/dk = 0$ when $k \rightarrow k_0$ and

$$q = u_f k - \left(\frac{u_f}{k_j} \right) k^2$$

then

$$0 = u_f - \left(\frac{u_f}{k_j} \right)^{2k_o}$$

and

$$k_o = \frac{k_j}{2} \tag{10.4}$$

It should be observed that this unique relationship is true only when the speed–density relationship is assumed linear. However, the procedures described above for relating optimum and jam density can be followed when nonlinear speed–density relationships are encountered.

The speed–flow relationship is shown directly to the right of the speed–density relationship in Figure 10.1 because of their common vertical scales. The equation for this relationship can be derived by substituting q/k for k [based on equation (10.2)] into equation (10.1), and solving for speed (u). The resulting speed–flow equation is

$$u = u_f - \left(\frac{u_f}{k_j} \right) \frac{q}{u} \tag{10.5}$$

Solving for q, equation (10.5) becomes

$$q = \frac{k_j}{u_f} (u_f u - u^2) \tag{10.6}$$

The upper limb of the speed–flow curve is described as the free-flow regime and the lower limb is referred to as the congested flow regime. Under free-flow conditions, the speed decreases as the flow level increases up to the maximum flow. Further speed reductions coupled with flow reductions are encountered when density exceeds optimum density. The lower limb of the curve depicts this congested flow situation. Note that densities can be represented on the speed–flow diagram as radial lines extending up to the right from the origin. Steeper-sloped lines represent lower densities; that is, a vertical line represents a density of zero while a horizontal line represents a density of infinity. The relationship between optimum and free-flow speed can be determined in the following manner.

Since

$$q = \frac{k_j}{u_f} (u_f u - u^2)$$

then

$$q_m = \frac{k_j}{u_f} (u_f u_o - u_o^2)$$

Substituting $u_o k_o$ for q_m and $2k_o$ for k_j gives

$$u_o k_o = \frac{2k_o}{u_f}(u_f u_o - u_o^2)$$

Solving for u_o yields

$$u_o = \frac{u_f}{2} \qquad \qquad (10.7)$$

Given equations (10.4) and (10.7), the maximum flow (capacity) can be determined from free-flow speed and jam density as follows. Since $q = uk$, then $q_m = u_o k_o$
   Since

$$u_o = \frac{u_f}{2} \quad \text{and} \quad k_o = \frac{k_j}{2}$$

then

$$q_m = \frac{u_f k_j}{4} \qquad \qquad (10.8)$$

Again, it is important to remember that equations (10.4), (10.7), and (10.8) are all based on a linear speed–density relationship.

The three diagrams shown in Figure 10.1 are redundant, for it is obvious that if any one relationship is known, the other two are uniquely defined. However, all three relationships are shown because each has a particular purpose and use. For example, the speed–density relationship is used in most theoretical work for two reasons. First, there is a single-valued speed for each single-valued density, which is not true in the other two relationships. Another reason for its theoretical use is that its form is similar to car-following theory formulations (i.e., speed as a function of space). Turning to the flow–density relationship, this relationship is often used as the basis for freeway control systems. Density (or percent occupancy) is used as the control parameter and flow (or productivity) is the objective function. For example, under low-density conditions no control is needed because all the demand is being satisfied at a high level of service. As density is observed to increase, increased control is required to maintain densities below the optimum density value. Finally, the speed–flow relationship is used in design primarily to identify the trade-off between level of service (speed) and the level of productivity (flow). The *Highway Capacity Manual* [17] uses this relationship with levels of service depicted on the vertical scale and service flow rates on the horizontal scale.

10.2 PRACTICAL CONSIDERATIONS

The discussion in this chapter so far has been limited to theoretical stream flow relationships without reference to field observations. Before proceeding to the next section which is devoted to the gradual evolution of theoretical traffic stream models, the importance of field location and some characteristics of speed–flow–density field measured relationships are discussed.

10.2.1 Importance of Field Location

To obtain meaningful speed–flow–density measurements, the data collection must be taken at the right location during a selected time period. Putting it in another way, the location and time period of field measurements significantly affect the resulting speed–flow–density measurements. Consider the directional roadway depicted at the top of Figure 10.2 with traffic moving from left to right. There are no entrances or exits along the roadway, and in the middle portion of the study section the capacity has been reduced due to a lane drop. To describe the example clearly, assume three-lane sections at the upstream and downstream portions and a two-lane section in the middle portion.

Measurement stations are established at locations A, B, C, and D. Speeds, flows, and densities are measured independently at each station. Station A is at the upstream end of the study section, away from any influence due to the lane-drop location. Station

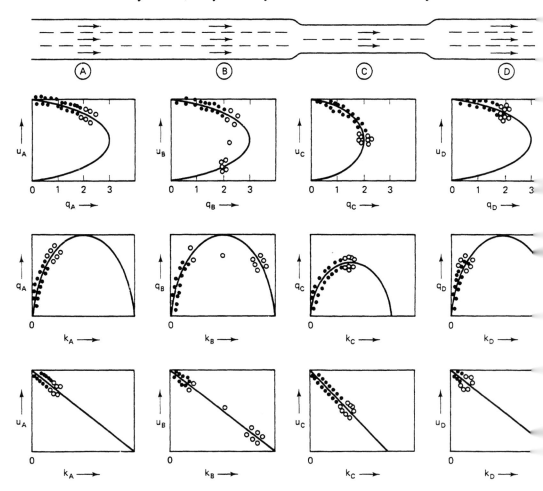

Figure 10.2   Importance of Location

B is in the three-lane section, but just a short distance upstream of the lane-drop location. Station C is in the two-lane section and station D is in the downstream three-lane section.

Theoretical speed–flow, flow–density, and speed–density curves are shown on Figure 10.2 directly below the four measurement stations. The curves are identical for stations A, B, and D; however, the curves for station C are different. For station C, the roadway capacity is only two lanes (two-thirds of the capacity of stations A, B, and D) and the roadway jam density is two-thirds of stations A, B, and D.

Assume that this is an inbound section of roadway carrying traffic to the central business district during the morning peak period. From 6:00 to 7:30 A.M. the traffic demand increases from a very small flow up to a flow equivalent to two lanes of capacity. The hypothetical collected data measurements are plotted on all diagrams of Figure 10.2 as solid dots. While stations A, B, and D are operating only at two-thirds of capacity and operating at relatively high speeds and low densities, station C has reached its capacity with significantly lower speeds and higher densities.

Now assume that the demand from 7:30 to 8:00 A.M. increases to a flow equivalent of two and one-half lanes of capacity. What data measurements would be observed at the four stations? Station A is fairly easy to predict since it is *not* influenced by the two-lane section and its capacity is greater than two and one-half lanes. The resulting data points are plotted as hollow circles. Now consider station B. Its data points would be identical to station A's data points until it is affected by the two-lane section. When this occurs, the flow at station B will be equal to the flow at station C (two lanes of flow) and station B would be congested with exhibited low speeds and high densities. The resulting data points are plotted as hollow circles.

Now consider station C. Although the traffic demand is equivalent to two and one-half lanes of traffic, its capacity is only two lanes, and thus the flow will remain fairly uniform right at capacity. Station C would be operating at optimum speed and optimum density. The resulting data points are plotted as hollow circles. Finally, consider station D. Its capacity of three lanes exceeds its demand of two and one-half lanes of traffic. However, station C can pass only two lanes of traffic, so that station D will exhibit uniform data points of two lanes of traffic and relatively high speeds and low densities. The resulting data points are plotted as hollow circles.

Now assume that the demand after 8:00 A.M. begins to decrease. What data measurements would now be observed at the four stations? The previously plotted hollow circles would revert back to solid dots at all stations and further data collection would result in new data points being superimposed on previously collected data points.

Now review the original theoretical curves with the hypothetical measured data points. Stations A, C, and D will only exhibit data points in the free-flow regime and only in the case of station C are measurements available over the complete free-flow regime. The data point configuration for station B does exhibit observations in the free-flow and congested-flow regimes. However, data points are not provided over the complete theoretical curves. In the free-flow regime, data points are obtained up to about two and one-half lanes of flow. In the congested-flow regime, data points are clustered at a congested flow rate approximately equivalent to two lanes of flow, which is the downstream capacity. Note that some observations may be recorded in the region between the free-flow regime and the congested-flow regime.

None of the data sets from the four measurem   stations covers the complete range of possible speed, flow, and density values. If ι   ɔadway at station B had a little lower capacity, say equivalent to two and one-half lanes of flow, the complete free-flow regime would be obtained. Even then, however, the measurements in the congested-flow regime would be lir ited to a relative small segment. The distribution of data points in the congested-flow regime could be improved if the capacity of the roadway at station C varied with time from a very small capacity to a capacity equivalent to two and one-half lanes of flow. Although in real life, capacity might change with time due to merging, weaving, and/or incidents, it is unlikely that a wide range in capacity changes would occur, particularly at the time of field measurements.

In summary, there are two important points to keep in mind with traffic stream models based on field observations. First, the location limits the range of flow, speed, and density values. A corollary is that given a set of flow, speed, and density measurements, the type of location can be determined. (Review the diagrams shown in Figure 10.2 without reference to the roadway sketch. Could these locations be described?) The second important point is that in validating traffic stream models, the data sets used may influence the results and the comparison between models.

10.2.2 Field Observations

The results of field studies that show the relationships between flow, speed, and density are important to observe before proceeding to theoretical traffic stream models. Four sets have been selected to show flow–speed–density data point patterns for a high-speed freeway [2], a freeway with a 55-mile per hour speed limit [1], a tunnel [3], and an arterial street [4]. Particular attention will be given to consistency of data point patterns; ranges of flow, speed, and density observed values;   nd to flow, speed, and density parameters.

Relationships between flow, speed, and density observed values for a high-speed freeway are shown in Figure 10.3. These data were obtained from the Santa Monica Freeway (detector station 16) in Los Angeles. This urban roadway incorporates high design standards and operates under nearly ideal conditions. A high percentage of the drivers are commuters who use this freeway on a regular basis. The data were collected automatically by the California Department of Transportation as part of their freeway surveillance and control project. Measurements are averaged over 5-minute periods. The speed–density plot shows a very onsistent data point pattern and displays a slight S-shaped relationship. The sample provides a rather uniform distribution of observed densities over the range from near zero to 130 vehicles per mile per lane, but no density values were observed over 130 vehicles per mile per lane. The free-flow speed is slightly over 60 miles per hour and the jam density cannot be estimated. The flow–density plot also shows a very consistent data point pattern, and while the free-flow portion appeared somewhat as a parabola, the congested-flow portion is relativeı flat with a tail to the right. The maximum flow or capacity appears to be just under 2000 vehicles per hour per lane, and the optimum density is approximately 40 to 45

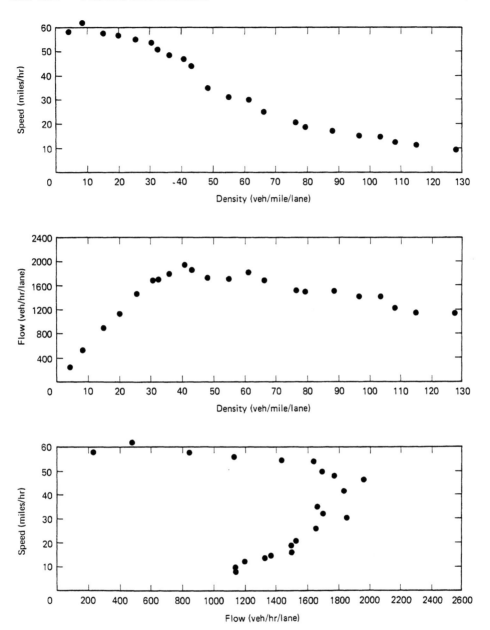

Figure 10.3   Highspeed Freeway Data (From Reference 2)

vehicles per mile per lane. The speed–flow plot shows a consistent data point pattern except when flows are over 1800 vehicles per hour per lane. It would appear that the optimum speed is not well defined and could lie anyway between 30 and 45 miles per hour.

Relationships between flow, speed, and density observed values for a freeway having a speed limit of 55 miles per hour are shown in Figure 10.4. These data were obtained from the Eisenhower Expressway near Harlem Avenue in Chicago. This roadway incorporates high design standards, operates under nearly ideal conditions, and is

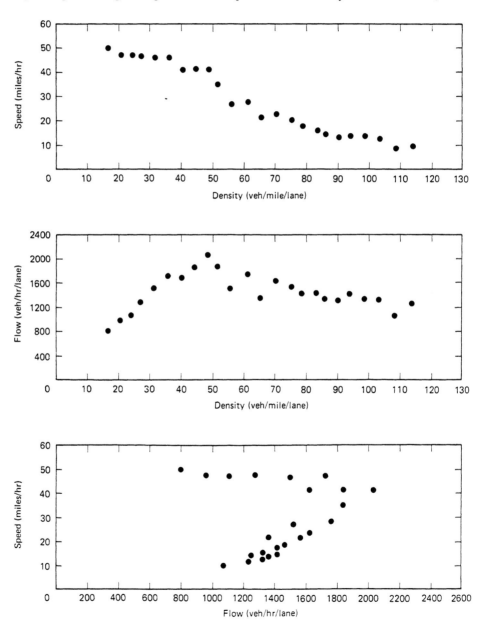

Figure 10.4   Freeway Data (55-mile/hr Speed Limit) (From Reference 1)

used by regular commuters. In comparison with the previous described data collection set, the Chicago data set was collected about 10 to 15 years earlier, and the speed limit of 55 miles per hour appeared to have a high degree of compliance. The data were automatically collected by the Illinois Department of Transportation as part of their freeway surveillance and control project. Measurements are averaged over 5-minute periods during the afternoon peak period. The speed–density plot shows a very consistent data point pattern and displays a S-shaped relationship with a discontinuity at a density of 50 to 55 vehicles per mile per lane. Although the sample provides a rather uniform distribution of observed densities over its range, observations were not recorded for density ranges below 15 or above 115 vehicles per mile per lane. The free-flow speed appears to be about 55 miles per hour, but the jam density cannot be estimated. The flow–density plot also shows a very consistent data point pattern. Like the first data set, while the free-flow portion is somewhat shaped like a parabola, the congested-flow portion is relatively flat with a tail to the right. The maximum flow or capacity appears to be about 2000 vehicles per hour per lane and occurs at an optimum density of about 50 vehicles per mile per lane. The speed–flow plot generally shows a consistent pattern except for two or three data points inside the enclosure curves and the absence of more than one data point at flows over 1850 vehicles per hour per lane. The optimum speed appears to be on the order of 40 miles per hour.

Relationships between observed values of flow, speed, and density in a tunnel are shown in Figure 10.5. These data were obtained in the Holland Tunnel in New York. This two-lane directional roadway is a tunnel underneath the Hudson River that connects New Jersey and New York. The design features are somewhat restrictive, with 11-foot lanes, no shoulders, and a typical tunnel alignment consisting of a downgrade followed by an upgrade. The data were automatically collected by the Port of New York Authority as part of their tunnel surveillance and control project. Measurements have been averaged over 5-minute periods. Drivers in the tunnel are regular commuters. The speed–density plot shows a very consistent data point pattern, but unlike the two earlier data sets, displays almost a linear configuration rather than a S-shaped relationship. Like earlier data sets, densities over 110 vehicles per mile per lane were not obtained. The free-flow speed appeared to be about 45 miles per hour, and the jam density could not be estimated. The shape of the flow–density plot is similar to the earlier flow–density plots with a parabola-type shape in the free-flow regime and a relative flat slope with a tail to the right in the congested-flow regime. The maximum flow or capacity was about 1350 vehicles per hour per lane and occurred at optimum densities of 50 to 60 vehicles per mile per lane. The speed–flow plot generally showed a consistent pattern in the free-flow regime, but the data points were clustered in the congested regime. There appears to be a bottleneck downstream which has a capacity on the order of 1100 to 1200 vehicles per hour per lane. Optimum speed appears to be about 25 miles per hour.

Relationships between observed values of flow, speed, and density for an arterial street are shown in Figure 10.6. This data set was obtained from a Newcastle University research team in the United Kingdom who are studying incident detection techniques on arterial streets. Data were collected from a detector located at the upstream

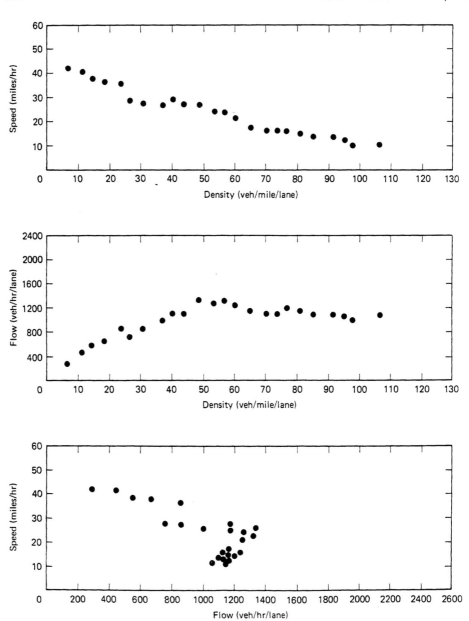

Figure 10.5   Tunnel Data (From Reference 3)

end of a link that had a signal at the downstream end. This detector was part of a traffic-responsive signal control system called "SCOOT." The data were collected for short periods, and unlike the earlier described controlled access freeway and tunnel sites, the environment around the detector site could be classified as uncontrolled. The

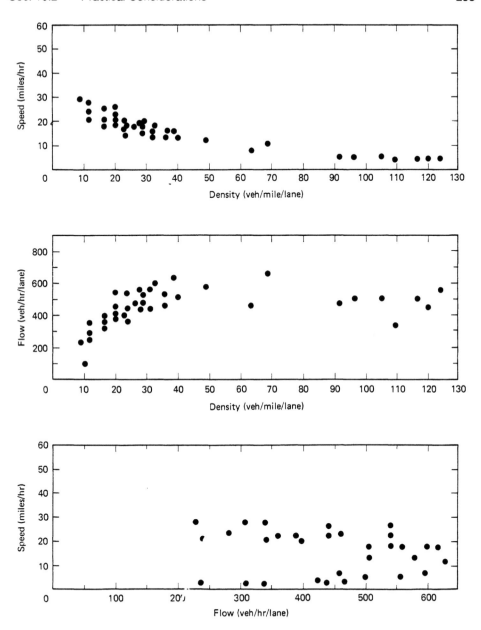

Figure 10.6   Arterial Street Data (From Reference 4)

speed–density plot shows a fairly consistent data point pattern except for the infrequent observations in the density range 40 to 90 vehicles per mile per lane. There appeared to be two distinctly different operational modes. One operational mode was when the site was unaffected by the downstream signal and free-flow conditions existed, and the other occurred when the queue from the signal backed into the detector site. Free-flow speeds were about 30 miles per hour and jam densities were not observed. The data point pattern was not S-shaped but exhibited a continuous decreasing slope with increased densities. The flow–density plot in the free-flow regime is similar to earlier flow–density plots with a parabola-type shape and the congested flow regime is relatively flat with a tail to the right. The maximum flow is between 600 and 700 vehicles per hour per lane. The scarcity of observations with densities between 40 and 90 vehicles per mile per lane provides suspicion that the site can never reach its capacity because of queue backups from the downstream signal. The speed–flow plot provides a less consistent pattern than that of the three earlier data sets. The highest flows can be identified, but the optimum speed is difficult to estimate.

The analysis of flow–speed–density relationships based on field measurements can be a very difficult task. Unique demand–capacity relationships over time of day and over length of roadway must be present. Even then the complete range of flow, speed, and density values will probably not be recorded. Parameter values of flow, speed, and density are often difficult to estimate and can greatly vary between sites. Many other factors affect flow–speed–density relationships, such as design speed, access control, presence of trucks, speed limits, number of lanes, and so on. Bridging the gap between theory and practice is a challenge to the theoretician and the professional. Having observed real-world flow–speed–density relationships, it is now appropriate to learn of proposed theoretical traffic stream models.

10.3 PROPOSED INDIVIDUAL MODELS

Over the years a number of traffic stream models have been proposed. The earlier models assumed a single regime phenomenon over the complete range of flow conditions includind free-flow and congested flow situations. Later models attempted to improve on the earlier models by considering two separate regimes (free-flow regime and congested-flow regime) and attempted to generalize by introducing additional parameters that could be used to distinguish between roadway environments.

10.3.1 Single-Regime Models

The first single-regime model was developed by Greenshields in 1934, based on observing speed–density measurements obtained from an aerial photographic study [5]. These measurements were presented in Figure 7.1 and Greenshields concluded that speed was a linear function of density. Section 10.1 was based on the assumption of a linear speed–density relationship, so all equations, illustrations, and derivations are for the Greenshields model. The model requires knowledge of the free-flow speed and jam density parameters in order to solve numerically for the speed–density relationship. The free-flow speed is relatively easy to estimate in the field and generally lies between the speed limit and the design speed of the roadway. On the other hand, the estimation and use of jam density present problems. Jam density values are difficult to obtain in the field, but a general value of 185 to 250 vehicles per mile per lane can be calculated assuming that a stopped vehicle occupies 21 to 28 feet of roadway space. The use of this jam density value also presents a problem because according to this model, optimum density is equal to one-half of the jam density value. This is not compatible with observed optimum density values on the order of 40 to 70 vehicles per mile per lane.

The Greenberg model was the second single-regime model that was proposed [6]. Observing speed–density data sets for tunnels, with particular attention to the congested portion such as shown in Figure 10.5, he concluded that a nonlinear model might be more appropriate. Using a hydrodynamic analogy he combined the equations of motion and continuity for one-dimensional compressible flow and derived the following equation:

$$u = u_o \ln \left(\frac{k_j}{k} \right) \qquad (10.9)$$

One of the important results of Greenberg's work was the bridge that was discovered between his proposed macroscopic model and the third General Motors car-following model. This bridge was described earlier in Chapter 6 and was the foundation for later discovery that almost all developed car-following theories could be related mathematically to most macroscopic traffic stream models [7, 8]. The Greenberg model requires knowledge of the optimum speed and jam density parameters. Like the Greenshields model, jam density is difficult to observe in the field, and estimating optimum speed is even more difficult than estimating free-flow speed. A crude estimate is that the optimum speed is approximately one-half of the design speed. Another disadvantage of this model is that free-flow speed is infinity. Later, Edie, recognizing this disadvantage, proposed a two-regime modeling approach with the Greenberg model being used for the congested regime [9].

The third single-regime model was proposed by Underwood as a result of traffic studies on the Merritt Parkway in Connecticut [10]. Underwood was particularly interested in the free-flow regime and was disturbed by free-flow speed going to infinity in the Greenberg model. Hence a new model was proposed as shown in the following equation:

$$u = u_f e^{-k/k_o} \qquad (10.10)$$

This formulation requires knowledge of the free-flow speed, which is fairly easy to observe, and the optimum density, which is difficult to observe and varies depending on the roadway environment. Another disadvantage of this model is that speed never reaches zero and jam density is infinity. Again Edie, recognizing this disadvantage, proposed a two-regime modeling approach with the Underwood model being used for the free-flow regime [9].

A fourth model was proposed by a group of researchers at Northwestern University when they observed that most speed–density curves appear as S-shaped curves [1]. The Northwestern group proposed the following equation:

$$u = u_f e^{-1/2(k/k_o)^2} \qquad (10.11)$$

This formulation appears related to the Underwood model in that knowledge of the free-flow speed and optimum density are required and also, speed does not go to zero when density goes to jam density.

Further development of single-regime models was directed toward the introduction of a parameter in the formulation which would provide for a more generalized modeling approach. For example, Drew proposed a formulation based on Greenshields' model, but with the introduction of an additional parameter n as shown in the following equation [11].

$$u = u_f \left[1 - \left(\frac{k}{k_j} \right)^{n+1/2} \right] \tag{10.12}$$

When the parameter n is set equal to 1, the formulation converts to the Greenshields model. However, varying the parameter n, a family of models can be developed. Drew suggested varying n from -1 to $+1$ and called these models a linear model ($n = +1$), a parabolic model ($n = 0$), and an exponential model ($n = -1$). Pipes-Munjal proposed a somewhat similar formulation that would provide a more generalized approach to single-regime models [12]. Their proposed formulation is shown by the following equation.

$$u = u_f \left[1 - \left(\frac{k}{k_j} \right)^{n} \right] \tag{10.13}$$

Again in this formulation when $n = 1$, the Greenshields model can readily be identified.

To study further the attributes of these single-regime models and to provide an opportunity to compare model characteristics, the initial four models were applied to the freeway data set (55-mile per hour speed limit) that was shown in Figure 10.4. The models could be applied in two different ways. The way the models were applied in this example was to determine the best regression fit for each model separately and then to observe the resulting flow parameter values. Another way would have been to select parameter values based on the inspection of the freeway data set, use these parameter values in applying each model, and observe how well each model fit the freeway data set by inspection and by calculating the mean deviations.

The resulting flow–speed–density relationships for each of the models based on the best regression fit are superimposed on the freeway data set in Figure 10.7. First, the resulting flow–speed–density relationships are discussed and then the model results are compared with the parameter values that appear to represent the field measured data set.

At density levels below 20 vehicles per mile per lane, the Greenberg and Underwood models overestimate speed. In the density range 20 to 60, all models underestimate speed and flow, and this is particularly disconcerting because of their estimation of capacity. As densities increase from 60 to 90, all models appear to track the field data reasonably well. At densities over 90 vehicles per mile per lane, the Greenshields model begins to deviate from the field data and at a density of 125, speeds and flows are predicted to approach zero.

Table 10.1 compares and summarizes the flow, speed, and density parameter values for each model, with estimates based on the field measured data set. All model

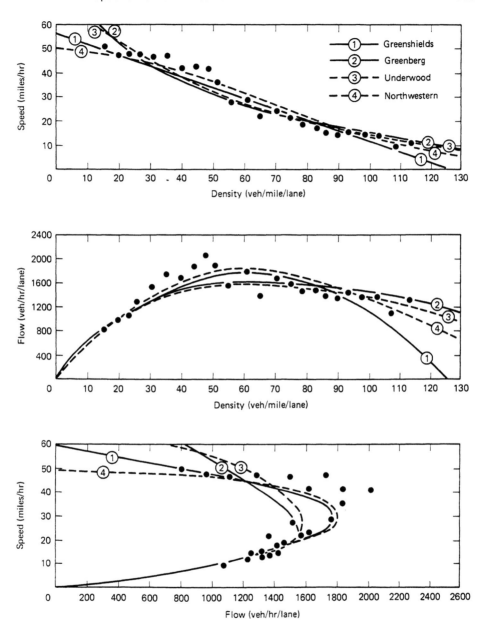

Figure 10.7   Single-Regime Models Superimposed on Freeway Data Set (From Reference 1)

predictions of maximum flow are less than the estimates from the data set. The Greenberg and Underwood models have the lowest predictions of maximum flow. In regard to free-flow speeds, the Greenberg and Underwood models predict much higher values

TABLE 10.1   Comparison of Flow Parameters for Single-Regime Models

Flow Parameter	Data Set	Single-Regime Models			
		Green-shields	Green-berg	Under-wood	North-western
Maximum flow, q_m	1800–2000	1800	1565	1590	1810
Free-flow speed, u_f	50–55	57	∞	75	49
Optimum speed, u_o	28–38	29	23	28	30
Jam density, k_j	185–250	125	185	∞	∞
Optimum density, k_o	48–65	62	68	57	61
Mean deviation	—	4.7	5.4	5.0	4.6

Source: Reference 1.

than would be expected from the data set. The Greenberg model appears to underestimate optimum speed. The Greenshields model significantly underestimates jam density, while the Underwood and Northwestern models predict jam densities of infinity. The Greenberg model slightly overestimates the optimum density. Finally, the Northwestern model exhibited the lowest mean deviation and the Greenberg model had the largest mean deviation.

In summary, four single-regime models have been described and then applied to a freeway data set. Each model had deficiencies over some portion of the density range. The most disconcerting feature of these models is their inability to track faithfully the measured field data near capacity conditions. One can observe a discontinuity in the flow–speed–density relationships as depicted by measured field data in the vicinity of capacity conditions. This has lead several researchers to propose two-regime models with separate formulations for the free-flow and congested-flow regimes.

10.3.2 Multiregime Models

Edie first proposed the idea of two-regime models in 1961 because of reservations of using car-following based models under free-flow conditions and his observation of the poor performance of the Greenberg model under free-flow conditions [9]. More specifically, Edie proposed the use of the Underwood model for the free-flow regime and the Greenberg model for the congested-flow regime. The equations for these two models were given earlier in the chapter as equations (10.9) and (10.10).

Supporting the idea of the use of multiregime models, a Northwestern University research team proposed three additional model formulations [1]. The first was the use of the Greenshields-type model for the free-flow regime and the congested-flow regime separately. The equation for the Greenshields model was shown as equation (10.1). The second proposed multiregime model suggested a constant-speed model for the free-flow regime and a Greenberg model for the congested-flow regime. The equation for the Greenberg model was shown as equation (10.9). The last proposed multiregime model suggested a three-regime model with the free-flow regime, transitional-flow regime, and congested-flow regime each being represented by the Greenshields formulation, as shown in equation (10.1).

The first difficulty in multiregime models is determining the breakpoint between regimes. The Northwestern researchers applied the work of Quandt on likelihood functions to identify the breakpoints between regimes for all four multiregime models [13, 14] using the earlier presented freeway data set. Then using regression analysis, the best model was selected for each regime. The resulting equations and breakpoints for the four multiregime models are presented in the Table 10.2.

TABLE 10.2   Equations and Breakpoints for Multiregime Models

Multiregime Model	Free-Flow Regime	Transitional-Flow Regime	Congested-Flow Regime
Edie model	$u = 54.9e^{-k/163.9}$ $(k \leq 50)$	—	$u = 26.8 \ln\left(\dfrac{162.5}{k}\right)$ $(k \geq 50)$
Two-regime linear model	$u = 60.9 - 0.515k$ $(k \leq 65)$	—	$u = 40 - 0.265k$ $(k \geq 65)$
Modified Greenberg model	$u = 48$ $(k \leq 35)$	—	$u = 32 \ln\left(\dfrac{145.5}{k}\right)$ $(k \geq 35)$
Three-regime linear model	$u = 50 - 0.098k$ $(k \leq 40)$	$u = 81.4 - 0.913k$ $(40 \leq k \leq 65)$	$u = 40.0 - 0.265k$ $(k \geq 65)$

Source: Reference 1.

To study the attributes of these multiregime models and to provide an opportunity to compare model characteristics, the equations and breakpoints shown in Table 10.2 are superimposed on the freeway data set in Figure 10.8. The multiregime models all track the freeway data set in a very reasonable manner and much better than any of the single-regime models. Table 10.3 compares and summarizes the flow, speed, and density parameter values for each model with estimates based on the field measured data set. In regard to maximum flow, the Edie model slightly overestimates while the other three models slightly underestimate. The linear two-regime slightly overestimates free-flow speed and the modified Greenberg slightly underestimates free-flow speed. The optimum speed is slightly overestimated by the Edie and three-regime linear model. All models underestimate jam density rather significantly. The three-regime linear model

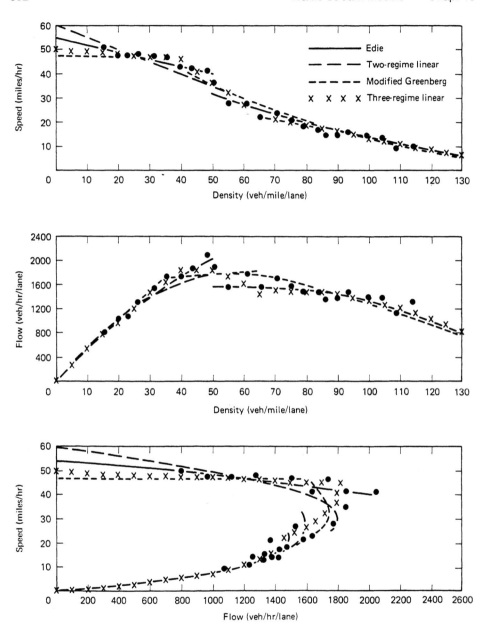

Figure 10.8   Multiregime Models Superimposed on Freeway Data Set (From Reference 1)

underestimates optimum density. All multiregime models have smaller mean deviations than those of the single-regime models. The Edie and three-regime linear models have the smallest mean deviations.

TABLE 10.3   Comparison of Flow Parameters for Multiregime Models

Flow Parameter	Data Set	Multiregime Model			
		Edie	Two-Regime Linear	Modified Greenberg	Three-Regime Linear
Maximum flow, q_m	1850–2000	2025	1800	1760	1815
Free-flow speed, u_f	50–55	55	61	48	50
Optimum speed, u_o	28–38	40	30	33	41
Jam density, k_j	185–250	162	151	146	151
Optimum density, k_o	48–65	50	59	54	45
Mean deviation	—	3.6	4.2	3.9	3.6

Source: Reference 1.

In summary, the multiregime models provide a considerable improvement over single-regime models. However, the multiregime models and particularly the single-regime models had different strengths and weaknesses. Further, each model appeared discretely different rather than as exhibiting a continuous spectrum of a family of models. Two earlier points set the stage for Section 10.4. Recall that the General Motors and Port of New York researchers found that the Greenberg model could be derived from the third car-following model [7, 8]. Is it possible that other traffic stream models are related to other car-following models? Also, recall that Drew [11] and Pipes [12] introduced the idea of generalizing traffic stream models into families of models by inserting the parameter n. By varying n a spectrum of traffic stream models could be formulated.

10.4 PROPOSED FAMILY OF MODELS

The sequential development of car-following theories was presented in Chapter 6 and the final generalized model is shown by the following expression [7, 8]:

$$\ddot{X}_{n+1}(t + \Delta t) = \frac{\alpha_{\ell,m}[\dot{X}_{n+1}(t + \Delta t)]^m}{[X_n(t) - X_{n+1}(t)]^\ell}[\dot{x}_n(t) - \dot{X}_{n+1}(t)] \qquad (10.14)$$

It was also shown in Chapter 6 that the Greenberg macroscopic model (discussed earlier in this chapter) could be derived from the third car-following model, in which the m and

ℓ values in equation (10.14) were 0 and 1, respectively [7, 8]. Figure 6.6 presented an m and ℓ matrix on which the various car-following models could be coordinated into a family of car-following models. The Greenberg macroscopic model can be placed on this matrix at m and ℓ values of 0 and 1, respectively.

It was hypothesized that if the Greenberg model was related to the third car-following model, then other macroscopic models might be related to other car-following models. If this was true, the macroscopic models could be connected into a continuous family of models through the m and ℓ matrix. The work of Gazis et al. [7, 8] and May and Keller [3, 15] demonstrated that almost all proposed macroscopic models were related to almost all car-following theories. This relationship is summarized in Figure 10.9 as an m and ℓ matrix. Both the earlier car-following models shown in Figure 6.6 and the four single-regime macroscopic models discussed earlier in this chapter are shown in Figure 10.9. Thus the Greenberg, Greenshields, Underwood, and Northwestern macroscopic models were equivalent to the generalized car-following model when m and ℓ values were 0 and 1, 0 and 2, 1 and 2, and 1 and 3, respectively. This exciting development not only provided a bridge between macroscopic and microscopic models but provided a framework for individually developed macroscopic models in a continuous spectrum of integer and noninteger models.

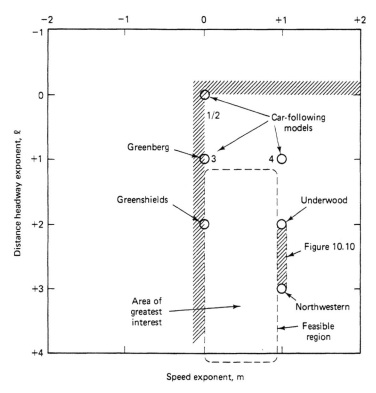

Figure 10.9   Car Following Theories and Macroscopic Models on m and ℓ Matrix

10.4.1 Single-Regime Models

As a simple illustration of this continuous spectrum of integer and noninteger models, consider the portion of the m and ℓ matrix where m is constant and equal to 1 while ℓ varies from 2 to 3 in steps of 0.2 [3]. By substituting these combinations of m and ℓ values into the generalized car-following model and then transforming them into the equivalent macroscopic forms, the resulting speed–density relationships are shown in Figure 10.10. The spectrum of speed–density relationships varies from the Underwood model ($m = 1, \ell = 2$) to the Northwestern model ($m = 1, \ell = 3$). Four additional noninteger models are also shown in Figure 10.10 ($m = 1$ with $\ell = 2.2, 2.4, 2.6,$ and 2.8).

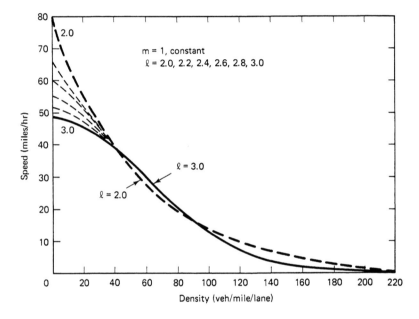

Figure 10.10   Influence of the Use of Non-Integer Exponents on the Speed–Density Relationship (From Reference 3)

In examining the m and ℓ matrix of Figure 10.9, one should recall from equation (10.14) that m is the exponent of the speed term in the numerator and ℓ is the exponent of the spacing term in the denominator. Since the speed term of the sensitivity component should remain in the numerator and the spacing term of the sensitivity component should remain in the denominator, then m and ℓ values should never be negative. Hence portions of the matrix shown in Figure 10.9 should not be considered.

Another limitation that can be imposed on the m and ℓ matrix is the desire for free-flow speed (u_f) and jam density (k_j) to have finite values and not go to infinity. For example, free-flow speeds go to infinity when $\ell \leq 1$ and jam densities go to infinity when $m \geq 1$. Therefore, considering these limitations and the nonnegative m and ℓ limitations in the preceding paragraph, the most feasible area on the m and ℓ matrix for consideration is when $0 \leq m < 1$ and $\ell > 1$. In this feasible region, the generalized speed–density equation can be shown to be

$$u^{1-m} = u_f^{1-m} \left[1 - \left(\frac{k}{k_j} \right)^{\ell-1} \right]$$

(10.15)

The Greenshields model lies within this feasible area $(m = 0, \ell = 2)$ and equation (10.15) becomes the well-recognized Greenshields model:

$$u = u_f \left(1 - \frac{k}{k_j} \right)$$

(10.16)

Researchers at the University of California have applied this m and ℓ matrix approach to a large number of data sets in order to select the *best* speed–density equation for each data set [2, 3, 15]. As an example of the approach undertaken, consider again the freeway data set shown in Figure 10.4. Each point (to the nearest 0.1 value) was analyzed in the m and ℓ matrix, and through regression analysis the *best*-fitting speed–density equation was determined and the mean deviation and resulting flow parameters calculated. With these results available for each m and ℓ point on the matrix, contour maps were constructed. These eight contour maps are shown as Figures 10.11 and 10.12.

To illustrate further the interpretation of these contour maps, consider again the Greenshields model $(m = 0$ and $\ell = 2)$. Inspection of these eight contour maps for the Greenshields model provides the following information.

- Free-flow speed, u_f 57 miles/hr
- Jam density, k_j 125 veh/mile/lane
- Optimum speed, u_o 29 miles/hr
- Optimum density, k_o 62 miles/hr
- Maximum flow, q_m 1800 veh/hr/lane
- Mean deviation   4.7 miles/hr
- Speed at $k = 20$ 49 miles/hr
- Speed at $k = 120$ 1 mile/hr

These are the same results as shown in Table 10.1.

Now two separate and sometimes conflicting criteria can be considered in selecting the *best* model to represent the freeway data set: best regression fit or acceptable flow parameters. For example, the best regression fit occurs when $m = 1.6$ and $\ell = 3.0$ and the mean deviation is slightly less than 4.2 miles per hour (see Figure 10.12b). However, with this best-fit model, some poor flow parameters result. For example, the jam density goes to infinity and the maximum flow would be less than 1700 vehicles per hour per lane. A compromise solution would be to select acceptable ranges of values in each contour map, overlay them, and see if any m and ℓ model penetrates all contour maps in an acceptable range. By inspection of the freeway data set shown in Figure 10.4, the following acceptable ranges might be selected for each contour map.

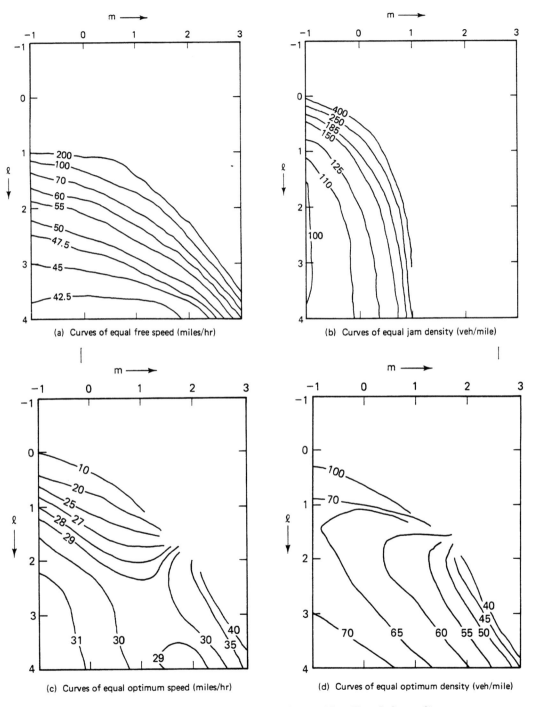

(a) Curves of equal free speed (miles/hr)

(b) Curves of equal jam density (veh/mile)

(c) Curves of equal optimum speed (miles/hr)

(d) Curves of equal optimum density (veh/mile)

Figure 10.11   Flow Parameter Contour Maps (From Reference 3)

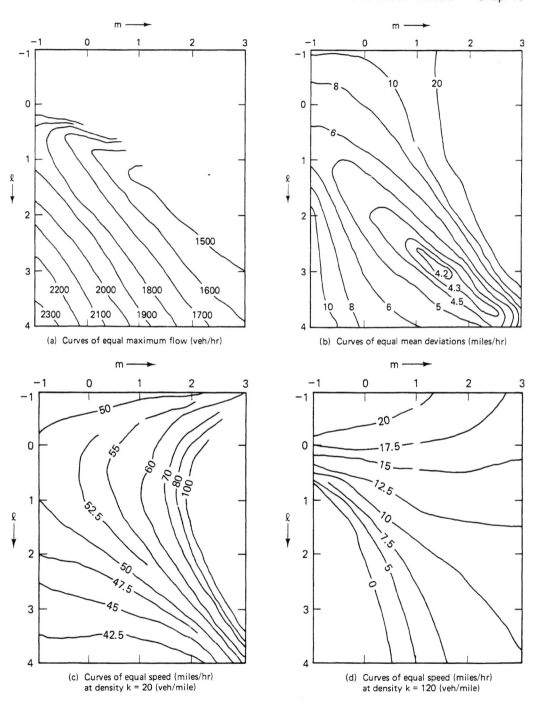

(a) Curves of equal maximum flow (veh/hr)

(b) Curves of equal mean deviations (miles/hr)

(c) Curves of equal speed (miles/hr) at density k = 20 (veh/mile)

(d) Curves of equal speed (miles/hr) at density k = 120 (veh/mile)

Figure 10.12   Additional Flow Parameter and Mean Deviation Contour Maps (From Reference 3)

- Free-flow speed u_f 50–55 miles/hr

- Jam density k_j 185–250 veh/mile/lane

- Optimum speed u_o 28–38 miles/hr

- Optimum density k_o 48–65 veh/mile/lane

- Maximum flow q_m 1800–2000 veh/hr/lane

- Mean deviation ≤ 4.5 miles/hr

- Speed at $k = 20$ 45–50 miles/hr

- Speed at $k = 120$ 6–12 miles/hr

Overlaying the contour maps, the only m and ℓ model that penetrates all contour maps in an acceptable range is when $m = 0.8$ and $\ell = 2.8$. The resulting speed–density equation is

$$u = u_f \left[1 - \left(\frac{k}{k_j} \right)^{1.8} \right]^5 \tag{10.17}$$

In later work at the University of California, the procedure described previously was applied to 32 data sets [2, 3, 15]. These data sets included the freeway and tunnel data sets shown earlier in this chapter and other freeway data sets from California, Pennsylvania, Virginia, and West Germany. The selected models for each of the 32 data sets are shown on the m and ℓ matrix of Figure 10.13. The selected models were positioned on the m and ℓ matrix between the Greenshields, Underwood, and Northwestern models. It was originally hoped that the position of a site data set on the m and ℓ matrix might be related to the site's characteristics. If so, in the future m and ℓ values could be selected on the basis of site characteristics. Unfortunately, the influence of site characteristics could not be clearly recognized.

Since acceptable ranges of the various flow parameters can be estimated for a site without too much difficulty while determining the best regression fit requires a data set for the site and considerable analysis, a nomograph procedure based on acceptable flow parameters was developed by Easa and May [16]. This nomograph is shown as Figure 10.14.

Another interesting application of the m and ℓ matrix approach to single-regime models is to determine the equation that best fits a unique specified flow–speed–density relationship. The 1985 *Highway Capacity Manual* [17] proposes a specific graphical flow–speed–density relationship for ideal freeway conditions, but without a specified equation. While not ideal, a single-regime model with m and ℓ values of 0.8 and 2.5, respectively, was fitted to the HCM graphs, which resulted in the equation

$$u = 60 \left[1 - \left(\frac{k}{260} \right)^{1.5} \right]^5 \tag{10.18}$$

This single-regime model is superimposed on the HCM graphs in Figure 10.15.

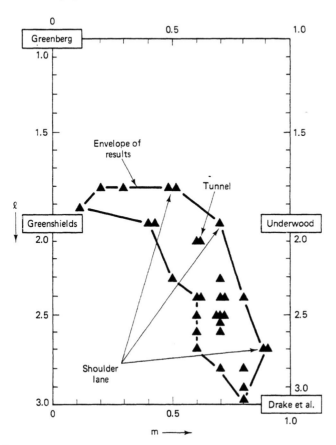

Figure 10.13   Selected Single-Regime Models on m and ℓ Matrix(From Reference 2)

The use of the m and ℓ matrix, and the concept of a spectrum of speed–density models are not limited to the single-regime approach. Such an approach for two-regime models is discussed in the next section.

10.4.2 Multiregime Models

The work on developing a family of single-regime models was extended to developing a family of two-regime models [3]. As discussed earlier and shown in the field-measured data sets, there appeared to be a discontinuity between the free-flow regime and the congested-flow regime. Further, even with the development of the family of single-regime models, the prediction of flows approaching capacity was not completely satisfying.

The procedures employed in this two-regime approach were identical to the previous single-regime approach, with the following exceptions. For the free-flow regime, only data points with density values less than 60 vehicles per mile per lane were included, and the jam density criterion of 185 to 250 vehicles per mile per lane was

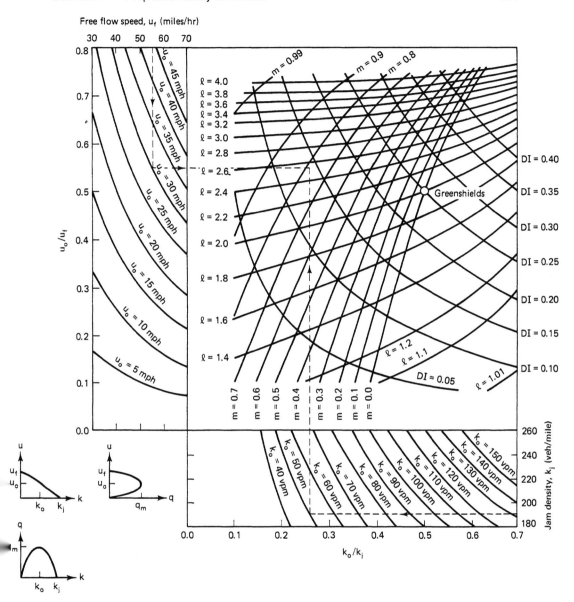

Figure 10.14   Nomograph for Selecting Single-Regime Models (From Reference 16)

removed. For the congested-flow regime, only data points with density values greater than 50 vehicles per mile per lane were included, and the free-flow speed and maximum flow criteria were removed. The overlapping breakpoints were selected based on visual inspection of the 32 data sets and the desire to have some overlap in the discontinuity zone.

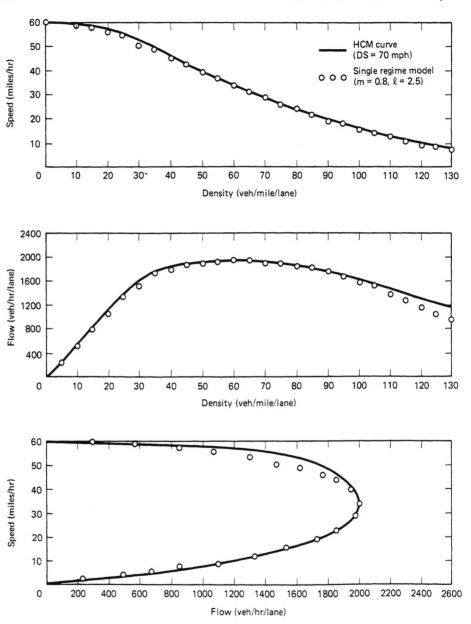

Figure 10.15   Single-Regime Model Superimposed on HCM Ideal Freeways Graphs
(Design Speed = 70 miles/hr)

The locations of the selected two-regime models in the m and ℓ matrix are shown in Figure 10.16. The free-flow models are located with ℓ values greater than 1.7 but fall into two groups in regard to the m value: some clustering around $m = 0$, while others are slightly less than $m = 1.0$. With the elimination of the free-flow speed and

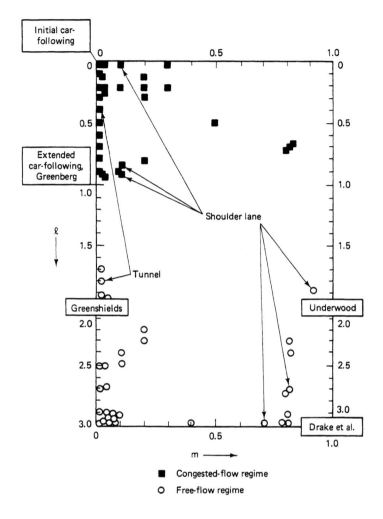

Figure 10.16   Selected Two-Regime Models on m and ℓ Matrix (From Reference 2)

maximum flow criteria, the congested-flow models clustered around m-values approaching zero and ℓ-values lying between 0 and 1. It is interesting to note that this portion of the m and ℓ matrix is located very close to the original car-following models.

As a summary to the discussion of single- and two-regime model families, the single-regime model and the two-regime model that were selected as the best ones for the freeway data set are shown in Figure 10.17, superimposed over the field-measured data set. As would be expected, the two-regime model represents the field-measured data set better than the single-regime model. However, there is still some concern about the accuracy of the predictions just below capacity flow. In hindsight, limiting the flow-free regime to densities values below 50 rather than 60 would have resulted in an improved free-flow regime model.

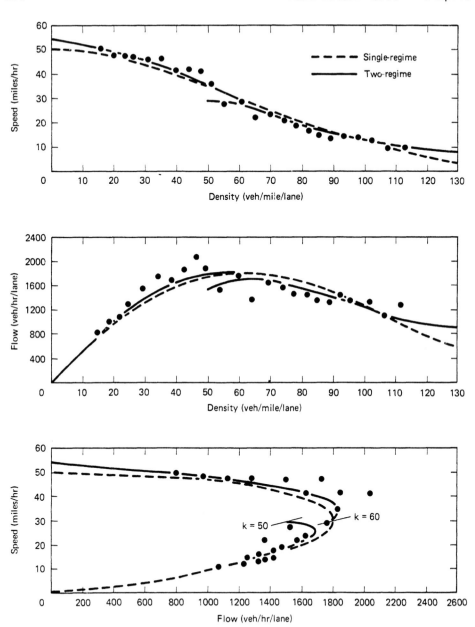

Figure 10.17   Selected Single-Regime and Two-Regime Models

Future research will extend current approaches and discover new approaches to the understanding of flow–speed–density relationships as incorporated into traffic stream models. Traffic stream models are essential for the planning, design, and operations of transportation facilities, particularly in theoretical investigations, control strategies, and capacity analysis.

10.5 PEDESTRIAN STREAM MODELS

An understanding of flow–speed–density relationships is also important in the planning, design, and operations of pedestrian facilities. There are a great many similarities between the flow phenomena of vehicles and pedestrians: Particles occupy so much space, require gaps between units, and their speeds depend on the open area in front of them. Behavior of particles in both vehicular and pedestrian streams vary among themselves as well as over time and space. Bottlenecks occur in both types of traffic streams, and their effects on upstream and downstream segments are similar.

The two major differences between vehicular and pedestrian stream flow are the numerical values of the flow characteristics and the use of lanes or width to define stream files. There are obviously numerical differences in characteristics, such as space occupied, gaps between particles, the influence of space on speeds, and the like. The other less obvious difference is that while on vehicular facilities there are well-defined integer numbers of lanes, on pedestrian facilities the lanes are less defined and the stream files may vary over time and flow situations. Perhaps the key description is that pedestrain streams can be compressed in the transverse dimension. Consequently, when dealing with pedestrian flows, the units employed for flow, speed, and density are different from vehicular flow as shown in Table 10.4. Hence, for pedestrian flow, the fundamental relationship that flow is equal to the product of speed and density becomes

$$v = S \times D \tag{10.19}$$

TABLE 10.4 Comparison of Vehicular and Pedestrian Units of Measure

Flow Characteristic	Units of Measure	
	Vehicular Streams	Pedestrian Streams
Flow	Vehicles per hour per lane (q)	Persons per minute per foot of width (v)
Speed	Miles per hour (u)	Feet per minute (S)
Density	Vehicles per mile per lane (k)	Persons per square foot (D)
Space	Feet per vehicle (on lane basis) (h_d)	Average square feet per person (M)

Since pedestrian densities (persons per square foot) are difficult to visualize and also are numerically very small fractions. space (*M*), which is defined as the reciprocal of density (*D*), is substituted into equation (10.19) and becomes

$$v = \frac{S}{M}$$

(10.20)

where v = flow (persons per minute per foot of width)

S = speed (feet per minute)

M = space (average square feet per person)

Most research into pedestrian stream models started with the assumption of a linear speed–density relationship but with different linear relationships for different types of pedestrians [18, 19]. The linear speed–density relationship is then converted to a speed–space relationship, and finally into flow–space and speed–flow relationships. These four relationships based on the examples given in the 1985 *Highway Capacity Manual* [18] are shown in Figure 10.18. Field studies have indicated a capacity value of approximately 25 persons per minute per foot of width regardless of pedestrian type.

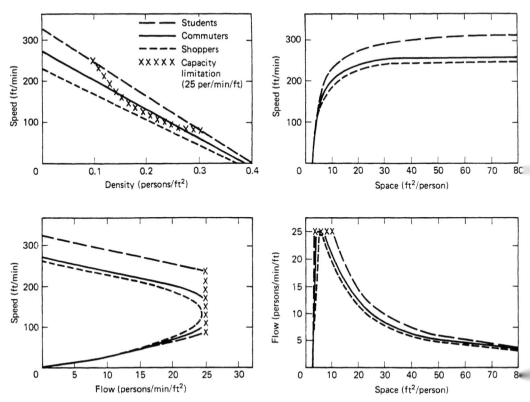

Figure 10.18   Pedestrian Stream Flow Diagrams

The relationships shown in Figure 10.18 have been modified to exhibit this capacity limitation. The resulting pedestrian stream equations and flow parameters for the three pedestrian types are summarized in Table 10.5.

TABLE 10.5 Equations and Flow Parameters for Pedestrian Stream Models

Characteristic	Student Pedestrians	Commuter Pedestrians	Shopper Pedestrians
Speed–density equation	$S = 320 - 800D$ ·	$S = 270 - 692D$	$S = 260 - 684D$
Speed–space equation	$S = 320 - \dfrac{800}{M}$	$S = 270 - \dfrac{692}{M}$	$S = 260 - \dfrac{684}{M}$
Flow–space equation	$v = \dfrac{320}{M} - \dfrac{800}{M^2}$	$v = \dfrac{270}{M} - \dfrac{692}{M^2}$	$v = \dfrac{260}{M} - \dfrac{684}{M^2}$
Speed–flow equation	$v = 0.4S - \dfrac{S^2}{800}$	$v = 0.39S - \dfrac{S^2}{692}$	$v = 0.38S - \dfrac{S^2}{684}$
Maximum flow (capacity)	32.0 person/min/ft (25 person/min/ft[a])	26.3 person/min/ft (25 person/min/ft[a])	24.7 person/min/ft
Free–flow speed	320 ft/min	270 ft/min	260 ft/min
Optimum speed	160 ft/min (85–235 ft/min)	135 ft/min (104–166 ft/min)	130 ft/min
Jam density	0.40 person/ft^2	0.39 person/ft^2	0.38 person/ft^2
Optimum density	0.20 person/ft^2 (0.11–0.29 person/ft^2 [a])	0.195 person/ft^2 (0.15–0.24 person/ft^2 [a])	0.19 person/ft^2
Jam space	2.50 ft^2/person	2.56 ft^2/person	2.63 ft^2/person
Optimum space	5.00 ft^2/person (3.4–9.4 ft^2/person[a])	5.12 ft^2/person (4.2–6.6 ft^2/person[a])	5.26 ft^2/person

[a]Assumed capacity limitation of 25 person/min/ft.

Source: Reference 18.

Considerable research has been directed to the development and application of pedestrian stream models. These investigations have considered such factors as the effect of width, grades, and pedestrian characteristics. The research has extended into such areas as stairways, escalators, and merging–crossing situations. The reader may wish to consult the literature in this field in order to understand more fully the complexities and challenges of pedestrian stream models [19].

10.6 SELECTED PROBLEMS

1. Determine flow–speed–density equations and flow parameters for the tunnel data shown in Figure 10.5. Visually select a linear speed–density curve from observing the data points.

2. Describe a location and circumstances when flow, speed, and density measurements could be obtained over their complete range of values.

3. Compare the two freeway data sets shown in Figures 10.3 and 10.4. Note differences and similarities in the flow–speed–density relationships and your estimates of their flow parameters.

4. Observe the presence and absence of discontinuities in the flow–speed–density relationships of the data sets shown in Figures 10.3 through 10.6. Identify discontinuities, and attempt to describe how they might have occurred.

5. Observe that the highest recorded density values in Figure 10.3 through 10.6 are on the order of 110 to 130 vehicles per mile per lane. On the other hand, jam density estimates are on the order of 185 to 250 vehicles per mile per hour. Why is it difficult to obtain observations in this density range? How would you design a field study to collect data in this density range?

6. Estimate flow parameters visually from the tunnel data set shown in Figure 10.5. Apply the Greenshields, Greenberg, Underwood, and Northwestern single-regime models to the data set. Show results in a form similar to Figure 10.7. Based on appearance, comment as to how well each of the single-regime models fits the tunnel data set.

7. Where feasible determine the optimum-speed/free-flow speed ratio and the optimum density/jam density ratio for the Greenshields, Greenberg, Underwood, and Northwestern single-regime models. These ratios sometimes are helpful in selecting the most appropriate model for a given data set.

8. The optimum speed and the optimum density for the high-speed freeway data set is assumed to be 40 miles per hour and 50 vehicles per mile per lane, respectively. Determine the equations for the Greenberg, Underwood, and Northwestern models, and construct flow–speed–density relationships in a fashion similar to Figure 10.7.

9. Apply the Edie two-regime model to the tunnel data set shown in Figure 10.5. Select what you consider to be appropriate flow parameters and breakpoint location between the free-flow regime and the congested-flow regime. Summarize your results in a fashion similar to Figure 10.8.

10. Continue work on Problem 9 by performing a sensitivity analysis of the effect of flow parameters and breakpoint location on the suitability of the resulting flow–speed–density relationships.

11. Derive the following macroscopic models from the appropriate car-following equations: (a) Greenshields, (b) Underwood, and (c) Northwestern. (*Hint:* Refer to derivation of Greenberg model in Chapter 6.)

12. Construct a diagram like Figure 10.10 except cover the portion of the m and ℓ matrix where $m = 0$ and l lies between 1 (Greenberg) and 2 (Greenshields). Select required flow parameters from inspection of freeway data set shown in Figure 10.4.

13. The feasible region on the m and ℓ matrix for single-regime models is shown in Figure 10.9. Reproduce Figure 10.9 and denote areas on the m and ℓ matrix which are feasible areas for the free-flow portion and the congested-flow portion of two-regime models.

14. Inspection of a freeway data set such as shown in Figure 10.3 reveals a free-flow speed of 60 miles per hour, a jam density of 180 vehicles per mile per lane, and a maximum flow of

2000 vehicles per hour per lane. Consider the area of the m and ℓ matrix when $0 \leq m < 1$ and $l > 1$. Construct two contour maps like those shown in Figure 10.11 and 10.12: one for optimum speed (u_o) and the other for optimum density (k_o). Considering the optimum speed and optimum density that result, what m and ℓ values appear to be the most appropriate by inspection of Figure 10.3?

15. Identify the advantages and disadvantages of selecting the Greenberg, Greenshields, Underwood, and Northwestern single-regime models to represent the freeway data set by interpreting the contour maps shown in Figures 10.11 and 10.12.

16. The traffic stream model with m and ℓ values of 0.8 and 2.8, respectively, was selected on the basis of the contour maps shown in Figures 10.11 and 10.12. Comment on the suitability of the resulting parameters. Which parameters are the most constraining? Which are the least constraining?

17. Using the nomograph shown in Figure 10.14, select the best m and ℓ models for the data sets presented in Figures 10.3 through 10.6. For selected models in the feasible region of Figure 10.9, solve for the resulting macroscopic equations and superimpose the curve on the data set figures.

18. A single-regime macroscopic model was fitted to the 1985 *Highway Capacity Manual* flow–speed–density relationship shown in Figure 10.15. Perform analysis to attempt to improve the macroscopic single-regime model that could be used to represent the *Highway Capacity Manual* relationships in Figure 10.15.

19. For capacity and level of service analysis, the free-flow portion of the flow–speed–density relationships shown in Figure 10.15 is of the greatest interest. Through analysis, select the macroscopic model for the free-flow regime that best fits the relationships shown in Figure 10.15. Superimpose your selected free-flow regime model on a figure similar to Figure 10.15.

20. An 8-foot-wide directional pedestrian-way for commuters connects two subway lines. Between trains the flow in the pedestrian-way is almost nonexistent, while the volume–capacity ratio increases to 0.9 shortly after a train arrival. Some subway reconstruction is planned which will reduce the capacity of the middle portion of the pedestrian-way to an effective 6-foot width. Plot numerical flow–speed–density relationships similar to Figure 10.2 for (a) station A; (b) station B; (c) station C; and (d) station D.

10.7 SELECTED REFERENCES

1. J. S. Drake, J. L. Schofer, and A. D. May, Jr., A Statistical Analysis of Speed Density Hypotheses, *in Third International Symposium on the Theory of Traffic Flow Proceedings*, Elsevier North Holland, Inc., New York, 1967.

2. Avishai Cedar and Adolf D. May, *Further Evaluation of Single- and Two-Regime Traffic Flow Models*, Transportation Research Board, Record 567, TRB, Washington D. C., 1976, pages 1–15.

3. A. D. May, Jr. and H. E. M. Keller, *Non-integer Car-Following Models*, Highway Research Board, Record 199, HRB, Washington D. C., 1967, pages 19–32.

4. Provided through personal correspondence from Dr. Michael Bell, Newcastle University, summer 1987.

5. B. D. Greenshields, A Study in Highway Capacity, *Highway Research Board, Proceedings*, Vol. 14, 1935, page 458.

6. H. Greenberg, An Analysis of Traffic Flow, *Operations Research,* Vol. 7, 1959, pages 78–85.

7. D. C. Gazis, R. Herman, and R. Potts, Car-Following Theory of Steady-State Traffic Flow, *Operations Research,* Vol. 7, 1959, pages 499–595.

8. D. C. Gazis, R. Herman, and R. W. Rothery, Nonlinear Follow-the-Leader Models of Traffic Flow, *Operations Research,* Vol. 9, 1961, pages 545–567.

9. L. C. Edie, Following and Steady-State Theory for Non-congested Traffic, *Operations Research,* Vol. 9, 1961, pages 66–76.

10. R. T. Underwood, *Speed, Volume, and Density Relationships, Quality and Theory of Traffic Flow,* Yale Bureau of Highway Traffic, New Haven, Conn., 1961, pages 141–188.

11. Donald R. Drew, *Traffic Flow Theory and Control,* McGraw-Hill Book Company, New York, 1968, Chapter 12.

12. L. A. Pipes, Car-Following Models and the Fundamental Diagram of Road Traffic, *Transportation Research,,* Vol. 1, No. 1, 1967, pages 21–29.

13. R. E. Quandt, The Estimation of the Parameters of a Linear Regression System Obeying Two Separate Regimes, *Journal of the American Statistical Association,* Vol. 53, 1958, pages 873–880.

14. R. E. Quandt, Tests of the Hypothesis That a Linear Regression System Obeys Two Separate Regimes, *Journal of the American Statistical Assocation,* Vol. 25, 1960, pages 324–330.

15. A. D. May, Jr. and H. E. M. Keller, Evaluation of Single-and Two-Regime Traffic Flow Models, *Fourth International Symposium on the Theory of Traffic Flow Proceedings,* Karlsruhe, West Germany, 1968.

16. S. M. Easa and A. D. May, *Generalized Procedure for Estimating Single- and Two-Regime Traffic-Flow Models,* Transportation Research Board, Record 772, TRB, Washington, D.C., 1980, pages 24–37.

17. Transportation Research Board, *Highway Capacity Manual,* Special Report 209, TRB, Washington D. C., 1985, Chapter 3.

18. Transportation Research Board, *Highway Capacity Manual,* Special Report 209, TRB, Washington D. C., 1985, Chapter 13.

19. John J. Fruin, *Pedestrian Planning and Design,* New York Metropolitan Association of Urban Designers and Environmental Planners, New York, 1971, 206 pages.

20. Daniel L. Gerlough and Matthew J. Huber, *Traffic Flow Theory—A Monograph,* Transportation Research Board, Special Report 165, TRB, Washington D. C., 1975, Chapter 4.

11

Shock Wave Analysis

Flow–speed–density states change over space and time. When these changes of state occur a boundary is established that demarks the time–space domain of one flow state from another. This boundary is referred to as a shock wave. In some situations the shock wave can be very mild, like a platoon of high-speed vehicles catching up to a slightly slower moving vehicle. In other situations the shock wave can be a very significant change in flow states, as when high-speed vehicles approach a queue of stopped vehicles.

An introduction to shock waves was presented in Chapter 7, and the reader may wish to review that material before proceeding further in this chapter. The introduction was limited to qualitative analysis, with primary attention given to flow discontinuities or shock waves demarking boundaries between free-flow regimes and congested-flow regimes. Two simple qualitative examples of shock waves were described: a signalized intersection and a lane-drop location. The introduction concluded with further description of shock waves and a classification of types of shock waves which included:

- Frontal stationary
- Backward forming*
- Forward recovery*
- Rear stationary
- Backward recovery*
- Forward forming*

In this chapter, attention will be directed toward quantitative analysis of all types of shock waves. After a brief historical perspective, the chapter begins with the

---

*Using more general terminology, the phrase "-moving" can be substituted for "-forming" and "-recovery."

derivation of shock wave equations. Then three examples are presented to illustrate the application of shock wave theory. The first example is for a traffic signal location and the second at the site where a slow-moving truck enters a highway. The last example is of a pedestrian-way in a bottleneck situation. The text of the chapter concludes with a discussion of shock wave analysis complexities and refinements. Selected problems and references are included at the end of the chapter.

11.1 AN HISTORICAL PERSPECTIVE

Richards [1] appears to have published the first paper on shock wave analysis applied to transportation facilities. In his paper he assumed a linear speed–density relationship and gave particular attention to the discontinuity of density. In the following year, Lighthill and Whitham [2] published their renowned thesis on a theory of traffic flow on long, crowded roads. It is still considered a monumental document and highly recommended. It covers a wide selection of topics, including discussions on flow–concentration curves, theory of bottlenecks, and flow at junctions. A research group at Airborne Instruments Laboratory concerned with traffic congestion studied backward waves [3].

In the late 1950s and early 1960s, several other researchers gave their attention to shock wave analysis. These included Greenberg, Edie, and Foote [4, 5] with their work in the New York tunnels; Franklin [6] with his proposed structure for traffic shock waves; and Pipes [7] with his fine summary paper reviewing previous work.

By the late 1960s several major textbooks were published. These included a Transportation Research Board publication entitled *An Introduction to Traffic Flow Theory* [8], which included the earlier Lighthill and Whitham [2] paper and the Pipes paper [7]. Wohl and Martin, in their textbook *Traffic System Analysis for Engineers and Planners* [9], discuss hydrodynamic analogies and their applications. In his textbook *Traffic Flow Theory and Control* [10], Drew reviewed the previous work on shock waves, proposed an approach to bottleneck control, and finally introduced the concepts of momentum-kinetic energy and internal energy.

Pipes [11, 12] continued his work with hydrodynamic theories describing wave phenomenon. Rorbech [13], concerned with spill-backs from signalized intersections, provided design procedures for the length of approach lanes. Gerlough and Huber prepared an excellent state-of-the-art chapter on hydrodynamic and kinematic traffic models in their monograph on *Traffic Flow Theory* [14].

The most recent research publications have been concerned with the application of shock wave analysis to signalized intersections and to highway facilities such as freeways. Michalopoulos and associates [15, 16] have reported on their research in applying shock wave analysis to signalized intersections, with particular concern with queue dynamics and the effects of signal control. Leutzbach and Kohler, Wirasinghe, and May have applied shock wave analysis to highways. Leutzbach and Kohler [17] have used shock wave analysis to estimate delay due to incidents. Wirasinghe [18] has applied shock wave analysis to estimate individual and total delays using graphical means. May and associates [19, 20] have developed and continuously improved a free-way simulation model called FREQ, which includes the modeling of congestion employing shock wave theory.

11.2 SHOCK WAVE EQUATIONS

Consider an uninterrupted segment of roadway for which a flow–density relationship is known. Such a flow–density relationship is shown in Figure 11.1a. For some period of time, a steady-state free-flow condition exists, as noted on the flow–density diagram as state A. The flow, density, and speed of state A are denoted as q_A, k_A, and u_A, respectively. Then, for the following period of time, the input flow is less and a new steady-state free-flow condition exists, as noted on the flow–density diagram as state B. The flow, density, and speed of state B are denoted as q_B, k_B, and u_B, respectively. Note that in state B, the speed (u_B) will be higher, and these vehicles will catch up with vehicles in state A over space and time.

To visualize this more clearly, two additional sketches are included in Figure 11.1. Figure 11.1b is an illustration of the flow states in a distance–time diagram. Note that the scales of distance and time are selected in such a manner that the rays representing speeds in the d–t diagram are parallel to the rays representing speeds in the q–k diagram. The heavy line identifies the shock wave or discontinuity between states A and B, or stated in simple terms, it is the distance–time trace where the higher-speed vehicles in flow state B join the lower-speed vehicles in flow state A. The symbol ω is used to represent the shock wave and the subscripts A and B denote that the shock wave lies between flow states A and B.

Figure 11.1c shows a slice of the roadway at time t which depicts the two flow states. Three speeds with motion to the right are indicated: u_B, the vehicular speed in flow state B; u_A, the vehicular speed in the downstream flow state A; and ω_{AB}, the shock wave speed between the two flow states. In this case, the direction of the shock wave speed is fairly clear, but in more complicated situations, where the direction may not be clear, the analyst should assume the shock wave as a forward-moving shock wave (moving in the direction as traffic). That is, assume a positive shock wave speed. If, upon completing the analysis, the shock wave speed becomes negative, the shock wave is a backward-moving shock wave and thus moving in the opposite direction as traffic.

At the shock wave boundary, the number of vehicles leaving flow condition B (N_B) must be exactly equal to the number of vehicles entering flow condition A (N_A) since no vehicles are destroyed nor created. The speed of vehicles in flow condition B just upstream of the shock wave boundary relative to the shock wave speed is ($u_B - \omega_{AB}$). The speed of vehicles in flow condition A, just downstream of the shock wave boundary relative to the shock wave speed, is ($u_A - \omega_{AB}$). Therefore, N_B and N_A can be calculated using the following equations:

$$N_B = q_B t = (u_B - \omega_{AB}) k_B t \tag{11.1}$$

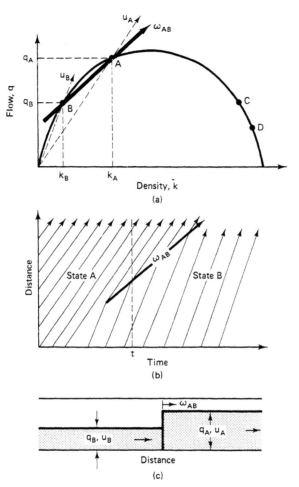

Figure 11.1   Shock Wave Analysis Fundamentals

$$N_A = q_A t = (u_A - \omega_{AB}) k_A t \tag{11.2}$$

Setting $N_B = N_A$ and solving for ω_{AB}, equations (11.1) and (11.2) become

$$(u_B - \omega_{AB}) k_B t = (u_A - \omega_{AB}) k_A t \tag{11.3}$$

and

$$\omega_{AB} = \frac{q_A - q_B}{k_A - k_B} = \frac{\Delta q}{\Delta k} \tag{11.4}$$

Therefore, the shock wave speed between two states is equal to the change in flow divided by the change in density. With flows expressed in vehicles per hour and densities expressed in vehicles per mile, the shock wave speed is in units of miles per hour. These units of measurement are confirmed by the sketches in Figure 11.1. It should also be noted that since the shock wave speed is equal to the ratio of the change in flow to the change in density, the shock wave can be depicted in the top sketch of Figure

11.1 as the slope of the line connecting flow conditions A and B. Since q_A is larger than q_B and k_A is larger than k_B, equation (11.4) results in a positive shock wave velocity. This can also be seen in the sketches of Figure 11.1 since the slope of the line representing the shock wave speed is up to the right, indicating a positive shock wave speed moving in the direction of traffic. Recall that a shock wave moving in the direction of traffic is called a forward-moving shock wave.

Now consider stationary and backward-moving shock waves in light of the sketches of Figure 11.1. In these two examples, flow state B will remain unchanged, while flow state A will be moved to the right side of the flow–density curve marked C and D. First consider flow condition C replacing flow condition A. In that situation, q_C is equal to q_B, and equation (11.4) indicates a shock wave speed of zero. This is confirmed by Figure 11.1a, since the line joining flow condition B with flow condition C is a horizontal line. This is called a stationary shock wave.

Now consider flow condition D replacing flow condition A. In this situation q_B is greater that q_D, while k_B is less than k_D and equation (11.4) indicates a negative shock wave. This is confirmed by Figure 11.1a, since the line joining flow conditions B and D is a downward sloping line. This is called a backward-moving shock wave.

The theory developed thus far assumes an instantaneous change in flow states and implies that even when the roadway is free of traffic downstream, the drivers of lead vehicles in a particular flow state will not increase their speeds. These limitations and possible refinements are discussed in the last section of this chapter.

11.3 SHOCK WAVES AT SIGNALIZED INTERSECTIONS

Shock wave analysis at signalized intersections is a common application because of the concern for the length of queues interfering with upstream flow movements. Examples include queues extending out of left-turn lanes into through traffic lanes and queues extending upstream to block adjacent intersections.

Shock waves at signalized intersections can be analyzed if a flow–density relationship is known for the approach to the signalized intersection and if the flow state of the approaching traffic is specified. For this example a flow–density curve and the approaching traffic flow state (A) are shown in Figure 11.2a. A distance–time diagram is shown in Figure 11.2b with distance and time scales selected so that a given slope in the two diagrams represent a specific vehicle or shock wave speed. The stop line on this approach to the signalized intersection is located at the traffic signal band, with green and red phases indicated as light and dark strips, respectively.

During time t_0 to t_1, the signal is green and traffic proceeds on the approach, through the intersection, and downstream under flow state A (q_A, u_A, and k_A). The trajectory of individual vehicles are shown as dashed lines. At time t_1, the traffic signal changes to red and the flow state immediately upstream of the stop line changes to state B while the flow state immediately downstream changes to state D. Three shock waves begin at time t_1 at the stop line: ω_{AD}, a forward-moving shock wave; ω_{DB}, a frontal stationary shock wave; and ω_{AB}, a backward-moving shock wave. The speeds of these

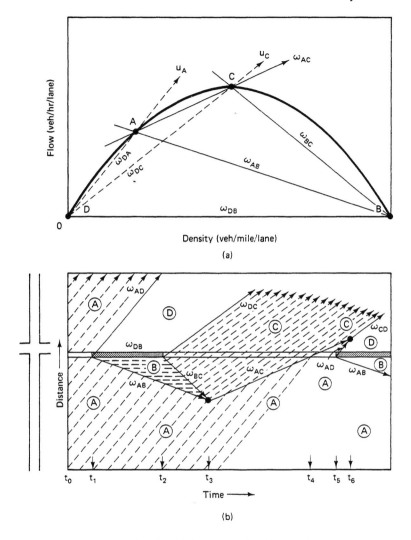

Figure 11.2 Shock Waves at Signalized Intersections

three shock waves are depicted on the flow–density diagram of Figure 11.2a and can be calculated using the following equations.

$$\omega_{DA} = \frac{q_D - q_A}{k_D - k_A} = + u_A \tag{11.5}$$

$$\omega_{DB} = \frac{q_D - q_B}{k_D - k_B} = 0 \tag{11.6}$$

$$\omega_{AB} = \frac{q_A - q_B}{k_A - k_B} = -\frac{q_A}{k_B - k_A} \tag{11.7}$$

These flow states of A, B, and D continue until time t_2 when the signal changes to green. A new flow state is introduced (flow state C) at time t_2 at the stop line when the flow at the stop line increases from 0 to saturation flow. This causes two new shock waves, ω_{DC} and ω_{BC}, while terminating shock wave ω_{DB}. The speeds of these two new shock waves can be graphically seen in the flow–density diagram of Figure 11.2a and calculated using the following equations:

$$\omega_{DC} = \frac{q_D - q_C}{k_D - k_C} = +u_C \tag{11.8}$$

$$\omega_{BC} = \frac{q_B - q_C}{k_B - k_C} = -\frac{q_C}{k_B - k_C} \tag{11.9}$$

The flow states of D, C, B, and A continue until ω_{AB} and ω_{BC} intercept at time t_3. The time interval between t_2 and t_3 can be calculated to be

$$t_3 - t_2 = r \left(\frac{\omega_{AB}}{\omega_{BC} - \omega_{AB}} \right) \tag{11.10}$$

where r is the effective duration of the red phase. The location of the queue dissipation at time t_3 can also be calculated by means of the equation

$$Q_M = \frac{r}{3600} \left[\frac{(\omega_{BC})(\omega_{AB})}{\omega_{BC} - \omega_{AB}} \right] \tag{11.11}$$

At time t_3 a new forward-moving shock wave ω_{AC} is formed, and the two backward-moving shock waves, ω_{AB} and ω_{BC}, are terminated.

The shock wave ω_{AC} is shown in Figure 11.2 and calculated using the equation

$$\omega_{AC} = \frac{q_A - q_C}{k_A - k_C} \tag{11.12}$$

The flow states of D, C, and A continue until time t_5; but first consider time t_4. At time t_4, the forward-moving shock wave ω_{AC} crosses the stop line, and the flow at the stop line goes from a maximum flow of q_C to the arrival flow of q_A. The period of time from the start of the green phase until the stop-line discharge rate drops below its maximum value (t_2 to t_4) can be calculated to be

$$t_4 - t_2 = \frac{r(\omega_{AB})}{\omega_{BC} - \omega_{AB}} \left(\frac{\omega_{BC}}{\omega_{AC}} + 1 \right) \tag{11.13}$$

At t_5, which is the beginning of the red phase, the shock wave pattern upstream of the signal begins to repeat itself. However, the shock wave pattern downstream of the signal deviates from the earlier pattern. Note that at the beginning of the red, shock wave ω_{AD} is formed, but it travels downstream only until it intercepts shock wave ω_{AC}. At time t_6, the ω_{AC} and ω_{AD} shock waves terminate, and a new shock wave, ω_{CD}, is created. As long as the traffic demand and signal timing plan remain unchanged, the shock wave pattern will repeat itself every signal cycle.

11.4 SHOCK WAVES ALONG A HIGHWAY

In their monograph entitled *Traffic Flow Theory* [14], Gerlough and Huber attributed the following example to Edie. A steady flow state existed when a slow-moving truck entered the roadway and proceeded to travel at a constant lower speed for some distance before turning off the roadway. It is assumed that passing is not possible. The objective of the example problem is to construct a distance–time diagram on which all shock waves are shown, as well as distance–time traces of selected vehicles.

A flow–density diagram and distance–time diagram are shown in Figure 11.3. Prior to the truck turning onto the highway, the flow condition is represented by flow state A. At time t_1 the truck enters the highway and travels at a reduced speed of μ_B, causing the flow state behind the truck to go to B and resulting in the flow state D occurring in front of the truck. Two forward-moving shock waves and the truck trajectory begin at time t_1 at the location where the truck enters the highway: ω_{AD}, ω_{AB}, and $u_B(\omega_{DB})$.

At time t_2 the truck leaves the highway, and now with the unrestricted flow conditions ahead, the traffic can operate at capacity (flow state C). Two new shock waves begin at time t_2 at the location where the truck leaves the highway: ω_{DC} and ω_{BC}. The shock wave, ω_{DC}, is a forward-moving shock wave, while ω_{BC} is a backward-moving shock wave.

At time t_3 the shock waves ω_{BC} and ω_{AB} collide and terminate, which results in another discontinuity between flow states C and A. This establishes a new forward-moving shock wave ω_{CA}.

The total picture of the trajectories of vehicles before, during, and after the truck travel is shown in the lower portion of Figure 11.3. The result is the formation of five forward-moving shock waves and one backward-moving shock wave. The truck not only affects the traffic stream when it is present on the roadway, but for sometime thereafter. Imagine what the driver of the vehicle denoted as "x" experiences. When this driver passes the point where the truck had entered, the truck had just exited from the highway. The driver slows down and travels at the truck speed without knowing why. A little later the driver speeds up some, but not as fast as his original speed. In summary, this driver experiences congestion at a time period after the truck has exited and without knowledge as to what caused the congestion.

Now consider some variations of this problem. Two situations will be considered. In the first situation, all flow states will be assumed to remain the same except that flow state A will vary from flow state D to flow state C. If flow state A were assumed to be near flow state D, the only significant change would be that shock wave ω_{AB} would approach the speed of the truck u_B and the flow states B and C would be significantly reduced. In other words, if the approach flow was extremely small, the truck would have little influence on roadway operations.

Now consider the flow state A moving up the free-flow portion of the flow–density curve until its flow is the same as the flow for flow state B ($q_A = q_B$). The shock wave ω_{AB} would become a stationary shock wave at the point where the truck had entered the roadway. Flow states B and C would be larger than shown in Figure

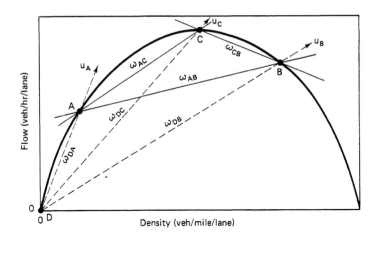

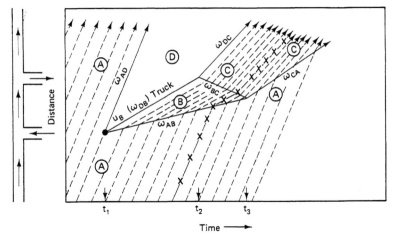

Figure 11.3   Shock Waves along a Highway

11.3. If the flow state A moved closer to the flow state C, the shock wave ω_{AB} would become a backward-moving shock wave, and the shock wave ω_{CA} would become a much slower forward-moving shock wave. The result would be a greatly increased space–time domain with flow states B and C. The extreme case would be if the roadway were operating at capacity when the truck entered, the shock waves ω_{AB} and ω_{BC} would be equal backward-moving shock waves and congested flow would never end.

Consider now the second situation. In this situation, all flow states will be assumed to remain the same except that flow state B (caused by the truck) will vary over the complete range of the flow density diagram. If the truck enters the roadway at a speed between free-flow speed and u_A, the truck will obviously not delay the other vehicles in the traffic stream. (In fact, the truck will be delayed by the other vehicles.)

If the truck enters the roadway at a speed between u_A and u_C, the distance–time diagram will be somewhat similar to Figure 11.3, except that ω_{DB} and ω_{AB} would be steeper, and ω_{BC} would be a slow-moving forward shock wave. The shock wave ω_{BC} would not exist if the truck speed was u_C.

Finally, consider that the truck enters the roadway at a speed close to zero. The shock wave ω_{AB} would be a rather fast-moving backward shock wave, and the space–time domain of flow states B and C would be significantly increased.

11.5 SHOCK WAVES ALONG A PEDESTRIAN-WAY

The application of shock wave analysis is not limited to highway vehicular traffic. Consider the discussion of pedestrian stream models in Chapter 10 and the following example of shock wave analysis along a pedestrian-way.*

A pedestrian-way for commuters connects two subway lines. Between trains, the flow in the pedestrian-way is almost nonexistent, while the flows increase almost to capacity after a train arrival. Some subway reconstruction is planned, which will reduce the capacity of the middle portion of the pedestrian-way. Given are the flow–density diagram shown in Figure 11.4a, the demand flow (flow state A) after a train arrival, and the reduced capacity (q_B) due to the subway reconstruction project. The objective of the example problem is to construct a distance–time diagram on which all shock waves are shown as well as selected distance–time projectories of pedestrians for that portion of the pedestrian-way upstream of the reconstruction area.

A subway train arrives and the first pedestrian reaches the reconstruction area at time t_1. The flow state just prior to t_1 is flow state D. The pedestrian arrival flow rate becomes q_A and immediately at t_1, the pedestrians begin to queue up going into the reconstruction area. Since the capacity of the reconstruction area is equivalent to q_B, the flow just upstream of the reconstruction area is limited to q_B, resulting in flow state B. The result is a flow discontinuity between flow states A and B, and a backward-moving shock wave (ω_{AB}) is established. This shock wave continues until the last passenger on the train reaches the upstream end of the congested area at time t_2. Thereafter, no further pedestrians arrive from this train and the last pedestrian proceeds to the reconstructed area at speed u_B, and this also marks the discontinuity or shock wave between flow states B and D. The last passenger from that subway train enters the reconstruction area at t_3. The queueing pattern repeats itself with the arrival of the first passenger from the second train at time t_4.

The analyst particularly would like to calculate the length of the maximum queue, the time of occurrence of the maximum queue length, and the time when queueing ceases. First consider the time of queueing ($t_3 - t_1$). If N persons leave the train and use the pedestrian-way, and the capacity of the reconstruction area is equivalent to q_B, the time of queue can be calculated as follows:

---

*The symbols that will be used for flow, speed, density, and shock wave speed are q, u, k, and ω. However, units for pedestrian flow characteristics are persons per minute per foot of width, feet per minute, persons per square feet, and feet per minute, respectively. All flows must be for the total width of the pedestrian-way.

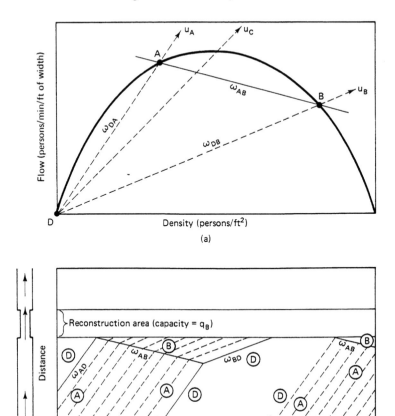

Figure 11.4   Shock Waves along a Pedestrian-Way

$$N = q_B (t_3 - t_1) \tag{11.14}$$

$$t_3 - t_1 = \frac{N}{q_B} \tag{11.15}$$

Now consider the time of the occurrence of the maximum queue length (Q_m). Since

$$\omega_{AB} = \frac{Q_m}{t_2 - t_1} \tag{11.16}$$

and

$$\omega_{BD} = \frac{Q_m}{t_3 - t_2}$$

then

$$\omega_{AB} (t_2 - t_1) = \omega_{BD} (t_3 - t_2) \tag{11.17}$$

since

$$t_3 - t_1 = (t_3 - t_2) + (t_2 - t_1) \tag{11.18}$$

By substitution,

$$t_2 - t_1 = \frac{N \omega_{BD}}{q_B (\omega_{AB} + \omega_{BD})} \tag{11.19}$$

The length of the maximum queue can be calculating as follows: since

$$\omega_{AB} = \frac{Q_m}{t_2 - t_1} \tag{11.20}$$

then

$$Q_m = \omega_{AB} (t_2 - t_1) \tag{11.21}$$

Finally, a very serious queueing situation would exist if the queue from the passengers on one train had not been discharged before the passengers from the next train had arrived at the reconstruction area. In fact, if this occurred, the maximum length of the queue would continue to increase with the arrival of every train. An important safeguard is to ensure that the following equation holds true, where h_t is the train headway.

$$h_t \geq \frac{N}{q_B} \tag{11.22}$$

11.6 COMPLEXITIES AND REFINEMENTS

The applications of shock wave theory presented earlier in this chapter are based on a number of simplifying assumptions and limitations. The purposes of this section are to identify these assumptions and limitations, describe their effects and added complexities, and discuss possible refinements.

The following is a list and brief description of some of the assumptions and limitations that are most significant.

- The capacity over the length of the study section is either constant or changes instantaneously to specific constant values at prespecified points along the study section.

- The capacity at a location over the entire time duration of the study is either constant or changes instantaneously to specific constant values at prespecified points in time.

- The demand for service over the length of the study section is constant and there is only one entrance and one exit.

- The demand for service over the entire time duration of the study is either constant or changes instantaneously to specific constant values at prespecified points in time.
- A single flow–density relationship is specified for the entire length of the study section.
- The selected flow–density relationship does not vary over the time duration of the study.
- Only a single bottleneck is studied and the possibility of queue collisions and queue splits are not considered.
- All vehicles travel at exactly the same speed for a specific condition on the flow–density relationship.
- Drivers do not anticipate changes in downstream flow conditions and are assumed to change their speeds instantaneously only at shock wave boundaries.

Consider the example shown in Figure 11.2 of shock waves at signalized intersections. The capacity is assumed to be constant over space and time except for the red periods at the signal. The demand is assumed to be constant over space and time. A single flow–density relationship is specified for the entire length and time duration of the study, and only a single bottleneck is considered. The time–space diagram clearly shows that vehicle speeds are identical for a specific flow regime condition. Now consider driver behavior patterns in the vicinity of the various shock waves.

The diagram indicates that vehicle speeds in flow regime A are the same as the speed of the shock wave AD, and hence no vehicles in flow regime A cross into flow regime D. Field observations suggest that vehicles at the rear of the platoon travel at slower speeds and the shock wave AD would most likely curve up and to the right. The frontal stationery shock wave DB is fairly realistic with some possible modifications due to stopping and starting delays.

Driver behavior in the vicinity of shock waves AB and BC indicate the lack of driver anticipation of downstream flow conditions. Field observations would suggest that drivers approaching a stopped queue of vehicles caused by a signalized intersection anticipate the need for deceleration and do so in advance, and hence change the position of the shock wave AB. In a like manner, stopped vehicles seeing the green signal indication and observing the line of vehicles ahead accelerating, would modify their behavior and thus change shock wave BC.

More disconcerning are the effects of the assumptions related to shock wave DC and the nonconvergence of shock wave DC with shock wave AC (and CD). Flow regime D demarks a space unoccupied by vehicles so that the lead vehicles being discharged from the signal into flow regime C have an empty road ahead of them. Undoubtedly, some of the vehicles will travel faster than speed denoted at capacity and the shock wave DC would be curvilinear and have a steeper slope. Related to this is the realization that these assumptions lead to nonconvergent forward-moving shock waves and result in nonreducing capacity flow conditions downstream of the signalized intersection.

Similar observations can be drawn from Figures 11.3 and 11.4 of shock waves along a highway and a pedestrian-way. The reader is encouraged to study these two figures and to assess the effects of the various assumptions and limitations on the resulting diagrams.

Limited literature is available to suggest possible refinements in shock wave analysis. Lighthill and Whitham [2, 8] provide some insights as to how to handle increased inflow situations, and shock wave analysis at bottlenecks and signalized intersections. Wohl and Martin [9] and a little later, Gerlough and Huber [14], reinforced the Lighthill and Whitham approaches. Stephanopoulos et al. [16] provided further insights into traffic queue dynamics at signalized intersections. The FREQ model [20] has computerized shock wave analysis at freeway bottlenecks and has the capability of handling interactive multiple bottlenecks, varying capacity over space and time, and varying demand over space and time.

Shock wave analysis is an important tool in the analysis of flow and queueing problems. Existing theory is available to approximate relatively simple problems. Further research is needed to refine existing theory in order to more realistically represent complex flow and queueing problems.

11.7 SELECTED PROBLEMS

1. Conduct a literature search on shock wave theory and application as applied to transportation for the period since 1980. Prepare a bibliography and classify into theory and applications. Further, subclassify applications into highway vehicular and other, and under highway vehicular subdivide into interrupted and uninterrupted flow situations. Comment on the extensiveness (or lack thereof) of the literature.

2. Rework the example problem shown in Figure 11.1 by reversing the flow states A and B. That is, have flow state B occur first, followed by flow state A. Draw diagrams like Figure 11.1 and derive equations (11.1) through (11.4).

3. Numerically solve the example problem shown in Figure 11.1 if flow states A and B are defined as follows: u_A and u_B are equal to 30 and 40 miles per hour, respectively, and k_A and k_B are equal to 48 and 24 vehicles per mile per lane, respectively. How many vehicles leave flow state B in a 1-hour period?

4. Use the flow–density diagram and combinations of the four flow states (A, B, C, D) of Figure 11.1a to draw distance–time diagrams (showing shock wave and vehicular trajectories) that result in the following types of shock waves: (a) frontal stationary, (b) backward forming, (c) forward recovery, (d) rear stationary, (e) backward recovery, and (f) forward forming.

5. Repeat Problem 4 with numerical solutions. Assume that the flow–density diagram is based on a linear Greenshields model, $u = 50 - 0.417k$, and the flows for states A, B, C, and D are 1440, 960, 960, and 600 vehicles per hour per lane, respectively.

6. Numerically, solve the example problem shown in Figure 11.2 if the speed–density relationship is linear and flow states A and C are defined as follows: u_A and u_C are equal to 40 and 26 miles per hour, respectively, and k_A and k_C are equal to 24 and 60 vehicles per mile per lane, respectively. Assume a 30-second green phase.

7. Plot a graph of the stopping shock wave (w_{AB}) in Figure 11.2 as a function of the arrival flow (q_A). Assume a linear speed–density relationship of $u = 50 - 0.417k$.

8. Plot a graph of the starting shock wave (w_{BC}) in Figure 11.2 as a function of the discharge flow (q_C). Assume a linear speed–density relationship $u = 50 - 0.417k$.

9. Calculate the minimum length of the cycle that will *not* result in queue spillover to the next signal cycle. Use the information provided in Problem 6.

10. Consider traffic flowing at 1000 vehicles per hour with a density of 20 vehicles per mile and a speed of 50 miles per hour. A truck with a speed of 12 miles per hour enters the traffic stream, travels for 2 miles, and then exits. No passing is possible and a dense flow state results with a density òf 100 vehicles per mile and a flow of 1200 vehicles per hour. Draw the resulting distance–time diagram showing shock waves and vehicle projectories. (Problem suggested by Edie.)

11. Using the information provided in Problem 10 and assuming a linear speed–density relationship, draw the distance–time diagram (showing shock waves and vehicle projectories) if the traffic was arriving at 1200 vehicles per hour.

12. By inspection of Figure 11.3, answer the following questions. (a) How fast would the truck have to travel to eliminate shock waves? (b) When would the shock wave (w_{AB}) become a rear stationary shock wave? (c) When would the shock wave (w_{AB}) become a backward forming shock wave? and (d) If traffic was arriving at capacity flow, when would shock waves w_{AB} and w_{BC} intercept?

13. The individual lanes on a long, tangent, two-lane directional freeway have identical traffic behavior patterns and each follows a linear speed–density relationship. It has been observed that the capacity is 2000 vehicles per hour per lane and occurs at a speed of 25 miles per hour. On one particular day when the input flow rate was 1800 vehicles per hour per lane, an accident occurred on the opposite side of the median which caused a gapers' block and caused the lane density to increase to 120 vehicles per mile. After 15 minutes the accident was removed and traffic began to return to normal operations. Draw the distance–time diagram showing shock waves and selected vehicle trajectories.

14. In Chapter 8 a directional freeway location was analyzed but queues were quantified in terms of numbers of vehicles. Apply shock wave analysis techniques to determine the length of queues at the end of each time period from 4:45 to 6:00 P.M.

15. In the problem graphically depicted in Figure 11.4, assume that the normal pedestrian-way for commuters is 8 feet wide and the reduced pedestrian-way in the reconstruction area is 6 feet wide. A subway train arrives and discharges 540 passengers who proceed to the pedestrian-way at a flow rate equivalent to a volume–capacity ratio of 0.9. Use appropriate pedestrian stream flow diagrams contained in Chapter 10. Plot resulting flow–density and distance–time diagrams similar to Figure 11.4.

16. Using the information given in Problem 15, plot the minimum width of the pedestrian-way in the reconstruction area that is needed as a function of the train headways. Assume no queue overflows from one train's arrival until the next.

17. In evaluating the operational consequences of the reconstruction work, the operator is very concerned that the queue due to the reconstruction does not extend upstream into the subway train platform area. Using the information given in Problem 15, plot the effect that the width of the pedestrian-way in the reconstruction area has on the length of the queue. Consider only one train arrival.

18. Trains are expected to arrive at a specific station every 5 minutes. Four hundred and twenty commuter passengers are predicted to be discharged onto the platform from every train during the peak period. From the platform, passengers are to be processed in a serial fashion through five subsystems, which are discussed below.

  1. From the platform to an escalator system, passengers will enter the first corridor which is 100 feet long and 4 feet wide at capacity.

  2. The escalator rises to the upper concourse level in 30 seconds over a horizon distance of 100 feet. The capacity of the escalator system is 70 passengers per minute.

  3. The passengers then proceed along a second corridor (identical to the dimensions of the first corridor).

  4. The passengers then pass through a turnstile system whose capacity is expected to be 50 passengers per minute.

  5. The passengers then proceed along a third corridor (identical to the dimensions of the first corridor) into an open sidewalk area.

There are two tasks. The first task is to draw a distance–time diagram showing the trajectories of the first and last passenger off the first two trains and indicating the resulting shock waves. The second task is to redesign the escalator and turnstile systems so that there is no queueing backing into the escalator or train platform area.

19. In developing and applying shock wave analysis techniques a number of simplifying assumptions were made. Identify these simplifying assumptions and discuss their implications to the analysis portrayed in Figures 11.1, through 11.4. Refine the four figures qualitatively to real-world conditions.

11.8 SELECTED REFERENCES

1. P. I. Richards, Shock Waves on the Highway, *Operations Research*, Vol. 4, No.1, 1956, pages 42–51.

2. M. H. Lighthill and G. B. Whitham, On Kinematic Waves: A Theory of Traffic Flow on Long Crowded Roads, *Proceedings of the Royal Society, Series A*, Vol. 229, 1957, pages 317-345. (Reprinted in Highway Research Board, Special Report 79, HRB, Washington, D.C., 1964, pages 7–35).

3. Airborne Instruments Laboratory, Backward Waves in Highway Traffic Jams, *Proceedings of IRE*, Vol. 45, No. 1, 1957, page 2A.

4. H. Greenberg, *A Mathematical Analysis of Traffic Flow-Tunnel Traffic Capacity Study*, Port of New York Authority, New York, 1958.

5. H. Greenberg, An Analysis of Traffic Flow, *Operations Research*, Vol. 7, No. 1, 1959.

6. R. E. Franklin, The Structure of a Traffic Shock Wave, *Civil Engineering Public Works Review*, Vol. 56, No. 662, 1961, pages 1186–1188.

7. L. A. Pipes, *Hydrodynamic Approaches—Part I: An Introduction to Traffic Flow Theory*, Highway Research Board, Special Report 79, HRB, Washington, D.C., 1964, pages 3–5.

8. D. L. Gerlough and D. G. Capelle, *An Introduction to Traffic Flow Theory*, Highway Research Board, Special Report 79, HRB, Washington, D.C., 1964, pages 3–35.

9. Martin Wohl and Brian V. Martin, *Traffic System Analysis for Engineers and Planners*, McGraw-Hill Book Company, New York, 1967, pages 338–345.

10. Donald R. Drew, *Traffic Flow Theory and Control*, McGraw-Hill Book Company, New York, 1968, pages 314–315 and 364–366.

11. L. A. Pipes, Wave Theories of Traffic Flow, *Journal of Franklin Institute*, Vol. 280, No. 1, 1965, pages 23–41.

12. L. A. Pipes, Topics in the Hydrodynamic Theory, *Transportation Research*, Vol. 2, No. 21, 1968, pages 143–149.

13. J. Rorbech, Determining the Length of the Approach Lanes Required at Signal-Controlled Intersections on through Highways, *Transportation Research*, Vol. 2, 1968, pages 283–291.

14. D. L. Gerlough and M. J. Huber, *Traffic Flow Theory*, Transportation Research Board, Special Report 165, TRB, Washington, D.C., 1975, pages 111–123.

15. P. G. Michalopoulos and G. Stephanopoulos, Oversaturated Signal Systems with Queue Length Constraint, *Transportation Research*, Vol. 11, 1977, pages 413–421.

16. Gregory Stephanopoulos, Panos G. Michalopoulos, and George Stephanopoulos, Modelling and Analysis of Traffic Queue Dynamics at Signalized Intersections, *Transportation Research*, Vol. 13A, 1979, pages 295–307.

17. W. Leutzbach and U. Kohler, Definitions and Relationships for Three Different Time Intervals for Delay Vehicles, *6th Symposium, Transportation and Traffic Theory*, Vol. 6, University of New South Wales, Sydney, Australia, 1974, pages 87–103.

18. S. Chandana Wirasinghe, Determination of Traffic Delays from Shock Wave Analysis, *Transportation Research*, Vol. 12, 1978, pages 343–348.

19. William A. Stock, Richard G. Blankenhorn, and Adolf D. May, *The FREQ3 Freeway Model*, University of California, Berkeley, Calif., ITTE Report 73-1, June 1973, 247 pages.

20. Tsutomu Imada and Adolf D. May, *FREQ8PE--A Freeway Corridor Simulation and Ramp Metering Optimization Model*, UCB-ITS-RR-85-10, University of California, Berkeley, Calif., June 1985, 263 pages.

21. Institute of Transportation Engineers, *Transportation and Traffic Engineering Handbook*, 2nd Edition, Prentice-Hall, Inc., Englewood Cliffs, N.J., 1982, pages 452–454.

22. William Nesbit, Making Waves on the Freeways, *American Way*, August 1979, pages 72–75.

12

Queueing Analysis

When demand exceeds capacity for a period of time or an arrival time headway is less than the service time (at the microscopic level) at a specific location, a queue is formed. The queue may be a moving queue or a stopped queue. Essentially, excess vehicles are stored upstream of the bottleneck or service area, and their departure is delayed to a later time period.

There are numerous examples of queueing processes in highway systems, such as at intersections, toll plazas, parking facilities, freeway bottlenecks, incident sites, merge areas, and behind slow-moving vehicles. Queueing processes occur in all transportation modes and in everyday situations, such as at grocery checkout counters, bank teller windows, and at restaurants.

Two analytical techniques can be employed in studying queueing processes: shock wave analysis and queueing analysis. Shock wave analysis (Chapter 11) can be employed when the demand–capacity process is deterministic, and is particularly well suited to evaluating the space occupied by the queueing process and to interacting queueing processes. Queueing analysis, which is covered in this chapter, can be employed for deterministic or stochastic processes, and the vehicles in the process are considered as being stored in a vertical queue.

This chapter is presented in three parts and is concluded with selected problems and references. The introductory part is devoted to specifying input requirements and to the development of a classification scheme for queueing analysis. The second portion of the chapter addresses deterministic queueing analysis and includes fundamental concepts and applications. The last part of the chapter is directed to stochastic queueing analysis and deals with single- and multichannel service systems, with a variety of probabilistic arrival and service distributions.

12.1 INPUT REQUIREMENTS AND CLASSIFICATION

The input requirements for queueing analysis include the following five elements.

- Mean arrival value
- Arrival distribution
- Mean service value
- Service distribution
- Queue discipline

The mean arrival value is expressed as a flow rate, such as vehicles per hour, or as a time headway, such as seconds per vehicle. The arrival distribution can be specified as a deterministic distribution or as a probabilistic distribution. The term, "demand" or "input," is sometimes substituted for the term "arrival."

The mean service value is expressed as a flow rate such as vehicles per hour, or as a time headway, such as seconds per vehicle. The service distribution can also be specified as a deterministic distribution or as a probabilistic distribution. The term "capacity," "departure," or "output" is sometimes substituted for the term "service."

The fifth and last element to be specified is queue discipline. The most common queue discipline encountered is referred to as "first in, first out." That is, the vehicles are served in the order in which they arrive, and often the symbol "FIFO" is used to designate this queue discipline. Other queue disciplines include "first in, last out" (FILO) systems encountered at elevators and "served in random order" (SIRO) systems.

A classification scheme is needed to assess the input characteristics in order to select the appropriate queueing analysis: deterministic queueing analysis or stochastic queueing analysis. If either the arrival distribution and/or the service distribution are probabilistic, the exact arrival and/or service time of each vehicle is unknown, and stochastic queueing analysis must be selected. On the other hand, if both the arrival and service distributions are deterministic, the arrival and service times of each vehicle are known and deterministic queueing analysis is selected. The next two major portions of the chapter are directed to deterministic queueing analysis and stochastic queueing analysis.

12.2 DETERMINISTIC QUEUEING ANALYSIS

Deterministic queueing analysis can be undertaken at two different levels of detail. The analysis can be at a macroscopic level, where the arrival and service patterns are considered to be continuous, or at a microscopic level, where the arrival and service patterns are considered to be discrete. The macroscopic level is normally selected when the arrival and service rates are high, while the microscopic level is often selected when the arrival and service rates are low. Primary attention is given in the next few sections to the macroscopic level of analysis, but the last section deals with the microscopic level of analysis.

12.2.1 Signalized Intersection Example

The signalized intersection serves as one of the best examples of deterministic queueing analysis at the macroscopic level because it is relatively simple and because of our personal experience with such queueing processes. Queueing on only one approach to a signalized intersection having only two signal phases for that approach is considered.

Only an undersaturated situation is considered in this example. That is, in each cycle the arrive demand is less than the capacity of the approach, no vehicles wait longer than one cycle, and there is no overflow from one cycle to the next. A number of simplifying assumptions are included with this first example of deterministic queueing analysis to emphasize the concept and reduce confusion with more complicated real-life situations. Later examples demonstrate techniques for handling more complicated real-life situations.

Figure 12.1a provides for all the input requirements needed to solve this problem. The arrival rate (λ) is specified in vehicles per hour and is constant for the study period. The service rate (μ) has two states: zero when the signal is effectively red and up to saturation flow rate (s) when the signal is effectively green. Note that the service rate can be equivalent to the saturation flow only when a queue is present. Otherwise, the service rate is equal to the arrival rate if the signal is green. The queue discipline is assumed to be a "first in, first out" (FIFO) system.

Directly below the flow rate versus time diagram, a cumulative vehicles versus time diagram is constructed. A horizontal line (Figure 12.1a) such as the arrival rate (λ) appears as a sloping line in Figure 12.1b with the slope equal to the flow rate. Thus the arrive rate goes through the origin and slopes up to the right with a slope equal to the arrival rate.

Now the service rate in (a) is transformed to (b). During the red period the service rate is zero, so the service is shown as a horizontal line in the lower diagram. At the start of the green period a queue is present, and the service rate is equal to the saturation flow rate (s). As shown in Figure 12.1b, the cumulative arrival line intersects the cumulative service line during the green period. At this point in time the queue is dissipated and the cumulative service line overlays the cumulative arrival line until the end of the green period. Then the pattern repeats itself, with the service rate varying again from zero to saturation flow rate to arrival flow rate.

A series of identical triangles are formed, with the cumulative arrival line forming the top side of the triangles and the cumulative service line forming the other two sides of the triangles. Each triangle represents one cycle length and can be analyzed to calculate a set of five measures of performance: time duration of queue (t_Q), number of vehicles experiencing queue (N_Q), queue length (Q), individual delay (d), and total delay (TD).

The time duration of queue is represented by the horizontal projection of the queueing triangle. It starts at the beginning of the red period and continues until the queue is dissipated. Its value varies between the effective red time and the cycle length, and is expressed in seconds. Two measures of performance are associated with queue time duration: time duration of queue (t_Q) and percent time queue is present (Pt_Q)

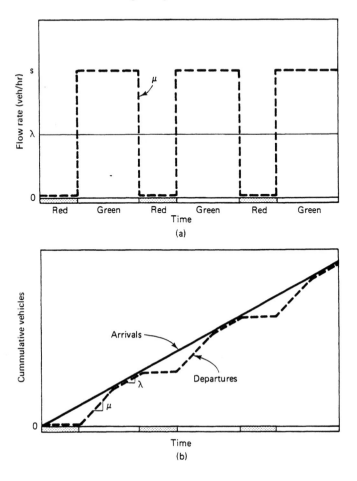

Figure 12.1   Queueing Diagram for Signalized Intersection

The equations for these two measures are:

$$\lambda t_Q = \mu \, (t_Q - r)$$

$$\lambda t_Q = \mu t_Q - \mu r$$

$$t_Q \, (\mu - \lambda) = \mu r$$

$$t_Q = \frac{\mu r}{\mu - \lambda} \tag{12.1}$$

$$P t_Q = \frac{100 t_Q}{C} \tag{12.2}$$

where λ = mean arrival rate (vehicles per hour)
    μ = mean service rate (vehicles per hour)
    t_Q = time duration of queue (seconds)

r = effective red time (seconds)

C = signal cycle time (seconds)

Pt_Q = percent time queue is present

The time duration of queue is helpful in understanding the storage and de-storage of vehicles in the queue and also in assessing the degree of saturation of the approach to the signalized intersection. Figure 12.2 is an extension of Figure 12.1 with additional information displayed on part (a).

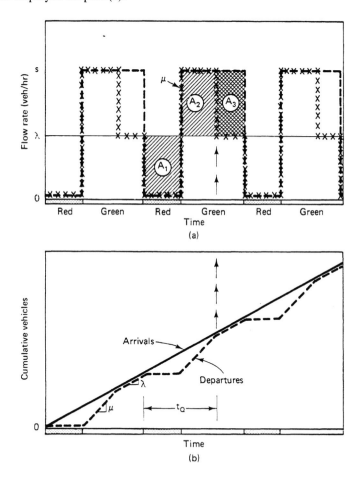

Figure 12.2   Extension of Signalized Intersection Queueing Diagram

Near the middle of the second green phase t_Q terminates, and the queue is dissipated. This point in time is projected onto Figure 12.2a, and three areas are denoted. The first area (A_1) represents the number of vehicles stored during the previous red phase. When the signal changes to green, the second area (A_2) begins to enlarge. The second area (A_2) denotes the number of vehicles that are de-stored. When the second area is equal to the first area, the queue is dissipated. The third area also has some significance, for it represents how many additional vehicles could be served during this signal cycle.

The number of vehicles experiencing queueing is represented by the vertical projection of the queueing triangle. The first vehicle experiencing the queue is the vehicle that arrives just after the signal turns red. All vehicles arriving during the red as well as vehicles arriving during the green but before the queue is dissipated experience the queueing process and are forced to stop or slow down significantly. Its value varies between λr and λC and is expressed in number of vehicles. Three measures of performance are associated with the number of vehicles experiencing queueing: number of vehicles queued (N_Q), number of vehicles per cycle (N), and percent of vehicles queued (PN_Q). The equations for these three measures are

$$N_Q = \frac{\lambda t_Q}{3600} \tag{12.3}$$

$$N = \frac{\lambda C}{3600} \tag{12.4}$$

$$PN_Q = \frac{100 t_Q}{C} \tag{12.5}$$

where N_Q = number of vehicles queued
$\quad\quad N$ = number of vehicles per cycle
$\quad PN_Q$ = percent of vehicles queued

The queue length is represented by the vertical distance through the triangle. At the beginning of red, the queue length is zero and increases to its maximum value at the end of the red period. Then the queue length decreases until the arrival line intersects the service line when the queue length is equal to zero. The queue length remains equal to zero until the end of the green period when the pattern repeats itself. Three queue length measures of performance are of primary interest: maximum queue length (Q_m), average queue length while queue is present $(\bar{Q}_Q)$, and average queue length $(\bar{Q})$. The equations for these three measures are

$$Q_M = \frac{\lambda r}{3600} \tag{12.6}$$

$$\bar{Q}_Q = \frac{Q_M}{2} = \frac{\lambda r}{7200} \tag{12.7}$$

$$\bar{Q} = \frac{Q_M t_Q}{2C} \tag{12.8}$$

where Q_M = maximum queue length (vehicles)
$\quad \bar{Q}_Q$ = average queue length while queue is present (vehicles)
$\quad\quad \bar{Q}$ = average queue length (vehicle)

Individual delay is represented by the horizontal distance across the triangle. The first vehicle to arrive after the beginning of red encounters the largest individual delay.

Each vehicle arriving thereafter experiences a smaller and smaller individual delay until the queue is dissipated. Vehicles arriving thereafter until the beginning of the next red encounter no individual delay. Three individual delay measures of performance are of primary interest: maximum individual delay (d_m), average individual delay while queue is present ($\bar{d}_Q$), and average individual delay ($\bar{d}$). The equations for these three measures are

$$d_M = r \tag{12.9}$$

$$\bar{d}_Q = \frac{r}{2} \tag{12.10}$$

$$\bar{d} = \frac{rt_Q}{2C} \tag{12.11}$$

where d_M = maximum individual delay (seconds)
$\quad \bar{d}_Q$ = average individual delay while queue is present (seconds)
$\quad \bar{d}$ = average individual delay (seconds)

The total delay per cycle is represented by the cross-sectional area of the queueing diagram triangle and is expressed in vehicle-seconds. Any one of the following equations can be used to calculate total delay (TD).

$$TD = \frac{N_Q r}{2} \tag{12.12}$$

or

$$TD = \frac{Q_M t_Q}{2} \tag{12.13}$$

or

$$TD = \bar{d}N \tag{12.14}$$

where TD is the total delay in vehicle-seconds.

12.2.2 Queueing Patterns

A variety of queueing patterns can be encountered and a classification of these patterns is proposed in this section. The classification scheme is based on how the arrival and service rates vary over time. For instance, the signalized intersection example presented in the preceding section has the characteristics of a constant arrival rate over time and a varying service rate over time.

A proposed classification scheme is shown in Figure 12.3, which is a 2 × 2 matrix resulting in four cells. Each cell can then be further subdivided into subclasses of queueing patterns.

Consider the upper left-hand cell that represents the pattern of a constant arrival rate and a constant service rate over time. Two rather less interesting subpatterns are encountered. If the arrival rate is less than the service rate, no queueing is ever

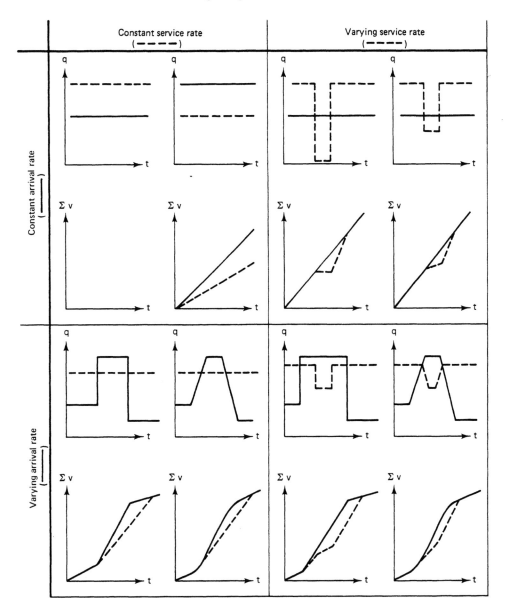

Figure 12.3   Deterministic Queueing Patterns

encountered. If on the other hand, the arrival rate is greater than the service rate, the queue has a never-ending growth with a queue length equal to the product of time and the difference between arrival and service rates.

The upper right-hand cell represents the pattern where the arrival rate is constant over time while the service rate varies over time. It should be noted that the service

rate must be less than the arrival rate for some periods of time, but greater than the
arrival rate for other periods of time. Again, this cell can be further subdivided into
subclasses of queueing patterns. Two are shown in Figure 12.3, but the service rate
does not have to be in the form of a square wave. That is, several changes in service
rates of different amounts can be encountered, and the changes do not have to occur
instantaneously but during transitional periods. The signalized intersection example fits
the diagram to the left in this cell, while the occurrence of an incident or accident would
result in a diagram similar to that shown to the right in this cell. In Section 12.2.3,
these particular queueing patterns are discussed further.

The lower left-side cell represents the pattern where the arrival rate varies over
time, while the service rate is constant over time. For queueing to occur and then be
dissipated, the arrival rate must be greater than the service rate for some periods of time
and less than the service rate during other periods of time. Two subclasses of queueing
patterns are shown in Figure 12.3. The one on the left in the cell indicates a square-
wave type of arrival rate, while the one on the right in the cell provides for transitional
periods during changes in arrival rates. Section 12.2.4 addresses this particular queueing pattern in greater detail.

The lower right-hand cell represents the more complex situation where both
arrival and service rates vary over time. For queueing to occur and then be dissipated,
the arrival rate must exceed the service rate and later be less than the service rate. Two
subclasses of queueing patterns are shown in Figure 12.3. The one on the left in the
cell indicates a square-wave type of arrival rate and an inverted square-wave type of
service rate. The diagram on the right side is an extension of the first one with transitional periods during changes in the arrival and service rates. Analysis of these more
involved queueing patterns are extensions of queueing patterns that are described in the
next two sections. Simulation is often employed when these more complex queueing
patterns are encountered particularly when sensitivity of parameter values is to be investigated. Such problems are addressed in Chapter 13.

12.2.3 Varying Service Rate Problems

Figure 12.3 and the accompanying text suggested two possible queueing patterns when
service rates vary over time while arrival rates are constant. Signalized intersections
and at-grade highway–railroad crossings are examples of one queueing pattern, while
an incident or accident site could cause the other queueing pattern. The essential difference between the two cases is whether the reduced service rate goes to zero or not.
Another less obvious difference is whether the service rate variation is repetitive, such
as at a fixed-time traffic signal, or whether it is not repetitive, such as at a
highway–railroad at-grade crossing, incident or accident site, or at a traffic–responsive
controlled traffic signal. Since the signalized intersection was addressed in Section
12.2.1, attention is now given to queueing analysis at the site of an incident.

The queueing diagram for the incident situation is shown in Figure 12.4. Figure
12.4a provides for all the input requirements needed to solve this problem. The arrival
rate (λ) is specified in vehicles per hour and is constant for the study period. The

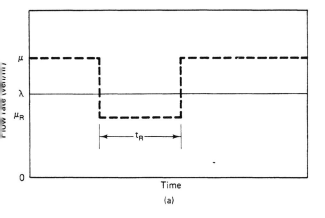

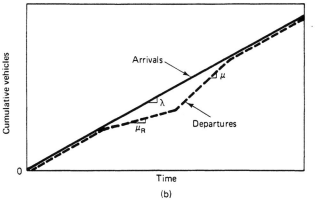

Figure 12.4   Queueing Diagram for
Incident Situation

normal service rate (without an incident) is indicated in the diagram as μ, and since it
exceeds the arrival rate, no queueing would normally exist. However, an incident
occurs that reduces the service rate to μ_R which is below the arrival rate, and this lower
service rate is maintained for t_R hours. As in the case for most highway situations, a
"first in, first out" (FIFO) queue discipline is assumed.

In Figure 12.4b, a cumulative vehicles versus time diagram is constructed. The
arrivals are shown as a straight line passing through the origin with a slope up and to
the right equivalent to the arrival rate (λ). For the first period of time the service line
follows the arrival line until the incident occurs. At that point in time the service rate
becomes equivalent to μ_R and maintains a flatter slope until the incident is removed.
Then the service rate increases to μ, and the service line has a steeper slope. This continues until the arrival line and service line intercept, at which time the service line once
again overlays the arrival line.

A triangle is formed with the cumulative arrival line forming the top side of the
triangle and the cumulative service line forming the other two sides of the triangle. The
triangle can be analyzed to calculate five sets of measures of performance. These are
the same measures of performance as discussed in Section 12.2.1, and the equations for
both situations are summarized in Table 12.1. Note that the time units identified with

TABLE 12.1   Queueing Performance Equations for Signalized
Intersections and Incident Situations

Performance Measures	Signalized Intersections	Incident Situations
Time duration in queue, t_Q (seconds/hour)[a]	$\dfrac{\mu r}{\mu - \lambda}$	$\dfrac{t_R(\mu - \mu_R)}{\mu - \lambda}$
Percent time queue is present, Pt_Q (percent)	$\dfrac{100 t_Q}{C}$	—
Number of vehicles queued, N_Q (vehicles)	$\dfrac{\lambda t_Q}{3600}$	λt_Q
Number of vehicles per cycle, N (vehicles)	$\dfrac{\lambda C}{3600}$	—
Percent of vehicles queued, PN_Q (percent)	$\dfrac{100 t_Q}{C}$	—
Maximum queue length, Q_M (vehicles)	$\dfrac{\lambda r}{3600}$	$t_R(\lambda - \mu_R)$
Average queue length while queue present, $\bar{Q}_Q$ (vehicles)	$\dfrac{\lambda r}{7200}$	$\dfrac{t_R(\lambda - \mu_R)}{2}$
Average queue length, $\bar{Q}$ (vehicles)	$\dfrac{\lambda r t_Q}{7200 C}$	—
Maximum individual delay, d_M (seconds/minutes)[a]	r	$\dfrac{60 t_R(\lambda - \mu_R)}{\lambda}$
Average individual delay while queue present, $\bar{d}_Q$ (seconds/minute)[a]	$\dfrac{r}{2}$	$\dfrac{30 t_R(\lambda - \mu_R)}{\lambda}$
Average individual delay, $\bar{d}$ (seconds)	$\dfrac{r t_Q}{2C}$	—
Total delay, TD (vehicle-seconds/vehicle-hours)[a]	$\dfrac{r t_Q \lambda}{2}$	$\dfrac{t_r t_Q(\lambda - \mu_R)}{2}$

[a]Time units: seconds for signalized intersections and minutes or hours for
incident situations.

signalized intersections are seconds and vehicle-seconds, while minutes and vehicle-hours are shown for the incident situations.

Consider a three-lane directional freeway with a capacity of 6000 vehicles per hour. Assume that during the middle of the day, the traffic demand is at 80 percent of capacity. Field studies have indicated that when incidents occur the capacity reduction is often equivalent to one lane of capacity and the reduction in capacity lasts for about 45 minutes. A sensitivity analysis study is to be undertaken to assess the effect of improvements in the incident detection and emergency service system for this freeway. Research has indicated that improvements in the system could decrease the duration of capacity reduction from 45 minutes down to 15 to 30 minutes. The queueing diagram for this problem is shown in Figure 12.5, and the resulting queueing measures of perfor-mance are summarized in Table 12.2. The queueing performance measures are all linear functions of incident duration (t_R) except for total delay, which varies with the square of the incident duration. For example, reducing the incident duration time from 45

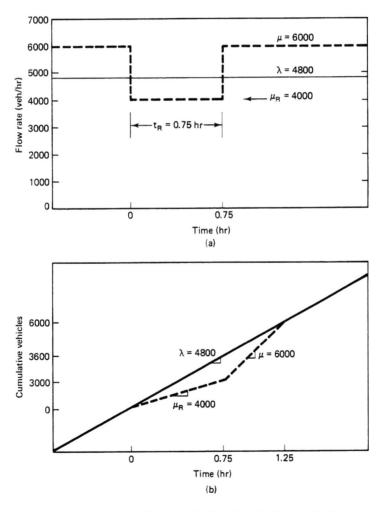

Figure 12.5   Queueing Diagram for Incident Situation Example Problem

TABLE 12.2 Performance Measures for Incident Situation Example Problem

Queueing Performance Measure	Equation	Modified Equation	Sensitivity Analysis			
			$t_R = 0.25$	$t_R = 0.50$	$t_R = 0.75$	$t_R = 1.00$
Time duration in queue, t_Q (hours)	$\dfrac{t_R(\mu - \mu_R)}{\mu - \lambda}$	$1.67t_R$	0.42	0.83	1.25	1.67
Number of vehicles queued, N_Q (vehicles)	λt_Q	$8000t_R$	2000	4000	6000	8000
Maximum queue length, Q_M (vehicles)	$t_R(\lambda - \mu_R)$	$800t_R$	200	400	600	800
Average queue length while queue is present, $\bar{Q}_Q$ (vehicles)	$\dfrac{t_R(\lambda - \mu_R)}{2}$	$400t_R$	100	200	300	400
Maximum individual delay, d_M (minutes)	$\dfrac{60t_R(\lambda - \mu_R)}{\lambda}$	$10t_R$	2.5	5.0	7.5	10.0
Average individual delay while queue is present, $\bar{d}_Q$ (minutes)	$\dfrac{30t_R(\lambda - \mu_R)}{\lambda}$	$5t_R$	1.2	2.5	3.8	5.0
Total delay, TD (vehicle-hours)	$\dfrac{t_R t_Q(\lambda - \mu_R)}{2}$	$666.67t_R^2$	42	167	375	667

minutes to 30 minutes through an improved incident detection and emergency service system would reduce most queueing performance measures by one-third and total delay by over 50 percent.

12.2.4 Varying Arrival Rate Problems

This portion of the chapter is based on a paper by May and Keller [3] in which a generalized deterministic queueing model was developed for a varying arrival rate problem. The queueing diagram for this generalized approach to the varying arrival rate problem is presented in Figure 12.6. The service rate (μ) is a constant vehicles per hour rate for the entire study period. The arrival rate (λ) takes on the form of a typical peak-period demand pattern, with a gradual increase in arrival rates in the early portion of the peak period and a gradual decrease in arrival rates in the later portion of the peak period.

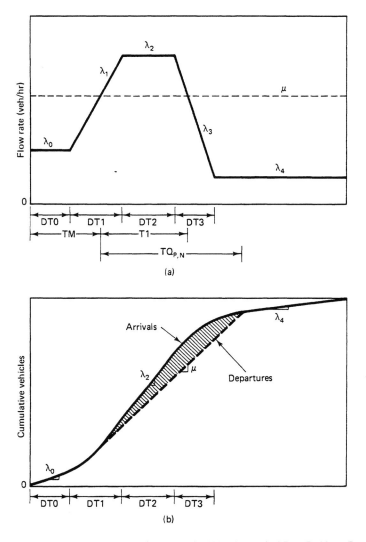

Figure 12.6   Queueing Diagram for Generalized Varying Arrival Rate Problem (From Reference 3)

More specifically, the arrival rate begins at a constant rate of λ_0 during time period DT0, which is less than the service rate (μ). During time period DT1, the arrival rate (λ_1) increases linearly from λ_0 to λ_2, and sometime during this period of time the arrival rate (λ_1) begins to exceed the service rate. During time period DT2 the arrival rate remains constant at λ_2. Then the arrival rate begins to decrease linearly from λ_2 to λ_4, and sometime during this period the arrival rate (λ_3) becomes less than the service rate. After time period DT3, the arrival rate remains at a constant rate (λ_4) for the rest of the study period.

The flow rates specified in Figure 12.6a are translated to the cumulative vehicle queueing diagram in Figure 12.6b. Note that λ_1 and λ_3 vary linearly on the top diagram and become curvilinear in the lower diagram.

This is referred to as a generalized approach since it will handle a great variety of arrival rate patterns. For example, if DT1 and DT3 are both set equal to zero, the arrival pattern will be rectangular and duplicates the arrival pattern shown in the left portion of the lower left-hand cell in Figure 12.3. On the other hand, if DT2 is set equal to zero, a triangle-shaped arrival pattern will result. Most other demand patterns encountered in real-life situations can be simulated by varying the various arrival rates (λ_0, λ_1, λ_2, λ_3, and λ_4) and duration times (DT0, DT1, DT2, and DT3).

The exact time that the arrival rate begins to exceed the service rate is not obvious by inspection and is denoted on Figure 12.6 as occurring TM time after the start of the study period. The equation for TM is

$$TM = DT0 + DT1 \left(\frac{\mu - \lambda_0}{\lambda_2 - \lambda_0} \right) \tag{12.15}$$

The exact time that the arrival rate becomes less than the service rate is also not obvious by inspection and is denoted on Figure 12.6 as occurring TI time after the end of the TM time period. The equation for TI can be found to be

$$TI = DT1 \left(\frac{\lambda_2 - \mu}{\lambda_2 - \lambda_0} \right) + DT2 + DT3 \left(\frac{\mu - \lambda_2}{\lambda_4 - \lambda_2} \right) \tag{12.16}$$

The duration of the queueing process (TQ) can be determined by investigating two cases. If the queue is dissipated during time DT3, the equation TQ is

$$TQ_P = TI + \left[\left(\frac{\lambda_2 - \mu}{\mu - \lambda_4} \right) (TI + DT2) \right]^{1/2} \tag{12.17}$$

On the other hand, if the queue is dissipated after the time period DT3, the equation for TQ would be

$$TQ_N = \frac{TI}{2} \left(\frac{\lambda_2 - \mu}{\mu - \lambda_4} + 2 \right) + \frac{DT2}{2} \left(\frac{\lambda_2 - \mu}{\mu - \lambda_4} \right) + \frac{DT3}{2} \left(\frac{\lambda_4 - \mu}{\lambda_4 - \lambda_2} \right) \tag{12.18}$$

The number of vehicles adversely affected by the bottleneck can be expressed as

$$N_Q = \mu(TQ) \tag{12.19}$$

The total delay TD in vehicle-hours is graphically shown in Figure 12.6b as the shaded area and can be computed by integration:

$$TD = \int_0^{TQ} [\lambda(T) - \mu(T)] \, dT \tag{12.20}$$

The solution of the integral gives the total delay as a function of the flow rates $\lambda_1(T)$, $\lambda_2(T)$, $\lambda_4(T)$, and $\mu(T)$ and the time intervals DT0, DT1, DT2, and DT3. If the queue ends after the decreasing input flow period, the total delay is

$$TD_N = \frac{\lambda_4 - \lambda_2}{2} \left[DT3 \left(\frac{1}{3} DT3 + DT2 + DT1(k) - TQ_N \right) \right.$$

$$\left. + DT2 \left(DT2 + 2(DT1)k - 2(TQ_N) \right) \right]$$

$$+ \frac{\lambda_4 - \mu}{2} [(DT1)^2 (k)^2 + (TQ_N)^2] - \frac{\lambda^4 - 2\mu}{2} [DT1(k)TQ_N] \qquad (12.21)$$

$$+ \frac{\lambda^2 - \mu}{2} \left[DT1(k) \left(TQ_N - \frac{2}{3} DT1(k) \right) \right]$$

where

$$k = \frac{\lambda_2 - \mu}{\lambda - \lambda_0}$$

If the queue ends during the decreasing input flow period, the total delay is

$$TD_P = \frac{\lambda_4 - \lambda_2}{6} \frac{1}{DT3} [TQ_P - DT1(k) - DT2]^3$$

$$+ \frac{\lambda_2 - \mu}{2} [TQ_P - DT1(k)]^2 \qquad (12.22)$$

$$+ \frac{\lambda_2 - \mu}{2} DT1(k) \left[TQ_P - \frac{2}{3} DT1(k) \right]$$

The maximum number of vehicles in the queue (Q_M) occurs at the end of the time interval TI and is given by

$$Q_M = \int_0^{TI} [\lambda(T) - \mu(T)] \, dT \qquad (12.23)$$

$$Q_M = \frac{\lambda_2 - \mu}{2} (TI + DT2) \qquad (12.24)$$

The average number of vehicles in the queue $\bar{Q}_Q$ can be computed from the total delay (TD) and the period of congestion (TQ) as follows:

$$\bar{Q}_Q = \frac{TD}{TQ} \qquad (12.25)$$

The maximum individual vehicle delay in hours (d_M) occurs to that vehicle which arrives at the end of the time interval TI.

$$d_M = \frac{\lambda_2 - \mu}{2\mu} (TI + DT2) \qquad (12.26)$$

The average individual vehicle delay in hours ($\bar{d}_Q$) can be obtained by dividing the total delay (TD) by the number of vehicles affected (N_Q).

$$\bar{d}_M = \frac{TD}{N_Q} \qquad (12.27)$$

The resulting equations presented in the previous several paragraphs are cumbersome to handle, and errors are likely if the equations are solved manually. Further, if a sensitivity analysis is to be performed, as the following example problem does, the manual effort is monumental. An alternative solution is to computerize the entire set of equations and provide input options to permit flexible use of the computer program. Such a computer program was developed and applied in the following example problem [3].

A freeway bottleneck exists during the afternoon peak period at a location where the capacity is 5500 vehicles per hour. The traffic demand before and after the peak period is 3000 vehicles per hour and the peak demand is 6600 vehicles per hour and lasts for 1 hour. The change in demand rate from 3000 to 6600 vehicles per hour extends over 1 hour, and similarly the change from 6600 to 3000 vehicles per hour extends over 1 hour. The queueing diagram for this freeway bottleneck problem is shown in Figure 12.7. It is desired to calculate the various queueing characteristics for

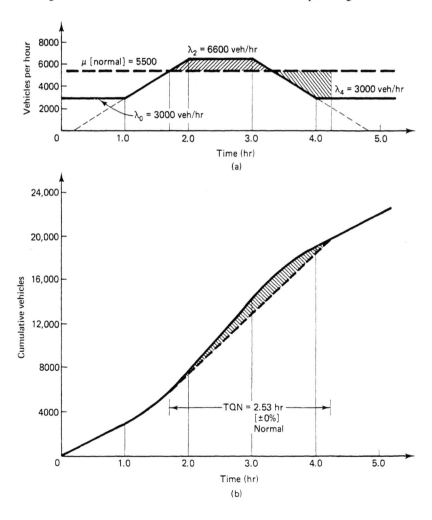

Figure 12.7   Queueing Diagram for Freeway Bottleneck Problem (From Reference 3)

the situation described above and then undertake a sensitivity analysis to determine the effect of increasing the bottleneck capacity on the queueing performance measures.

The results obtained from the computer program are shown in Figure 12.8. Part (a) displays the numerical results and part (b) indicates the percent reductions due to capacity increases.

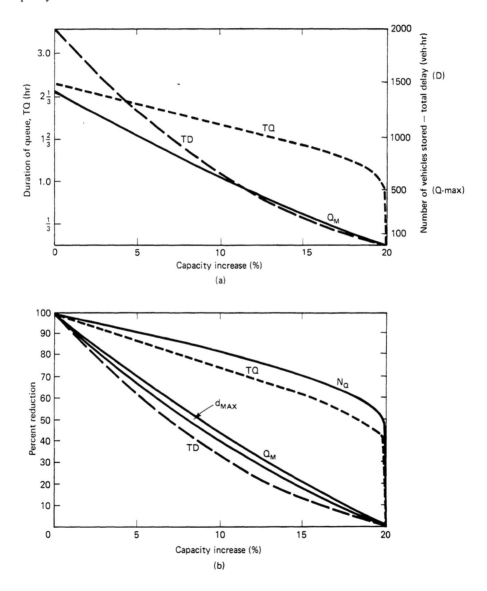

Figure 12.8   Effect of Increasing Capacity on Queueing Characteristics (From Reference 3)

12.2.5 Microscopic Analysis

Deterministic queueing analysis can also be undertaken at the microscopic level. The microscopic level of analysis requires that the arrival times and departure times of individual vehicles be known. Two examples will be presented: a signalized intersection and a rail transit line.

The queueing performance measures are to be determined for left-turning vehicles at a signalized intersection that has a separate left-turn lane and signal phase. Two location or stations are established in the field and the time of passage of every vehicle at each station is to be recorded. The upstream station is located at the beginning of the left-turn lane and upstream of any queueing process, while the downstream station is located at the centerline of the cross street.

The times of passage for each vehicle at each of the two stations is plotted in Figure 12.9. The uninterrupted travel time between stations is estimated to be 10 seconds. The arrival time curve is shifted to the right by 10 seconds and an arrival time curve is constructed for the downstream station. The shaded areas between the constructed arrival time curve and the departure time curve for the downstream station can be used to estimate the queueing performance measures.

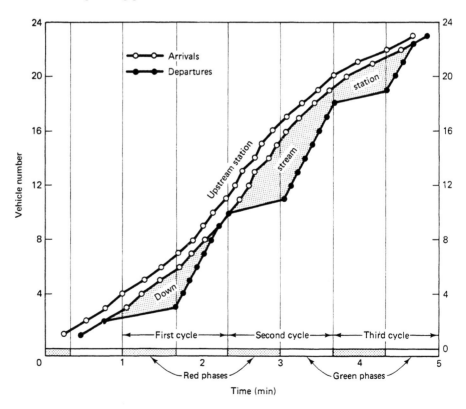

Figure 12.9   Microscopic Queueing Diagram for Signalized Intersection

The first shaded area is triangular in shape, which would be expected at an under-saturated signalized intersection. Vehicles 3, 4, and 5 arrive during the red phase and encounter delay. Vehicles 6, 7, and 8 arrive during the green phase but are delayed due to the queue. Vehicles 9 and 10 also arrive during the green phase but are not delayed since the previous queue was dissipated. The various queueing performance measures identified in Section 12.2.1 on signalized intersections can be estimated from the results shown in Figure 12.9. These performance estimates are summarized in Table 12.3.

TABLE 12.3 Performance Measures for Signalized Intersection

Performance Measure	First Cycle	Second and Third Cycles Combined
Time duration in queue, t_Q (seconds)	64	130
Percent time queue is present, Pt_Q (percent)	80	81
Number of vehicles queued, N_Q (vehicles)	6	12
Number of vehicles per cycle, N (vehicles)	8	13
Percent of vehicles queued, PN_Q (percent)	75	92
Maximum queue length, Q_M (vehicles)	3	6
Average queue length while queue present, $\bar{Q}_Q$ (vehicles)	2.0	3.2
Average queue length, $\bar{Q}$ (vehicles)	1.5	3
Maximum individual delay, d_M (seconds)	37	46
Average individual delay while queue is present, $\bar{d}_Q$ (seconds)	19.2	27.6
Average individual delay, $\bar{d}$ (seconds)	14.4	25.5
Total delay, TD (vehicle-minutes)	1.9	5.5

The second shaded area appears to have the shape of two connected triangles. This is due to the second cycle being oversaturated and its excess demand being transferred to the third cycle. Vehicles 11 through 15 arrive during the red phase and encounter delay. Vehicles 16, 17, and 18 arrive during the green phase but are delayed due to the queue. Vehicle 19 encounters the largest individual delay, for it arrives during one green phase but cannot be served until the next. Vehicles 20 and 21 arrive during the next red phase and encounter delay. Vehicle 22 arrives during the green phase and has some delay, while vehicle 23 arrives during the green phase and has no delay. The various queueing performance measures are estimated for the combined second and third cycles, and the results are summarized in Table 12.3.

Another example of deterministic queueing analysis at the microscopic level is a directional rail line in which the arrival and departure times of each train at stations along the route are automatically recorded [42]. A time–space diagram of train trajectories is presented in Figure 12.10. The trajectories of 24 trains are shown for a period of time between 7:00 and 9:00 A.M.

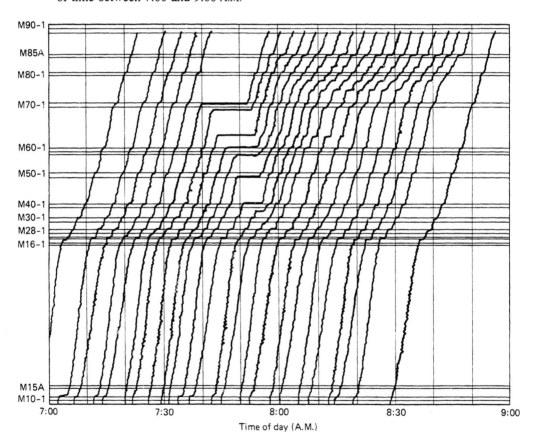

Figure 12.10   Time–Space Diagram for Train Trajectories (From Reference 42)

Close inspection of the time–space diagram indicates that there are two bottlenecks where queueing processes occur and train delays are encountered. The first bottleneck occurs at station M70 and begins at 7:40 A.M. It appears to be caused by a temporary track failure that lasts for about 12 minutes. After the track failure has been cleared, trains at closely spaced headways are released at station M70, which begin to overload station M85 at about 8:00 A.M. Station M85 then becomes a bottleneck and the queueing continues until about 8:50 A.M.

Queueing analysis permits an assessment of performance measures at the bottlenecks. A queueing diagram is shown in Figure 12.11 based on the time–space diagram information depicted in Figure 12.10. Three arrival–departure curves are constructed and plotted as denoted as M16 station arrivals, M70 station departures, and M85 station departures. In this way, the two bottlenecks can be analyzed separately. The next step is to plot the arrival time curves at the two bottleneck stations (M70 and M85) as if queueing had not existed. This is accomplished by determining the travel times between stations M16 and M70, and between stations M70 and M85, during the study period when the bottlenecks and hence queueing do not exist. These travel times are determined to be 14 and 4 minutes, respectively. Then the M16 station arrival curve is shifted to the right by 14 minutes and a new curve is constructed to represent the M70 station arrivals as if no queueing exists. In a similar manner the M70 departure curve is shifted to the right by 4 minutes and a new curve is constructed to represent the M85 station arrivals as if no queueing exists. The shaded area between the arrivals and departures curve for the M70 station and the shaded area between the arrivals and

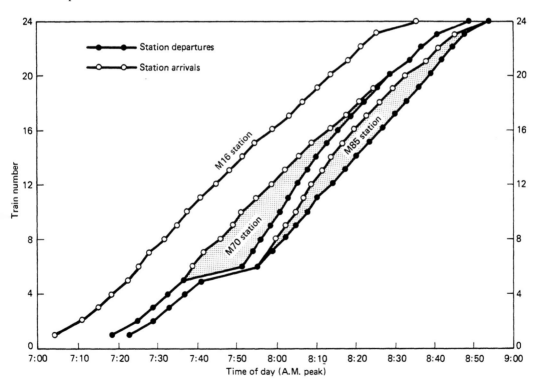

Figure 12.11   Microscopic Queueing Diagram for Rail Transit Line

departures curves for the M85 station capture the essence of the queueing performance measures at the two bottlenecks.

The queueing duration time at the first bottleneck due to a track failure of 12 minutes lasts for 47 minutes (7:38 to 8:25 A.M.) and delays 14 trains. As many as five trains are in queue at one time, and the average train delay for those trains in the queue was on the order of 6 minutes. The total delay is estimated to be 84 train-minutes of delay.

The queueing duration time at the second bottleneck due to the overload from delayed departures at M70 station lasts for 48 minutes (8:00 to 8:48 A.M.) and delayed 16 trains. As many as three trains are in the queue at one time and the average train delay for those trains in the queue was on order of 5 minutes. The total delay is estimated to be 80 train-minutes of delay.

12.3 STOCHASTIC QUEUEING ANALYSIS

A flowchart depicting various queueing situations is shown in Figure 12.12. If the arrival distribution and/or the service distribution is probablistic, the exact arrival and/or service time of each vehicle is unknown and stochastic queueing analysis rather than deterministic queueing analysis must be selected.

To use stochastic queueing analysis, the traffic intensity (ρ) must be less than 1.* Traffic intensity is

$$\rho = \frac{\lambda}{\mu} \qquad (12.28)$$

where ρ = traffic intensity
λ = mean arrival rate (vehicles per time interval)
μ = mean service rate per channel (vehicles per time interval)

The arrival and service rates can be expressed in a variety of ways. For example, the time interval may be specified in seconds, hours, and so on, but the arrival and service rates must employ the same time units. In some cases, mean arrival times and/or mean service times are given and require conversion to arrival and service rates as shown below:

$$\lambda = \frac{3600}{\bar{h}} \qquad (12.29)$$

$$\mu = \frac{3600}{\bar{s}} \qquad (12.30)$$

---

*This is true for single-channel systems, and the following material in this section is based on single-channel systems. In mutually shared multichannel systems, discussed in Section 12.3.3, an utilization factor (U_f) is introduced and is expressed as the ratio of the total arrival rate to the product of service rate per channel and the number of channels. For mutually shared multichannel systems, the utilization factor (U_f) must be equal to or less than 1.

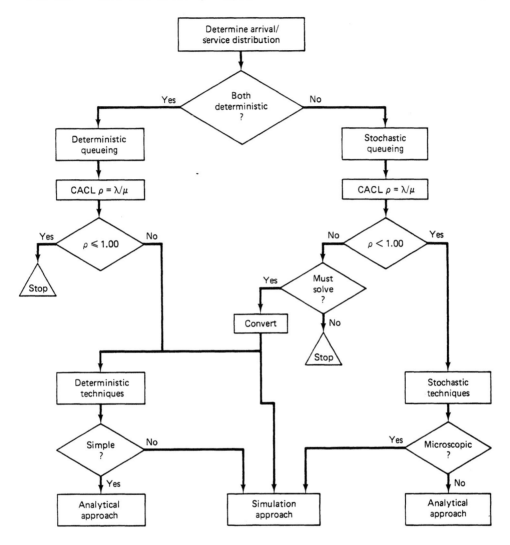

Figure 12.12   Flowchart of Queueing Analysis Approaches

where $\bar{h}$ = mean arrival time (seconds per vehicle)

$\bar{s}$ = mean service time (seconds per vehicle)

3600 = a constant to convert arrival and service rates to hourly rates

If the traffic intensity is greater than 1, the only possible solution approaches are to convert the queueing process to a deterministic queueing problem or to introduce multitime slice, varying mean arrival rates, and mean service rates and solve using microscopic simulation techniques. Deterministic queueing analysis was discussed in Section 12.2, and microscopic simulation techniques are discussed in Chapter 13.

Within stochastic queueing analysis there are two approaches. If the analyst wishes to study the stochastic queueing process macroscopically, an analytical approach based on sets of mathematical equations can be utilized. On the other hand, if the analyst wishes to study the stochastic queueing process microscopically, microscopic simulation techniques are required. The remainder of this chapter is devoted to the analytical approach of solving stochastic queueing problems.

There are many types of probability distributions of arrival and service rates. A classification scheme based on a few of the more common distributions is shown in Table 12.4. If the arrival and service distributions both have constant mean values, a deterministic queueing approach is employed. The letter D denotes a constant mean value, while M, E, and G represent random, Erlang, and generalized forms of probability distributions. An unlimited number of probability distributions can be utilized, but sets of mathematical equations have been derived for only a few of the more common probability distributions. For the development of sets of mathematical equations for other probability distributions, the reader may wish to consult queueing theory textbooks [33, 38–41].

TABLE 12.4   Classification Scheme of Probability Distributions
Used in Stochastic Queueing Problems[a]

Arrival Distribution	Service Distribution			
	Constant	Random	Erlang	Generalized
Constant	Deterministic queueing approach	D/M	D/E	D/G
Random	M/D	M/M	M/E	M/G
Erlang	E/D	E/M	E/E	E/G
Generalized	G/D	G/M	G/E	G/G

[a] D, constant mean value distribution; M, random distribution; E, Erlang distribution; G, generalized distribution.

Referring to Table 12.4, each cell is denoted by two letters (i.e., M/D). This represents a stochastic queueing situation in which the arrivals are randomly distributed while the service times are constant. This code identification scheme can be further extended by identifying the number of service channels, queue limits, and the queueing discipline characteristics. For example,

$$M/D/1 \ (\infty, FIFO)$$

identifies the same stochastic queueing problem described above and in addition indicates that there is a single service channel with queue discipline characteristics of permitted queue length of infinity with no diversion and a "first in, first out" service system. If only a code of M/D is given, it is understood to represent M/D/1 (∞, FIFO).

Three stochastic queueing situations based on the analytical approach are presented in the next three sections. The first two sections deal with single-channel services, with random arrival distributions and with random or constant service distributions [M/M/1 (∞, FIFO) and M/D/1 (∞, FIFO)]. The final section is devoted to multichannel systems [M/M/N (∞, FIFO)].

12.3.1 Random Service Problem

The set of equations used for a stochastic queueing problem in which both the arrival and service times are randomly- distributed is shown in Table 12.5. The discussion in this section is devoted to a single-service channel with no limit on queue length and "first in, first out" service.

TABLE 12.5   Queueing Performance Equations for Random
Arrival–Random Service Single-Channel Systems

Queueing Characteristics		M/M/1 (∞, FIFO)
Symbol	Definition	Equation
ρ	Traffic intensity	$\dfrac{\lambda}{\mu}$
$P(0)$	Probability of empty system	$1 - \rho$
$P(n)$	Probability of exactly n units in system	$\rho^n(1 - \rho)$
$E(m)$	Average number waiting to be served	$\dfrac{\rho^2}{1 - \rho}$
$E(m/m > 0)$	Average number waiting when queue is present	$\dfrac{1}{1 - \rho}$
$E(n)$	Average number in system (waiting and service)	$\dfrac{\rho}{1 - \rho}$
Var(n)	Variance of $E(n)$	$\dfrac{\rho}{(1 - \rho)^2}$
$E(v)$	Average time in system	$\dfrac{1}{\mu(1 - \rho)}$
$E(w)$	Average waiting time only	$\dfrac{\rho}{\mu(1 - \rho)}$

The traffic intensity, ρ, is the ratio of the mean arrival rate to the mean service rate and can vary from zero to 1. As ρ increases the various queueing performance measures increase at an increasing rate until when $\rho = 1$, the various measures go to infinity. The probability of one or more vehicles being in the system at any point in time is equal to the traffic intensity. Therefore, the probability of an empty system is equal to $(1 - \rho)$. If the analyst is interested in more specific information about the number of vehicles in the system, Gerlough and Huber [24] have shown that

$$P(n) = \rho^n (1 - \rho) \qquad (12.31)$$

where $P(n)$ = probability of exactly n vehicles in the system
ρ = traffic intensity
n = number of vehicles in system

The probabilities of n-or fewer vehicles in the system as a function of traffic intensity are shown in Figure 12.13. For example, if the arrival rate is 100 vehicles per hour and the service rate is 150 vehicles per hour, the traffic intensity is equal to 0.67. From Figure 12.13, the probability of exactly n vehicles in the system and the cumulative probabilities can be obtained. For example, the probability of exactly 0, 1, 2, 3, 4, and 5 vehicles in the system is 0.33, 0.22, 0.15, 0.10, 0.07, and 0.04, respectively. Note that as ρ approaches 1, the probabilities of a large number of vehicles in the system increases very rapidly.

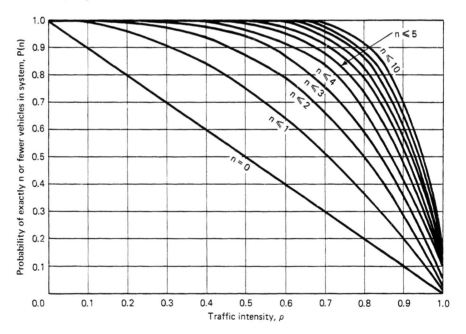

Figure 12.13 Probability of Exactly n or Fewer Vehicles in System as Function of Traffic Intensity

The queue length characteristics as a function of traffic intensity are shown in Figure 12.14. Using the previously calculated traffic intensity value of 0.67 and the equations shown in Table 12.5, the queue length characteristics would be [2]

$$
\begin{aligned}
E(M) &= 1.4 \text{ vehicles} \\
E(n) &= 2.0 \text{ vehicles} \\
E(m/m > 0) &= 3.0 \text{ vehicles} \\
\mathrm{Var}(n) &= 6.2 \text{ vehicles}
\end{aligned}
$$

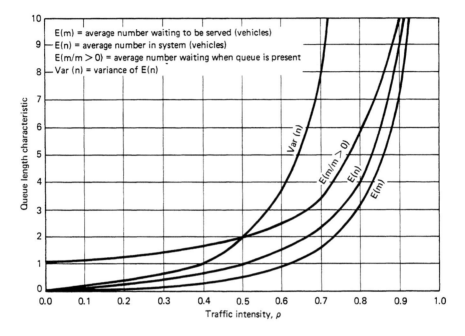

Figure 12.14   Effect of Traffic Intensity on Queue Length Characteristics

Without careful thought one might assume that the difference between the average number in the system $[E(n)]$ and the average number waiting to be served $[E(m)]$ should be equal to one for a single-channel server. However, there is a bias because of the probability of an empty system. In fact, it can be shown that

$$
E(n) = \rho + E(m) \tag{12.32}
$$

Thus the difference varies from zero (when $\rho = 0$) to 1 (when $\rho = 1$).

Special consideration should also be given to the relationship between the average number waiting when a queue is present $[E(m/m > 0)]$ and the average number waiting to be served $[E(m)]$. Contrary to initial intuition, the greatest difference occurs under low traffic intensity, and this difference approaches zero when traffic intensity approaches 1 because the probability of a empty system approaches zero.

In reference to the average number in system $[E(n)]$ and its variance $[\mathrm{Var}(n)]$, they are approximately equal under low traffic intensity, while the ratio becomes 1 to 10 when the traffic intensity equals 0.9.

The effects of arrival and service rates on average waiting time and total waiting time are shown in Figure 12.15. The two figures are constructed in a similar manner, with the mean arrival rate plotted on the vertical axis and the mean service rate on the horizontal scale. Two sets of contour lines are constructed on the figures. The traffic intensity contour lines appear as radial lines extending up and to the right from the origin. Steeper slopes represent higher traffic intensities. The second set of contour lines represent average system waiting time in Figure 12.15a and total system waiting time in Figure 12.15b. The two diagrams clearly show how waiting times increase rapidly as traffic intensity approaches unity.

12.3.2 Constant Service Problem

The set of equations used for a stochastic queueing problem in which the arrivals are randomly distributed and the service rate is constant is shown in Table 12.6. The discussion in this section is devoted to a single-service channel with no limit on queue length and "first in, first out" service.

As an example of queueing performance in single-channel systems with random arrivals and constant service, Figure 12.16 demonstrates the effects of arrival and service rates on average number and average time in the system. The contour lines in

TABLE 12.6   Queueing Performance Equations for Random
Arrival–Constant Service Single-Channel Systems

| Queueing Characteristics | | M/D/1 (∞, FIFO) |
Symbol	Definition	Equation
ρ	Traffic intensity	$\dfrac{\lambda}{\mu}$
$P(0)$	Probability of empty system	—
$P(n)$	Probability of exactly n units in system	—
$E(m)$	Average number waiting to be served	—
$E(m/m > 0)$	Average number waiting when queue is present	—
$E(n)$	Average number in system (waiting and service)	$\dfrac{2\rho - \rho^2}{2(1 - \rho)}$
$\text{Var}(n)$	Variance of $E(n)$	—
$E(v)$	Average time in system	$\dfrac{2 - \rho}{2\mu(1 - \rho)}$
$E(w)$	Average waiting time only	$\dfrac{\rho}{2\mu(1 - \rho)}$

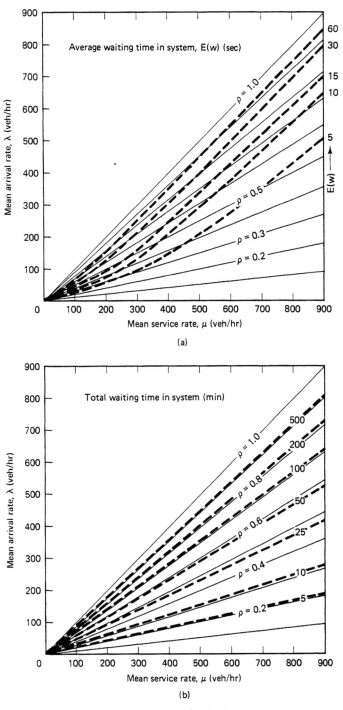

Figure 12.15   Effects of Arrival and Service Rates on Waiting Times

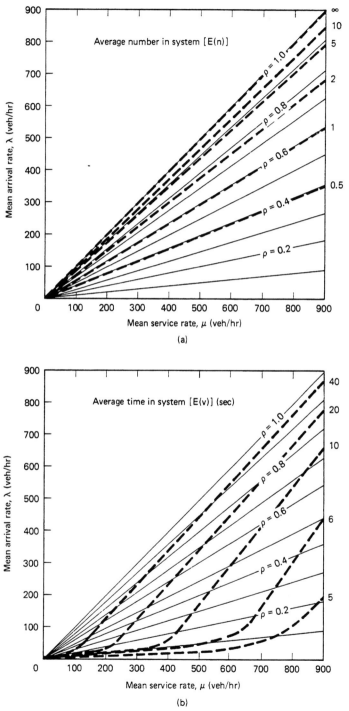

Figure 12.16   Effects of Arrival and Service Rates on Average Number and Average Time in System

Figure 12.16a represent the average number of vehicles in the system and extend up and to the right from the origin. Traffic intensity (ρ) can also be shown as contour lines in the same diagram. This diagram clearly shows the rapid increase in the average number in the system as traffic intensity approaches a value of 1.

Figure 12.16b shows contour lines representing average vehicle time in the system. This relationship is a little more complex since average time is a function of traffic intensity and mean service rate. This diagram clearly shows the rapid increase in the average time in the system as traffic intensity approaches a value of 1 and as mean service time approaches zero.

12.3.3 Multichannel Problem

In Sections 12.3.1 and 12.3.2 the problem of a single file of arriving vehicles being served by a single channel was addressed. If a multichannel server is provided (such as several toll gates at a toll plaza), two types of entry control could be implemented. One type would be to introduce as many entry channels as service channels and divide the arrival demand equally between channels. That is, the arrival rate for each channel would be equal to the total arrival rate divided by the number of channels. The service rate for each channel would be equal to the total service rate divided by the number of channels. The traffic intensity for each individual channel or for the combination of all channels would be

$$\rho = \frac{\lambda/N}{\mu/N} = \frac{\lambda}{\mu} \qquad (12.33)$$

where N is the number of channels. The key feature of this approach is that once the arriving traffic divides equally between channels serving the various service areas, the traffic cannot change channels even if their channel is queued and a parallel channel is empty. This causes inefficient use of the service area and results in greater delays and longer queues.

Another type of entry control can overcome this inefficient use of service area by combining the arrival traffic into one channel and then selecting the lead vehicle which is waiting, to go to the first available empty service channel. This will keep the service area busy as long as a queue is present. The difficulty is the practical implementation of such an entry control scheme. Probably the most successful examples of such an entry control scheme are in banks and at airport check-in counters.

An example problem is presented to demonstrate the application of the two entry control schemes and to provide a comparison of their performance. In this example, assume a total arrival rate of 800 vehicles per hour with the arrivals randomly distributed. Two toll gates are available for service, and each toll gate has a service time of 6 seconds, which is also randomly distributed. There is no limit on length of queue. The difference is the entry control scheme, that is, how the vehicles are served.

Consider the first entry control scheme in which the arriving vehicles are distributed equally between the two toll gates. In concept, there is a barrier between the two toll gates, and once a vehicle is assigned to a toll gate service channel, it cannot change to the other channel. Therefore, each channel is independent of the other, and the arrival rate and service times per channel are calculated and analyzed as a single-channel problem, as discussed in Section 12.3.1. In essence this multichannel problem is converted to a single-channel problem, M/M/1(∞, FIFO). The arrival rate per channel is 400 vehicles per hour and the service rate per toll gate is 600 vehicles per hour. The traffic intensity (ρ) can be calculated to be 0.67. The various queueing performance characteristics can be calculated using the equations shown in Table 12.5. These results and a comparison of results with the other entry control scheme are discussed next.

Now consider the other entry control scheme, where a single arrival queue is formed and the lead vehicle in the queue is always selected to go to the first available empty service channel. This is a multichannel problem that is analyzed as a multichannel problem and classified as M/M/2 (∞, FIFO). The total arrival rate (λ) is 800 vehicles per hour, the service per channel (μ) is 600 vehicles per hour, and the total service rate (μN) is 1200 vehicles per hour. Note that while λ denotes the total arrival rate, μ denotes the service rate per channel and μN denotes the total service rate. The queueing performance equations for random arrival–random service multichannel systems (where $N > 1$) are summarized in Table 12.7. The traffic intensity (ρ) can be calculated to be

TABLE 12.7   Queueing Performance Equations for Random Arrival–Random Service Multichannel Systems

Queueing Characteristics		M/M/N (∞, FIFO)
Symbol	Definition	Equation
ρ	Traffic intensity	$\dfrac{\lambda}{\mu}$
U_f	Utilization factor	$\dfrac{\rho}{N}$
$P(0)$	Probability of empty system	$\dfrac{1}{\displaystyle\sum_{n=0}^{N-1}\left(\dfrac{\rho^n}{n!}\right)+\dfrac{\rho^N}{N![(1-\rho)/N]}}$
$P(n)$	Probability of exactly n units in system ($n \le N$ or $n \ge N$)	$\dfrac{P(0)}{n!}$ or $\dfrac{\rho^n P_0}{N^{n-N}N!}$
$E(m)$	Average number waiting to be served	$\dfrac{P(0)\rho^{N+1}}{N!N}\left(\dfrac{1}{[(1-\rho)/N]^2}\right)$
$E(m/m > 0)$	Average number waiting when queue is present	$\dfrac{1}{(1-\rho)}/N$
$E(n)$	Average number in system (waiting and service)	$\rho + E(m)$
$E(v)$	Average time in system	$\dfrac{E(n)}{\rho\mu}$
$E(w)$	Average waiting time only	$E(v) - \dfrac{1}{\mu}$

1.33. A new term, the utilization factor (U_f), is introduced, which is the ratio of the total arrival rate divided by the total service rate, which in this example is equal to 0.67. While ρ can exceed 1.0, U_f cannot.

The queueing performance results of the two entry control schemes are summarized in Table 12.8.

TABLE 12.8 Queueing Performance Results of Two Entry Control Schemes

Queueing Characteristics		M/M/2(∞, FIFO) Result	
Symbol	Definition	Independent Channels	Shared Channels
ρ	Traffic intensity	0.67	1.33
U_f	Utilization factor	0.67	0.67
$P(0)$	Probability of empty system	0.33	0.20
$P(n)$	Probability of exactly 1, 2, and 3 units in system	0.22 0.15 0.10	0.27 0.18 0.12
$E(m)$	Average number waiting to be served	1.36	1.08
$E(m/m > 0)$	Average number waiting when queue is present	3.00	3.00
$E(n)$	Average number in system (waiting and service)	2.00	2.41
$E(v)$	Average time in system (seconds)	18.0	10.9
$E(w)$	Average waiting time only (seconds)	12.0	4.9

12.4 SELECTED PROBLEMS

1. Shock wave analysis or queueing analysis is employed when transportation systems are congested. Give examples when each analytical technique is most appropriate.

2. Queueing problems in transportation systems can be addressed using deterministic or stochastic queueing analysis techniques. Give examples when each technique is most appropriate.

3. Queue discipline is an input requirement for queueing analysis. The most common queue discipline is "FIFO." Identify other types of queue disciplines that could be encountered in transportation systems and give examples of each.

4. Consider the unique situation when the queue on an approach to a signalized intersection is dissipated exactly at the end of the effective green period. Modify equations (12.1) through (12.14) to represent this unique situation.

5. Considering the unique situation described in Problem 4, construct an illustration like Figure 12.2. Select scales so that crosshatched areas in part (a) are realistic.

6. Select a field location at an approach to a signalized intersection that has considerable queueing. With limited field measurements construct an illustration like Figure 12.2 that approximates the field conditions.

7. Solve equations (12.1) through (12.14) and construct a figure like Figure 12.2 for the following numerical example:

   - Vehicles approach at a uniform flow rate of 600 vehicles per hour.
   - Vehicles are served during the green period (when queue is present) at a uniform flow rate of 1800 vehicles per hour.
   - The queue discipline is a "first in, first out" (FIFO) system.
   - The effective green time is 40 seconds while the cycle length is 60 seconds.

8. Considering the numerical example described in Problem 7, plot a diagram of measures of performance on the vertical scale versus the mean arrival rate and volume–capacity ratio on the horizontal scale. Vary the mean arrival rate from 0 vehicles per hour up to the mean arrival rate that causes the approach to be loaded to capacity. The measures of performance should include t_Q, N_Q, Q_M, $\bar{d}$, and TD. Comment on these resulting relationships.

9. Convert the incident situation equations shown in Table 12.1 to the corresponding signalized intersection equations shown in the same table.

10. Consider the incident situation example problem shown in Figure 12.5. Rather than doing a sensitivity study of the effect of the time duration of the capacity reduction period, perform a sensitivity study of the effect of the reduced capacity value on resulting performance measures. Consider reduced capacities of 1/2 to 2 lanes in steps of 1/2-lane intervals. Construct a table similar to Table 12.2.

11. Conceptualize a freeway bottleneck situation in the field where the number of lanes on the freeway is reduced. Address the following issues:

   (a) What measures and/or observations would you make at what locations in the vicinity of the bottleneck to confirm that there is a freeway bottleneck?

   (b) Identify three different approaches that you could employ in order to construct a figure similar to Figure 12.6 for your bottleneck. (*Hint:* Focus on Figure 12.6a and on vertical and horizontal components of crosshatched area in Figure 12.6b.)

12. A queueing diagram for a freeway bottleneck problem is shown in Figure 12.7, and the effect of increasing capacity on queueing characteristics is presented graphically in Figure 12.8. Verify the no-capacity-increase numerical results.

13. Select a field location at an approach to a signalized intersection. With appropriate field measurements plot a figure similar to Figure 12.9 and estimate performance measures similar to Table 12.3.

14. A flow control strategy is to be implemented at station M16 as soon as a queue is detected downstream. The objective of the strategy is to eliminate interactions between trains downstream of station M16 while attempting to maintain the highest line capacity possible. Demonstrate the results of the flow control strategy in graphical form as depicted in Figures 12.10 and 12.11.

15. The classification scheme presented in Table 12.4 includes only four probability distributions. Review the selected references listed at the end of this chapter and identify other probability distributions that have been used in queueing analysis.

16. The numerical example discussed in Section 12.3.1 is based on an arrival rate of 100 vehicles per hour and a service rate of 150 vehicles per hour. Assuming that the arrival rate is reduced to 75 vehicles per hour, calculate the queueing performance using equations given in Table 12.5 and check your results with corresponding results shown in Figures 12.13, 12.14, and 12.15.

17. Solve Problem 16 using equations given in Table 12.6 and check your results with corresponding results shown in Figure 12.16. Compare the effect of random and constant service by comparing results of this problem with the results of Problem 16.

18. Plot two curves on a diagram similar to Figure 12.14. Both curves would represent average number in system $[E(n)]$, but one curve would be based on random service while the other would be based on constant service rate. Observe the effect of service distribution under different traffic intensity levels.

19. Verify the numerical results shown in Table 12.8.

20. Construct a summary table that combines the contents of Tables 12.5, 12.6, and 12.7. Through a literature search (**a**) attempt to complete the summary table where equations are missing; (**b**) attempt to extend the summary table by adding additional queueing characteristics equations; and (**c**) attempt to extend the summary table by adding additional columns representing other arrival and service time distribution combinations for single- and multichannel situations.

12.5 SELECTED REFERENCES

1. Wolfgang S. Homburger and James H. Kell, *Fundamentals of Traffic Engineering*, Institute of Transportation Studies, Berkeley, Calif., 1984, pages 4-3 to 4-4 and 4-8 to 4-9.

2. Adolf D. May, Traffic Flow Theory—The Traffic Engineer's Challenge, *Institute of Traffic Engineers World Traffic Engineering Conference, Proceedings,* Washington D.C., 1965, pages 290–303.

3. Adolf D. May and Hartmut E. M. Keller, A Deterministic Queueing Model, *Transportation Research,* Vol. 1, No.2, August 1967, pages 117–128.

4. Gordan A. Sparks and Adolf D. May, *A Mathematical Model for Evaluationg Priority Lane Operations on Freeways,* Highway Research Board Record, 363, HRB, Washington, D.C., 1971, pages 27–42.

5. Tsutomu Imada and Adolf D. May, *FREQ8PE—A Freeway Corridor Simulator and Ramp Metering Optimization Model,* UCB-ITS-RR-85-10, University of California, Berkeley, Calif., 2nd Edition, June 1985, 263 pages.

6. Institute of Transportation Engineers, *Transportation and Traffic Engineering Handbook,* Prentice-Hall, Inc., Englewood Cliffs, N.J., 1982, pages 460–469 and 502–503.

7. M. D. Raff, *A Volume Warrant for Urban Stop Signs,* Eno Foundation for Highway Traffic Control, Saugatuck, Conn., 1950, pages 62–75.

8. R. M. Oliver, Distribution of Gaps and Blocks in a Traffic Stream, *Operations Research,* Vol. 10, No. 2, 1962, pages 197–217.

9. W. R. Adams, Road Traffic Considered as a Random Series, *Journal of Institute of Civil Engineers,* Vol. 4, 1936, pages 121–130.

10. J. C. Tanner, The Delay to Pedestrians Crossing a Road, *Biometrika*, Vol. 38, 1951, pages 383–392.

11. A. J. Mayne, Some Further Results in the Theory of Pedestrian and Road Traffic, *Biometrika*, Vol. 41, 1954, pages 375–389.

12. G. H. Weiss and A. A. Maradudin, Some Problems in Traffic Delay, *Operations Research*, Vol. 10, No. 1, 1962, pages 74–104.

13. R. Herman and G. H. Weiss, Comments on the Highway Crossing Problem, *Operations Research*, Vol 9, No. 4, 1961, pages 828–840.

14. N. G. Major and D. J. Buckley, Entry to a Traffic Stream, *Australian Road Research Board Proceedings*, Vol. 1, 1962, pages 206–228.

15. R. Ashworth, The Capacity of Priority-Type Intersections with a Non-uniform Distribution of Critical Acceptance Gaps, *Transportation Research*, Vol. 3, No. 2, 1969, pages 273–278.

16. R. E. Allsop, Design at a Fixed-Time Signal I: Theoretical Analysis, *Transportation Science*, Vol. 6, No.3, 1972, pages 260–285.

17. M. Beckmann, C. B. McGuire, and C. B. Winsten, *Studies in the Economies of Transportation*, York University Press, York, England, 1956.

18. F. V. Webster, *Traffic Signal Settings*, Road Research Technical Paper No. 39, Great Britain Road Research Laboratory, Crowthorne, England, 1958.

19. G. F. Newell, Statistical Analysis of Flow of Highway Traffic through a Signalized Intersection, *Quarterly Applied Mathematics*, Vol. 13, 1956, pages 353–369.

20. T. P. Hutchinson, Delay at a Fixed Time Signal II: Numerical Comparisons of Some Theoretical Expressions, *Transportation Science*, Vol. 6, No. 3, 1972, pages 286–305.

21. D. R. McNeil, Growth and Dissipation of a Traffic Jam, *Transportation Research*, Vol. 3, 1969, pages 115–121.

22. Great Britain Road Research Laboratory, *Research on Road Traffic*, Her Majesty's Stationery Office, London, 1965.

23. D. L. Gerlough and D. G. Capelle, *An Introduction to Traffic Flow Theory*, Highway Research Board, Special Report 79, HRB, Washington, D.C., 1964, pages 49–96.

24. Daniel L. Gerlough and Matthew J. Huber, *Traffic Flow Theory—A Monograph*, Transportation Research Board Special Report 165, TRB, Washington, D.C., 1975, pages 137–173.

25. L. C. Edie, Traffic Delays at Toll Booths, *Operations Research*, Vol. 2, No. 2, 1954, pages 107–138.

26. K. Moskowitz, Waiting for a Gap in a Traffic Stream, *Highway Research Board Proceedings*, Vol. 33, 1954, pages 385–395.

27. Don Drew, *Traffic Flow Theory and Control*, McGraw-Hill Book Company, New York 1968, pages 223–254.

28. Martin Wohl and Brian V. Martin, *Traffic System Analysis for Engineers and Planners* McGraw-Hill Book Company, New York, 1967, pages 356–373.

29. G. F. Newell, Approximation Methods for Queues with Application to the Fixed-Cycle Traffic Light, *SIAM Review*, Vol. 7, No. 2, April 1965, pages 223–240.

30. Avishai Ceder and Philippe Marguier, Passenger Waiting Time at Transit Stops, *Traffic Engineering and Control*, Vol. 26, No. 6, June 1985, pages 327–329.

31. R. M. Kimber and P. N. Daly, Time-Dependent Queueing at Road Junctions: Observation and Prediction, *Transportation Research*, Vol. 20B, No. 3, June 1986, pages 187–203.

32. E. B. Lieberman, A. K. Rathi, and G. F. King, *Congested Based Control Scheme for Closely Spaced, High Traffic Density Networks,* KLD and Associates, Huntington Station, N.Y., 1986.

33. Alan J. Mayne, *Comprehensive Formulae for Queues and Delays: Some New Approximations,* University College, London, 1978.

34. Masao Kuwahara and Gordon F. Newell, Queue Evolution on Freeways Leading to a Single Core City During the Morning Peak, *Tenth International Symposium on Transportation and Traffic Theory Proceedings,* 1987, pages 21–40.

35. Margaret C. Bell, *Measures of Queueing Performance for a Traffic Network,* University of Newcastle upon Tyne, Newcastle, England, 1980, 26 pages.

36. Margaret C. Bell, *Queues at Junctions Controlled by Traffic Signals,* University of Newcastle upon Tyne, Newcastle, England, 1977.

37. Margaret C. Bell, *A Queueing Model for TRANSYT 7,* University of Newcastle upon Tyne, Newcastle, England, 1980, 27 pages.

38. Alec M. Lee, *Applied Queueing Theory,* Macmillan Publishing Co., Inc., New York, 1966, 244 pages.

39. Walter Helly, *Urban Systems Models,* Academic Press, Inc., New York, 1975, pages 103–132.

40. Frank A. Haight, *Mathematical Theories of Traffic Flow,* Academic Press, Inc., New York, 1963, 242 pages.

41. Gordon Newell, *Approximate Behavior of Tandem Queues,* Lecture Notes in Economics and Mathematical Systems 171: Operations Research, Springer–Verlag, Berlin, 1979, 410 pages.

42. Masami Sakita, *Use of Time–Space Diagram and Cumulative Arrival–Departure Diagram in Rail Mass Transit Operations Analyses,* University of California Lecture Notes, University of California, Berkeley, Calif., December 1987, 24 pages.

13

Computer Simulation Models

Computer simulation models can play a major role in the analysis and assessment of the highway transportation system and its components. Often they incorporate the other analytical techniques, such as demand–supply analysis, capacity analysis, traffic stream models, car-following theory, shock wave analysis, and queueing analysis, into a framework for simulating complex components or systems of interactive components. These components may be individual signalized or unsignalized intersections, residential or central business district dense networks, linear or network signal systems, linear or corridor freeway systems, or rural two-lane or multilane highways systems.

In addition, computer simulation models have been developed and applied to mode transfer locations such as at parking facilities, transit stations, and airports. Computer simulation model applications are not limited to highway transportation systems but have been used in all forms of transportation, such as bus systems, transit rail systems, railroad systems, air transport systems, waterway systems, ferry systems, pedestrian-ways, and elevator systems. In fact, computer simulation model techniques have no limits other than the creativity and resources of the developer and have been used in many fields outside of transportation.

This chapter has four major emphases. The first section of the chapter is an introduction to simulation which provides a starting point for the last three sections. The remaining three sections have been written so that each is self-contained, has a unique emphasis, and can be read based on the interest of the reader. The second section describes in qualitative terms the step-by-step process for developing a computer simulation model. In the third section a simulation model is developed and applied in a step-by-step fashion as described in Section 13.3. The final section contains an inventory and brief description of most currently used computer simulation models for various highway systems. The chapter concludes with a summary, selected problems, and selected references.

13.1 INTRODUCTION TO SIMULATION

This portion of the chapter is intended to give a brief introduction to simulation and provide a starting point for the remaining three sections. This introduction includes definitions of simulation, strengths and weaknesses of simulation techniques, and a historical perspective of simulation modeling.

Computer simulation modeling means different things to different people. In the very broadest sense simulation could be represented by such things as a physical model, a verbal description, a painting, photograph, or an equation. For purposes of this chapter, simulation is defined as follows:

- Simulation is a numerical technique for conducting experiments on a digital computer, which may include stochastic characteristics, be microscopic or macroscopic in nature, and involve mathematical models that describe the behavior of a transportation system over extended periods of real time.

It is interesting to review the many textbooks on simulation and observe how different authors interpret what is meant by simulation. One author proposes that simulation is a technique that permits the study of a complex system in the laboratory rather than in the field [1]. Another suggests that simulation is a dynamic representation achieved by building a model and moving it through time [2]. Mize and Cox propose that it is a process of conducting experiments on a model of a system [3]. Wohl and Martin describe simulation as an imitation of a real situation by some form of a model that assumes the appearance without the reality [4]. In a somewhat more involved definition, Naylor et al. suggest that a simulation of a system is the operation of a model that is a representation of the system, is amenable to manipulations, and from which properties concerning the behavior of the actual system can be inferred [5]. The reader may wish to consult other textbooks for additional definitions of simulation [6–10].

Computer simulation modeling can be a controversial subject because of successful and unsuccessful applications in the past. Transportation specialists have many analytical tools at their disposal, and the key to successful analysis is the selection of the correct tool for the problem at hand. Because of the complex nature of simulation and the extensive time commitments normally required, simulation should be considered as the technique of last resort. That is, select simulation only after all other analytical techniques have been considered.

Some of the strengths of simulation are outlined in Table 13.1. In addition, some proponents of simulation argue that there are higher levels of benefits not only in the model itself but by the learning process it provides to the developers and users. The developer must rigidly specify relationships and structure to the system being simulated and by so doing better understands the system and how it works. The user begins to identify important control parameters and trade-offs between measures of effectiveness.

TABLE 13.1   Simulation Model Strengths

- Other analytical approaches may not be appropriate.
- Can experiment off-line without using on-line trial-and-error approach.
- Can experiment with new situations that do not exist today.
- Can yield insight into what variables are important and how they interrelate.
- Time and space sequence information provided not only means and variances.
- System can be studied in real time, compressed time, or expanded time.
- Potentially unsafe simulation experiments can be conducted without risk to system users.
- Can replicate base conditions for equitable comparison of improvement alternatives.
- One can study the effects of changes on the operation of a system: What if . . . , what happens?
- Can handle interacting queueing processes.
- Can transfer unserved queued traffic from one time period to the next.
- Demand can be varied over time and space.
- Unusual arrival and service patterns can be modeled which do not follow more traditional mathematical distributions.

Some of the reservations about using simulation are outlined in Table 13.2. In addition, those individuals with the greatest concern about the role of simulation identify a set of situations which have led to unsuccessful uses of simulation models beyond those listed in Table 13.2. It is argued that there are simulators who see simulation as the only approach to every problem. Experiences show that in some cases the simulation model itself becomes the end product rather than the use of the simulation model. In other cases the model builder concentrates too much effort in small, less significant aspects of the model and loses sight of the total model.

In conclusion, it becomes clear that simulation is not a cure-all solution but one of many analytical techniques available to the transportation specialist. Careful study is

TABLE 13.2   Simulation Model Reservations

- There may be easier ways to solve the problem. Consider all possible alternative ways.
- Simulation is time consuming and expensive. Do not underestimate time and cost.
- Simulation models require considerable input characteristics and data, which may be difficult or impossible to obtain.
- Simulation models require verification, calibration, and validation, which, if overlooked, makes the model useless.
- Development of simulation models requires knowledge in a variety of disciplines, including traffic flow theory, computer programming and operation, probability, decision making, and statistical analysis.
- Simulation is not possible unless the developer fully understands the system.
- The simulation model may be difficult for nondevelopers to use because of lack of documentation or unique computer facilities.
- Some users may apply simulation models and treat them as black boxes and really do not understand what they represent.
- Some users may apply simulation models and not know or appreciate model limitations and assumptions.

required before selecting simulation as the appropriate approach to a specific problem. Simulation is usually more time consuming and costly than other approaches and requires greater knowledge of the system and more creativity. But it has the potential of providing a better working knowledge of a transportation system and discovery of improved ways to plan, design, and operate such systems.

From a historical perspective, computer simulation modeling has a short but intensive history. Work on digital computers began in the 1930s and was first used in computer simulation in the 1940s. By 1950, there were approximately 100 noncommercial digital computers in the United States. The first commercial digital computers became available in the United States in 1954. So in several decades, the digital computer has grown of age and with it, computer simulation.

The first acknowledged digital computer simulation was undertaken in the late 1940s by von Neumann and Ulam. The computer simulation dealt with nuclear-shielding problems which were too expensive and dangerous for experimental solution, and too complicated for analytical treatment. The *Simulation Journal* began publication in 1953.

The earliest computer simulation work in highway transportation occurred in the 1950s. Intersection simulation was undertaken by the Road Research Laboratory in the United Kingdom in 1951. The first simulation work in the United States was published in 1953 and reported on intersection and freeway models developed at the University of California at Los Angeles. This was quickly followed by intersection simulation work at the University of Michigan, major arterial simulation at Philco, bus terminal and car-following simulation by the Port of New York Authority, and freeway interchange and ramp merging simulation at Midwest Research Institute.

The development of simulation models grew rapidly during the 1960s and 1970s, and bibliographies were published devoted exclusively to computer simulation models developed for the highway system.

In the United States for example, Gerlough and Capelle authored a monograph *An Introduction to Traffic Flow Theory* in 1964 [11]. A chapter of this monograph was devoted to simulation of traffic flow. In 1975, the 1964 monograph was updated and the coverage of simulation of traffic flow was increased [12].

Fox and Lehman published a state-of-the-art article in the *Traffic Quarterly* in 1967 [13]. In 1975 [14], and updated in 1977 [15], the University of California at Berkeley published a bibliography identifying selected references of applications of computer simulation to transportation systems.

In Europe, two bibliographies were published during this period. Wigan prepared an annotated bibliography in 1969 while at the Road Research Laboratory in the United Kingdom [16]. In 1972 the Organization for Economic Cooperation and Development (OECD) published a control systems document that gave extensive coverage to computer simulation [17].

By 1981, traffic simulation modeling in the United States had developed to such an extent that a special three-day conference was held on this subject. Some 75 persons representing researchers, developers, and users attended this conference, which was conducted by the Transportation Research Board and sponsored by the U.S. Department of

Transportation. The conference had two objectives. The first was to inform the user community about the availability of models: developed and planned. The second objective was to obtain feedback from the users' community on their experiences and future needs. The conference conclusion was that computer simulation was strong at that time and it was anticipated that this field would gain further strength in the future [18].

The last portion of this chapter contains an up-to-date inventory and brief description of most currently used computer simulation models for various highway systems. The 1980s have exhibited an interesting transitional period from model development to model improvements. These model improvements include model refinements such as adding optimization submodels, additional measures of effectiveness, integration of simulation models, and modeling traveller responses. The model improvements have also included conversion to personal computers, user-friendly interactive graphical pre- and postprocessors, improved user manuals, and educational programs. However, before turning attention to current models, consider the steps necessary in developing a computer simulation model.

13.2 STEPS IN DEVELOPING A SIMULATION MODEL

The intent of this section is to introduce the reader to the procedural elements in developing a simulation model. An overview is presented followed by a qualitative description of each procedural element. Space does not permit an exhaustive treatment of simulation model development and the reader may also wish to follow through the actual development of a computer simulation model in Section 13.3 or to consult simulation textbooks for greater detail [2, 3, 5–10].

A flowchart of procedural elements is shown in Figure 13.1. Eleven elements are identified in this figure and each is discussed in the following paragraphs. Of course, the magnitude of the tasks and elements vary between simulation models, and since model development is an art as well as a science, one could structure the development of a model in different ways. Consider this as a qualitative guide.

The first element in the suggested procedure is problem identification. The objective is to state the problem explicitly so that alternative analytical techniques can be evaluated in the next procedural element. What are the desired outputs, and what inputs affect the outputs? What are the time and space boundaries of the problem? Are there important stochastic elements in the process, and do individual particles need to be modeled? Is queueing involved, and are there interacting queueing processes? Do traffic conditions vary over time, and are arrivals/departures uniquely different from the classical mathematical distributions? These are some of the questions that need to be addressed.

The second element is to answer the question: Is simulation the appropriate analytical technique for the problem at hand? Several more detailed questions need to be addressed. How could the problem be solved without simulation? Why is simulation a better solution? How have others solved this type of problem? Does one have the time and resources for a simulation approach? Can the problem really be solved?

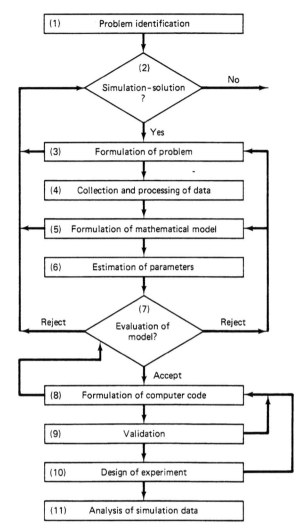

Figure 13.1   Flowchart of Procedural Elements

The third element is the formulation of the problem. This is accomplished by developing a first-level flowchart of the model consisting of three connected activities: input, process, and output. Particular attention is given to the input and output requirements. The objective of this element is to identify all needed input data for the model and associated output performance for the next data collection step. The input requirements normally include facility design elements, traffic demand patterns, operational rules and/or controls, and environmental conditions. The output requirements obviously vary depending on the type of problem but might include such measures of effectiveness as travel time, delay, fuel consumption, accidents, pollution, noise, queue lengths, and stops, etc.

The fourth element is the collection and processing of data based on the previously identified input and output requirements. It is often desirable simply to observe the traffic system to be studied and initially collect only a sample of data. Literature should be reviewed to reveal how and what data have been collected in previous simulation and related studies. Having designed the data collection plan, the field data are collected to meet the input requirements and associated output performance. Sample sizes must be adequate for both calibration and validation. The final task is the processing of the collected data so that it is compatible with simulation model requirements. This data processing may consist of calculating means and variances, plotting distributions and relationships, performing regression analysis, and unit conversions.

The fifth element is the formulation of the mathematical model. This is accomplished by developing second- and third-level flowcharts and is one of the most time-consuming and critical elements. The first-level flowchart described earlier is the starting point and attention is focused on the process activity that connects the input to the output. The second-level flowchart identifies all major subroutines and indicates their connectivity. The input requirements and anticipated output for each subroutine are noted. The third and most detailed set of flowcharts are flowcharts for each individual subroutine. The steps in each subroutine are identified, described, and connected.

The sixth element is the estimation of all required parameters in the model. Some parameters may be deemed to be deterministic, while others are stochastic. The deterministic parameters may be constant for all situations, may take on one of a set of constant values depending on the situation, or may vary over a range in a continuous manner through some form of regression. The stochastic parameters each require a mean and a variance as well as the identification of its distribution form. A stochastic parameter may be constructed in a relatively simple manner or can become extremely complex. The balance between realism and simplicity is often tested at this stage in the simulation modeling effort.

The seventh element is a manual evaluation of the current state of the model. The first task is to perform hand-calculation solutions of a variety of problem types so as to identify incomplete loops or closed loops, to check flexibility of data input and range of values, and to check reasonableness of intermediate and final outputs. In this process judgment decisions are required, such as adding, changing, and deleting variables; modifying deterministic and stochastic parameters; and modifying structure of the model. At some point in this process a decision is made to accept or reject the modified developed model. If the model is rejected, the developer reevaluates earlier established elements, and either modifies the earlier elements or terminates the effort to develop a simulation model.

Assuming that the model has been accepted in the previous element, the next element is the formulation of the computer code. The difficulty of this task depends on the quality of the three levels of flowcharts. With good flowcharts for relatively simple simulation models the computer code can almost be written on top of the flowcharts themselves. An important step is the selection of the computer language and computer facilities. While FORTRAN is most widely used and most transferable between computer facilities, special simulation languages should be considered. Factors to be considered include developers' knowledge of simulation languages, availability of compilers on selected computer facilities, compatibility of unique model features with simulation languages, and the extent to which the simulation model will be used by other investigators on other computer facilities. If the simulation model is expected to be used by others and there is an expectation that the model will be modified in the future, the computer code should be modular in structure with many comments inserted in the program. The final step is debugging the computer code which can be more time consuming than writing the code. A few tips follow.

- Debug one subroutine at a time and then begin to cluster them.
- Initially use known deterministic data rather than stochastic data.
- Perform manual calculations to check.
- Add intermediate output results temporarily.
- Look for more than one error at a time.
- Remember that computer code represents the flowcharts now, not necessarily real life.

The ninth element is validation. Actually, this element includes verification, calibration, and validation. The relationship of these three tasks are depicted graphically in Figure 13.2. Verification is a more exhaustive form of the earlier task of debugging code. The objective is to ensure that the computer code is doing exactly what is specified in the flowcharts. At this point no attention needs to be given to the real-life situation. The next task is calibration, and the difficulty of this task is dependent on the quality of field measurements, the comprehensiveness of the computer model, and the complexity of the system being studied. Only a portion of the collected input data and output performance can be used for calibration; the remainder of the collected data must be reserved for later validation. Reasonable adjustments are made in the model to match the model's output with real-life observations. The final task is validation. The unused portion of the field collected data input is placed in the model, and the model predictions are compared with the related unused portion of the field-observed real-life observations. No adjustments are made to the model and the differences record how well the completed model represents real life under the conditions tested. If the results of the validation effort are acceptable, the computer simulation model is ready for application. If not, further calibration and validation are required.

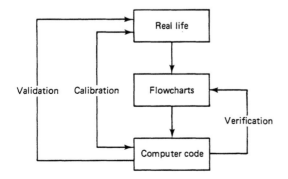

Figure 13.2   Verification, Calibration, and Validation Process

The tenth element is the formulation of the design of experiment. All applications of computer simulation models require a design of experiment, but its comprehensiveness depends on the size and flexibility of the model as well as the complexity and variety of situations to be evaluated. Assuming that the system performance for a multisurfaced set of control variables is desired, the following suggestions should be considered.

- Identify control variables.
- Select limits or bounds for each control variable.
- Select size of mesh (or interval) for each control variable.
- Select primary control variable(s) and hold secondary control variable(s) constant.
- Consider building loops within the computer code to vary primary control variable(s) automatically.
- Consider incorporating a search procedure within the computer code to determine the optimum conditions automatically.

Special attention is required for stochastic entities within the model. Replication of investigations for each cell is required and the number of replications needs to be determined. Each stochastic entity should have its independent generated random number sequence and flexibility in selecting the initial random number seed.

The eleventh and last element includes the production runs, analysis of results, and documentation. A log should be kept of all production runs, and the computer code should include the capability for labeling each output. The analysis of the results may reveal the need to modify the computer code or use auxiliary programs to plot graphical results, to perform statistical tests on results, and/or print results in report form. The documentation should not only report on the study results but should also include a user's manual and a computer systems operator's manual. All variables should be listed and defined; three-level final flowcharts included; computer code listed, including comments; and a sample problem containing a listing of sample input and associated sample output.

This section of the chapter has attempted to describe qualitatively a step-by-step process for developing a computer simulation model. There are obviously many variations, depending on the skills of the developer, the complexity of the problem, and the intended use of the model. To reinforce this qualitative discussion, a computer simulation model is developed and applied in Section 13.3.

13.3 AN EXAMPLE OF DEVELOPING A SIMULATION MODEL

The steps taken in developing this example computer simulation model are in parallel with the procedural elements described in Section 13.2. The example problem is based on a simulation model developed at the University of California in Berkeley and has the acronym "FREQ."

Ten generations of the FREQ model have now been developed, and later versions contain many special features, such as control and design improvement optimization, spatial and modal traveler responses, fuel and emission measures of effectiveness, incident and reconstruction investigation options, and many others. However, for purposes of demonstrating the development of a computer simulation model, the earlier and simpler FREQ3 model was selected. The FREQ3 model is a macroscopic, deterministic, and simulation-only model of a directional freeway [19].

13.3.1 Problem Identification

An analytical tool is needed to analyze and assess the traffic performance along a directional freeway. The primary input to the model will probably include freeway design features and origin–destination demand patterns. The model should be designed to study a peak traffic time period over a reasonable length of freeway. Only one direction of traffic needs to be considered, and only the directional freeway and its associated ramps need to be included initially. Because of the anticipated expansion of the model in a number of significant ways and the extensive time–space domain of the intended model, a deterministic macroscopic-type model is envisioned. Because of the variation of traffic demands over time, a multitime slice approach is anticipated. Interacting queueing processes will occur, and modeling this interacting queueing process is judged to be extremely important. There were misgivings at the beginning that the macroscopic nature of the proposed model and the lack of stochastic entities will limit the model, and efforts should be made in the model to compensate for this deficiency as much as possible.

13.3.2 Is Simulation the Solution?

Almost all analytical techniques are expected to be required for this problem, including demand–supply analysis, capacity analysis, traffic stream models, shock wave analysis, and queueing analysis. A computer simulation model appeared to be best to integrate these various techniques into one analytical framework. If simulation were not selected, shock wave analysis would probably be the best alternative since queueing is involved and there is concern for the spatial and temporal positioning of the queue. The development of the simulation model is expected to require considerable time and resources, and probably will exceed current allocations. A stage development of the model is most likely, which focuses attention on the modularity construction of the model. In the final analysis, it is concluded that developing a simulation model is feasible and is the preferred analytical approach to the problem at hand.

13.3.3 Formulation of the Problem

A first-level flowchart is shown in Figure 13.3. The anticipated input requirements include freeway and ramp capacities, origin and destination demands, and traffic stream models. The anticipated output requirements include freeway travel times, ramp delays,

Input   Output

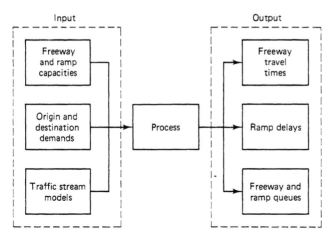

Figure 13.3   First–Level Flowchart of FREQ Model (From Reference 19)

and freeway and ramp queues. There was a tendency to consider expanding the input requirements and the output performances and thereby broadening the scope and power of the model. However, with concern for the limited resources at the very outset, the inputs and outputs were held to a minimum.

13.3.4 Data Collection and Processing

The first-level flowchart developed previously was used as a guide in determining the data to be collected. A nearby site, Route I-80 eastbound through the Berkeley area, was selected and a geometric layout of the freeway lanes and ramps constructed. The data collection field site is shown in Figure 13.4 and the freeway lanes and ramps are indicated. The field site was divided into 30 subsections and there were 11 on-ramps and 15 off-ramps. Floating car test runs were made along the freeway during the

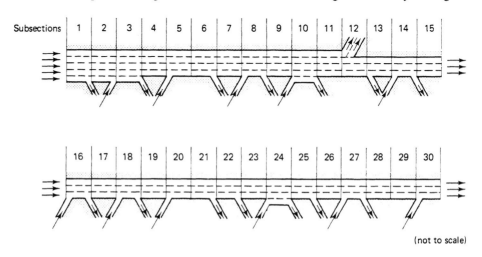

(not to scale)

Figure 13.4   Data Collection Field Site

afternoon peak period to ensure that the field site exhibited desirable characteristics such as queueing on the freeway, ramp delays, and varying traffic demands and performance. The following field studies were undertaken and the results processed as indicated.

- Freeway and ramp capacities were calculated using the *Highway Capacity Manual* and bottleneck capacities checked in the field.
- Origin and destination demands were obtained by a postcard survey of entering freeway traffic. *O–D* tables were constructed for each 15-minute period from 4:30 to 6:00 P.M. which included the pre-peak period, peak period, and post-peak period.   -
- Photographic studies were undertaken at several locations along the freeway to develop appropriate traffic stream models. The results were plotted in the form of speed versus volume/capacity ratio curves and equations developed.
- Freeway travel times were obtained by floating car runs during the study period. Average travel times were then calculated for each 15-minute period.
- The occurrence of queues on ramps were observed in the floating car study. Field studies were conducted at ramps where queues had been observed and delays estimated based on queue lengths and discharge rates. Average delays were then calculated for each 15-minute period.
- The occurrence of queues on the freeway were assumed to exist wherever the freeway speeds were observed to be less than 30 miles per hour in the floating car runs. A freeway queueing map was constructed on a time–space diagram.

13.3.5 Formulation of the Mathematical Model

A second-level flowchart was developed and is depicted in Figure 13.5. The program was envisioned as containing a main executive program calling other subroutines, which would do the actual work of the program. A brief description of each subroutine that was called is listed below. As each subroutine completed its tasks, it would return control to the main program, which in turn would call the next subroutine.

- READV   would read all input data and would store the data in appropriate format and location for later use by other subroutines.
- XRAMP   would prepare a special array for later merge analysis.
- RESET   would be called each time a new time slice is to be analyzed. The purpose of this subroutine would be initializing all computational arrays.
- SLICE   would retrieve the previously stored data for the time slice to be analyzed.
- RAMPQ   would be called to perform the on-ramp queueing analysis.
- VOLUM   would be called to calculate the demands for each freeway subsection.
- RAMPS   would be called to perform the merge analysis at each on-ramp.

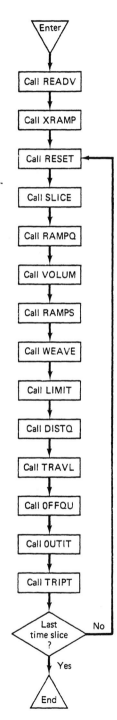

Figure 13.5   Second–Level Flowchart of FREQ Model (From Reference 19)

- WEAVE   would be called to reduce the capacity of any subsection in which weaving is encountered.

- LIMIT   would be called to compare subsection demands and capacities and to limit volumes to capacity limits.

- DISTQ   would be called to further limit subsection volumes due to upstream mainline queues.

- TRAVL   would be called to calculate travel times and travel distances.

- ØFFQU   would be called to perform off-ramp queueing analysis.

- ØUTIT   would be called to prepare a final printout of the basic results of the simulation.

- TRIPT   would be called to prepare an optional output of a travel time trip table; control would then be transferred to RESET to process the next time slice or to terminate the simulation if all time slices had been processed.

Third-level flowcharts were prepared for each subroutine. The flowchart for the ØFFQU subroutine is presented in Figure 13.6 as an example of a third-level flowchart. The ØFFQU subroutine was designed to check on adequacies of off-ramp capacities to handle off-ramp demands. Each off-ramp was checked in each time slice except for the last off-ramp, which is the mainline output (this is checked in the freeway subsection capacity–demand investigations). This subroutine is called by the main program after freeway travel times and travel distances are calculated (TRAVL) in each individual time slice. The variable NDMI is initialized and set equal to the number of freeway destinations minus one (accounts for mainline output destination). If NDMI is equal to zero, the off-ramp check does not need to be called. If NDMI is greater than zero, the variable ID is initialized starting with the first off-ramp and going to the last non-mainline off-ramp. The excess demand variable (EXØFFV) is calculated by subtracting the off-ramp capacity from the off-ramp demand. If there is no excess demand in this time slice, a check is made to see if there was an excess demand in the last time slice which was transferred to this time slice. The next calculation is to determine the unsatisfied transferred queue for the next time slice. If there was no transfer, there is no off-ramp queue and the next off-ramp is investigated. If there was a transfer, a check is made to see if all the queue can be discharged. If the queue can be discharged, the queue delay can be calculated and no off-ramp queue is transferred to the next time slice. If the queue cannot be discharged in this time slice, control is passed to the EXØFFV > 0 side of the subroutine.

Returning to the EXØFFV > 0 check, if EXØFFV > 0, the queue delay is calculated including any unsatisfied transferred queue from the last time slice. If the last off-ramp has not been analyzed, the next off-ramp is checked. If the last off-ramp has been analyzed, the control is returned to the main program.

The development of flowcharts can be a very time consuming and iterative process. However, the flowcharts are the bridge between the real world and the ultimate computer code, and comprehensive well-prepared flowcharts can greatly ease and improve the computer coding effort.

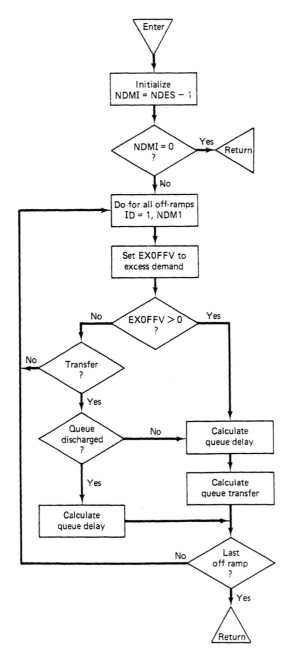

Figure 13.6   Third–Level Flowchart for ØFFQU Subroutine of FREQ Model (From Reference 19)

13.3.6 Estimation of Parameters

The macroscopic and deterministic nature of the proposed model significantly simplified the estimation of parameters. The freeway and ramp capacities were calculated based on the *Highway Capacity Manual.* Only the calculated weaving capacity varied over time and then only in discrete time slices. Consideration was given initially to including some stochastic variations to the time slice origin and destination demands. One approach considered was to develop demand distributions for each origin and destination, and to select the demands to be used by a stochastic process such as generalizing random numbers and selecting demand values from a normal distribution. An alternative approach which was ultimately selected was to suggest that a set of computer runs be made using lower-than-average, average, and higher-than-average origin and destination demands. The traffic stream models consisted of speed versus volume–capacity ratio relationships selected based primarily on the design speed of the subsection.

In conclusion, the decision to develop the FREQ model as a macroscopic and deterministic model greatly simplified this task of estimating parameters and in the large picture the entire development and application of the model. However, the cost of these simplications is the lack of ability to track individual vehicles in the system and to model any variations in system performance due to stochastic phenomena.

13.3.7 Evaluation of the Current State of the Model

The current state of the model was evaluated using hand calculation procedures following the flowcharts developed. The first checks considered only undersaturated conditions because of the greater ease in making manual calculations and also because only the simpler portions of fewer subroutines were utilized. Prior to checking the entire program, numerical checks were made of each subroutine. Later, oversaturated freeway situations were checked.

These checks on the flowcharts of the model were not trivial efforts. Considerable time and effort were required for each investigation and numerous modifications were made in the flowcharts. It was fortunate that the two-person team that developed the model, a traffic engineer and a computer programmer, worked well together and shared their expertise. After many trials and errors, the revised flowcharts were accepted and work began on the computer coding.

13.3.8 Formulation of Computer Code

The simulation language selected for the FREQ model was FORTRAN. The developers had more experience with FORTRAN and the model was designed at the outset for use on a wide variety of computer systems in various locations in the world. It was also a requirement of the contract to test and demonstrate the program on two governmental facilities elsewhere in the country.

The computer code for each subroutine flowchart was written and tested. Comments were added to the code and each line of code identified by subroutine and line number. Then the subroutines were clustered and further tested. Finally, the entire model was coded and sample problems developed to test the computer code starting with simple problems and continuing into more comprehensive and complex problems.

13.3.9 Verification, Calibration, and Validation

Verification was an extension of the debugging task described in the preceding paragraph. The model was applied to the data collection field site shown earlier in Figure

13.4. Spot checks were made to confirm that the computer program gave the same results as one would obtain by manually solving the problem following the flowcharts explicitly. However, the complete solution of the problem solved manually was extremely time consuming even for just one peak traffic period. Unique traffic situations were not completely verified, and all combinations of features of the model were not fully engaged because of limited resources. Over the years, several bugs were found when unique traffic situations were encountered, and this was the price for not undertaking more exhaustive verification.

The FREQ model was calibrated for the data collection field site shown in Figure 13.4. The criteria for calibration included bottleneck identification, trip travel times, and system performance. The model was considered calibrated when the following three conditions were met:

- Each freeway bottleneck was identified, and the predicted start and end of congestion occurred within 15 minutes of the field-observed conditions.
- The trip travel time along the freeway for each time slice predicted by the model was within plus or minus 10 percent of the field observed trip travel times.
- The system performance (in terms of total passenger-hours) for the entire freeway during the entire peak traffic peak predicted by the model was within plus or minus 2 percent of the field observed system performance.

The calibration procedure primarily was directed to adjusting bottleneck capacities. The traffic performance at these bottlenecks overshadowed all other traffic performance. Small changes in bottleneck capacities on the order of 1 to 2 percent changed location and occurrence of other bottlenecks, modified travel times by 10 or more percent, and significantly affected total system performance. The calibration procedure also included the modification of the speed versus volume–capacity ratio relationship. In fact, this calibration experience led to the development and incorporation into the FREQ model of empirically derived speed versus volume–capacity ratio relationships for California freeways.

A second day of peak period traffic data was used for the validation effort. The input data for the second day was put into the FREQ model and the freeway performance predicted by the model without adjustments. This predicted freeway performance was then compared with the field-measured freeway performance. The criteria identified earlier in the calibration process was used to accept or reject the validity of the model. For the selected second day there was agreement between the predicted and field-measured performance. It should be noted, however, that the data obtained for several other days did not meet the validation requirements. This was due to a variety of reasons, including the occurrence of freeway incidents and adverse weather conditions.

The conclusions of the verification, calibration, and validation process were as follows:

- In the verification process it is impossible to check the interaction of all elements in such a comprehensive and complex model. Undoubtedly, future applications will reveal flaws in the model when applied in very special circumstances.

- The calibration process revealed that the required accuracy for estimating bottleneck capacities exceeded present techniques for predicting capacities. It was concluded that calibration will be required whenever the model is applied to a new site.

- The validation process indicated that the model could reasonably represent a study site that had previously been calibrated provided that "normal" traffic conditions were encountered.

13.3.10 Formulation of the Design of Experiment

One of the first applications of the FREQ model was to investigate the effect that design improvements would have on freeway performance. A branch-and-bound manually derived search procedure was developed, the structure of which is shown in Figure 13.7, which became the design of experiment for this investigation.

The first level of investigation consisted of making one simulation run, which represented existing conditions without any design improvements. The total system travel time is recorded, and all freeway bottlenecks are identified.

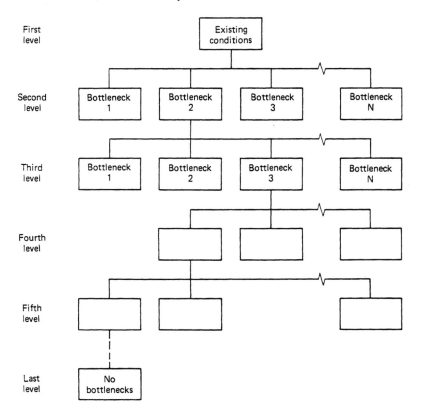

Figure 13.7   Design of Experiment for Design Improvement Investigations

The second level of investigation consisted of making one simulation run for each bottleneck site identified but with the capacity of the bottleneck increased to represent one additional lane of capacity. The reduction of total system travel time due to each individual bottleneck improvement is recorded and converted to an annual travel time savings. The annual cost for each bottleneck improvement is calculated and a cost-effectiveness ratio is determined for each bottleneck improvement. The bottleneck improvement that exhibited the lowest cost-effectiveness ratio was selected for improvement. The output of the selected run was inspected to identify all remaining freeway bottlenecks.

The third level of investigation consisted of making one simulation run for each bottleneck site identified in the second level of investigation, but with the capacity of the bottleneck increased to represent one additional lane of capacity. Similar to the second level of investigation, the annual travel time savings, annual improvement costs, and the cost-effectiveness ratio are calculated for each. The bottleneck improvement with the lowest cost-effectiveness is selected and the remaining bottlenecks identified. The fourth and later levels of investigations are continued in the same manner until no freeway bottlenecks remain.

In the original design of experiment only one "normal" set of inputs were used in the FREQ model. One could consider a set of initial conditions that could be represented by additional planes in Figure 13.7. The initial conditions could have represented weekend traffic, lower weekday traffic, higher weekday traffic, unusual weather situations, incident situations, future demand growth, and the like. Since no stochastic processes exist in the model, replications due to stochastic variations were not required.

13.3.11 Analysis of Simulation Data

Production runs were made with the FREQ model following the design of experiment described in previous paragraphs. A total of 46 production runs were made and the results are shown in Figure 13.8. A log was kept of these runs and each output was labeled in accordance with the design improvement being investigated (i.e., a code "22-04-07") indicated that a lane had been added to subsections 22, 04, and 07 in that order.

The final summary of the completed design of experiment was the graph shown in Figure 13.9. Based on the graph, the decision maker could select the desired improvement plan considering system performance, cost-effectiveness evaluation, and budget limitations.

The experiences gained in this investigation led to the development of the FREQ3D model, which combined the FREQ3 model with the branch-and-bound search program and automatically searched through the alternatives in one computer pass [20]. Extended versions of the FREQ3 model have also included FREQ3C, FREQ3CP FREQ4CP, FREQ5PE, FREQ6PL, FREQ6PE, FREQ7PE, FREQ8PL, FREQ8PE FREQ8PC, FREQ9PE, and FREQ10PC. [21–31].

It is difficult in a few pages or with one simulation example to convey the art and science of computer simulation modeling. In the final analysis one learns to simulate

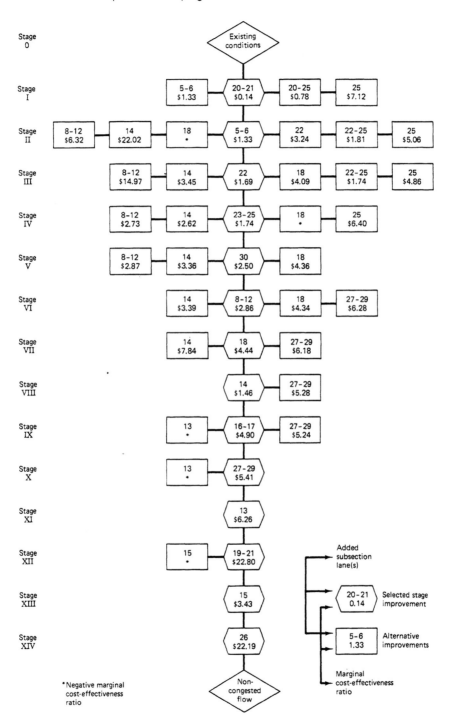

Figure 13.8   Results of Design Improvement Investigations (From Reference 20)

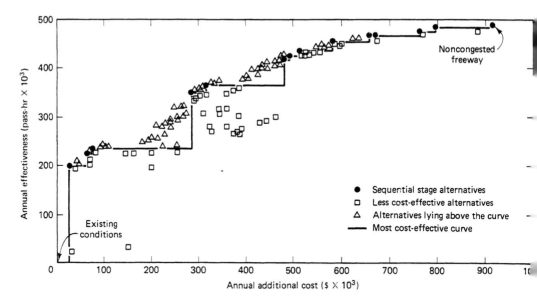

Figure 13.9   Cost-Effectiveness Diagram of Design Improvement Investigations (From Reference 20)

by actually developing a simulation model. In fact, simulation is a never-ending, learning experience.

The next section describes briefly some of the current computer simulation models in operation today. Learning about these current models may give the reader further insights into the development and application of computer simulation models.

13.4 SOME COMPUTER SIMULATION MODELS

Many computer simulation models are available today for analyzing various operating environments of the highway system. These operating environments include signalized intersections, arterial networks, freeway corridors, and rural highways. Both microscopic and macroscopic computer simulation models have been developed for each of the operating environments identified above. Selected models in each category will be identified and a brief description of one such model will serve as an example in each category.

Several recent publications have provided the basis for the inventory of available computer simulation models [18, 32–34]. The inventory is not complete, and newer versions of existing models as well as new models will be developed in the future. However, it is intended that this inventory and brief descriptions will give the reader further insights into the development and use of computer simulation models for the highway transportation system.

13.4.1 Signalized Intersections

Both microscopic and pseudo-macroscopic computer simulation models have been developed and are currently available for application at individual signalized intersections. The microscopic models include the single-intersection version of NETSIM [35], SIGSIM [36], and the TEXAS [37] models. The TEXAS model will be described as an example of this type of model. The macroscopic models are referred to as pseudo-macroscopic simulation models because they have attributes of both simulation and analytical models. Many macroscopic models of this type are currently available, a few of which are the CALSIG [38], CAPCAL [39], CAPSSI [40], POSIT [41], SIDRA [42], SIGNAL-85 [43], and SOAP-84 [44]. The CALSIG model is described as an example of this type of model.

The microscopic computer simulation model TEXAS was developed at the University of Texas beginning in the mid-1970s, and refinements and enhancements have continued through the 1980s. The TEXAS model is a microscopic, stochastic, time-scanning computer simulation model designed to predict intersection traffic performance as a function of the design, demand, and control of an existing or proposed intersection. The predicted performance includes delay, percent of vehicles stopped, travel time, and queue length for each approach and the whole intersection. The intersection design options are extensive, including the number of approaches, the number of lanes per approach, and the use made of individual lanes (i.e., exclusive turn lanes, mixed movement lanes, and through lanes). Turning movements and up to 15 classes of vehicles can be modeled to represent the intersection demand input. The model can simulate many types of intersection control, including stop sign control, yield sign control, pretimed signal control, and actuated signal control.

The components of the traffic simulation package are shown in Figure 13.10. The package consists of a geometry processor, a driver–vehicle processor, and a traffic simulation processor. The geometry processor calculates the geometric paths of vehicles on the approaches and in the intersection, identifies points of conflict between vehicle paths, and determines the minimum available sight distance along each inbound approach. The driver–vehicle processor characterizes the traffic stream to be simulated by generating queue-in time and other random descriptors for individually characterized driver–vehicle units. The traffic simulation processor processes each driver–vehicle unit through the intersection system, and gathers and reports a large selection of performance statistics.

The TEXAS program was written in FORTRAN IV and originally was operational on an IBM/CMS 4341 computer system. More recently, a PC version of the program has been developed and user-friendly enhancements incorporated. The model has proven to be a useful and effective method for predicting traffic performance at existing and proposed intersections.

Attention is now turned to the CALSIG model. The CALSIG model was developed at the University of California at Berkeley in the mid-1980s. The CALSIG model is a macroscopic, deterministic model designed to predict intersection traffic performance as a function of the design, demand, and control of an existing or proposed

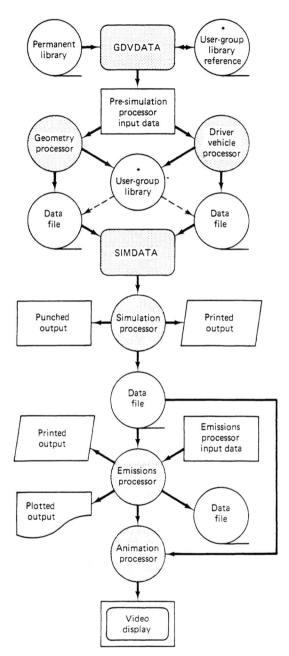

Figure 13.10   Components of the
TEXAS Traffic Simulation Package (From
Reference 37)

signalized intersection.   The predicted performance includes delay, percent vehicles
stopped, volume–capacity ratio, and queue status for each lane group, each approach
and the whole intersection. The lane configuration at the intersection can be specified
by the user or generated by the model. The turning movements and classification o

vehicles must be specified by the user. The signal control can be pretimed or actuated, and can be specified by the user or generated by the model. A unique feature of the model is that it can be used to perform up to four levels of analysis, ranging from planning analysis to intermediate analysis to volume–capacity analysis to operations analysis. The model is compatible with the 1985 *Highway Capacity Manual.*

The components of the CALSIG program are shown in Figure 13.11. The user inputs the turning movement demand and either specifies the geometrics or requests the program to generate the geometrics. A preliminary or planning-type analysis is performed and the user may further modify the geometrics and rerun, terminate the run, or request the next level of analysis. In the intermediate analysis, the user specifies the signal phase design and saturation flows or may request the program to generate them. An intermediate analysis is performed and the user may further modify signal phase design/saturation flows and rerun, terminate the run, or request the next level of analysis. In the comprehensive analysis (*v/c* analysis), the user specifies the signal control parameters or may request the program to generate them. A comprehensive analysis is performed and the user may further modify signal control parameters and rerun, terminate the run, or request the next level of analysis. In the final or operations analysis, the user requests the delay calculations and the final performance analysis is provided as well as suggestions for possible improvement.

The CALSIG program was written in Turbo-Pascal (Version 3) language for a personal computer. The CALSIG software package is for PC-DOS and MS-DOS, and requires at least one double-sided disk drive and a minimum of 260k memory. The program is designed as a user-friendly interactive program with accompanying graphics. The model is in its early stages of application but users indicate that it is an easy-to-use and effective model.

13.4.2 Arterial Networks

Both microscopic and macroscopic computer simulation models have been developed and are currently available for arterial network applications. The NETSIM [35, 45] model is the only microscopic computer simulation model available for arterial networks and it is described later in this section. There are a relatively large number of macroscopic models available, including MAXBAND [46], PASSER II [47], PASSER III [48], SIGOP III [49], SPAN [50], SSTOP [51], and TRANSYT [52]. The TRANSYT model is described as an example of this type of model.

The NETSIM network simulation model, formerly called UTCS-1, performs a microscopic simulation of an urban traffic network. It is designed to be applied by the traffic engineer and researcher as an operational tool for the purpose of evaluating alternative network control and traffic management strategies.

The input data requirements are rather extensive and include network supply features, traffic demand patterns, and signal timing plan. The network is made up of directional links and nodes, and physical features of each link must be specified. The traffic demands are entered as input and output network flows, and with specified turning movement patterns and traffic composition.

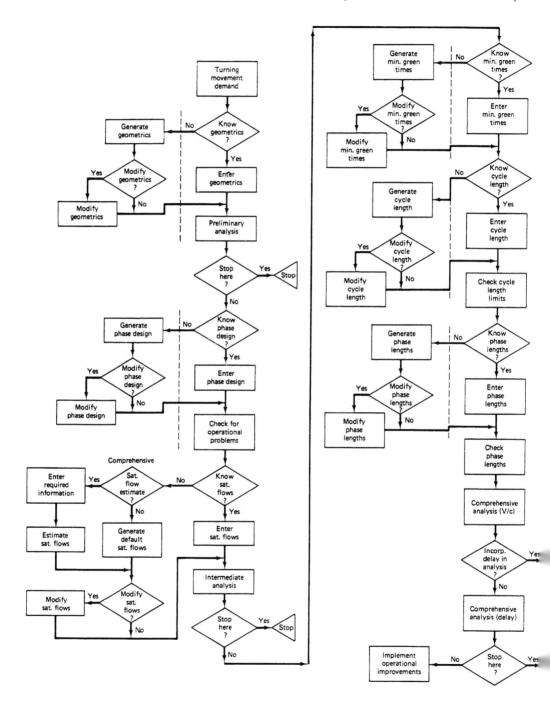

Figure 13.11   Procedural Flowchart for the CALSIG Model (From Reference 38)

A flowchart of the NETSIM model system is shown in Figure 13.12 and is divided into three major components: NETSIM pre-processor, NETSIM traffic simulator, and NETSIM post-processor.

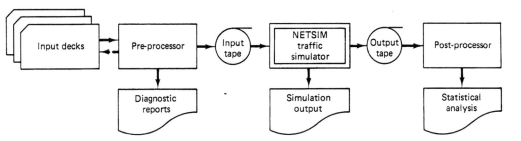

Figure 13.12   NETSIM Model Flowchart (From Reference 45)

The NETSIM pre-processor is designed to simplify the process of preparing and checking data inputs. It includes a comprehensive set of automatic "diagnostic checks" which are performed on all data inputs. It also provides for the convenient packaging of successive runs based on sequential modification of input conditions. The pre-processor may be operated either independently or may be integrated directly with the main program.

The NETSIM simulator contains the main simulation program. It consists of 60 separate routines, which may be linked together in a variety of optional configurations depending on the requirements of the user. The simulator requires as input a coded description of a street network, together with a prespecified control plan and a set of input volumes. Its output includes a set of standard measures of traffic performance, expressed as both link-specific and network-wide values.

The NETSIM post-processor consists of a set of standard data mainpulation and evaluation routines designed to operate on the outputs of the main simulation program to compare the results of two or more simulation runs, construct a "historical" data file summarizing their results, and subject the resultant data set to a set of standard statistical analysis.

The output consists of the input echo and network traffic performance. The performance includes travel time, number of stops, delay, fuel consumption, and emissions.

The NETSIM program is written in FORTRAN and is operational on mainframe computers such as the IBM 4341 computer [35]. Consideration is being given to developing a personal computer version [45].

The TRANSYT model has had a long history of continuous improvements from the late 1960s. It is probably the most used traffic simulation model in the world. The early versions of the model were developed in England with Robertson as the principal author. There have been nine English versions (TRANSYT1 through TRANSYT9). In the 1970s, Americans began to use TRANSYT and develop improved versions for American conditions. These American versions have included TRANSYT6C, TRANSYT6N, and TRANSYT7F, among others. The TRANSYT7F has continuously been enhanced in the United States during the 1980s and is the version most widely used. The TRANSYT model or variations of it have been used in many countries.

The TRANSYT model is a macroscopic, deterministic single-time period simulation and optimization model. The simulation submodel is used to calculate the performance of an arterial network for a specified signal timing plan. The optimization submodel is a hill-climbing optimization process which ultimately determines the near-optimum signal timing pian. The simulation submodel can be recalled to calculate the performance of the near-optimum signal timing plan. The input requirements include a user-specified signal timing plan, the network configuration and saturation flow rates on each approach to each intersection, and the intersection turning movements. A simplified flowchart for the TRANSYT model is shown in Figure 13.13.

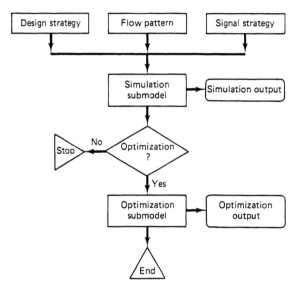

Figure 13.13   Simplified Flowchart for the TRANSYT Model (From Reference 52)

The output of the TRANSYT7F program includes an input data report; traffic performance table for each link, including calculations of travel time, delay, stops, queues and fuel consumption; and a summary table of network performance. In addition, the user may request signal controller tables, flow profiles, and time–space diagrams.

The TRANSYT7F program is written in FORTRAN and is operational on mainframe and personal computers. Many TRANSYT short courses have been held, and TRANSYT workbooks are readily available. A self-study user manual is also available.

13.4.3 Freeway Corridors

Both microscopic and macroscopic computer simulation models have been developed and are currently available for freeway corridors. The INTRAS [53] model is the only microscopic computer simulation model available for freeway corridors, and it i described later in this section. There are several macroscopic models available, including CORQ [54], FREQ [19–31], FRECON2 [55] (this model is based on earlier model called MACK and FREFLO), and KRONOS [56] models. The FRECON2 model is described as an example of this type of model.

The microscopic computer simulation model INTRAS was developed by KLD and Associates in the late 1970s, and refinements and enhancements have continued through the 1980s. The INTRAS model is a microscopic, stochastic, vehicle-specific, time-stepping computer simulation model designed to predict traffic performance for a directional freeway and surrounding surface street environment based on user-specified design, demand, and control. The predicted performance is unique in several ways. In addition to predicting normal traffic performance measures, such as speeds, distances, travel times, and fuel consumption, a variety of graphical plots are available, such as distance–time traces and density contour maps. The predicted performance is also exhibited as expected traffic measurements that would be derived from detectors placed along the freeway. The network design includes specifications for each link as to its link type, number of lanes, and connectivity to other links in the network. The expected flow rate on each link is specified by vehicle classification and lane usage. The control can include ramp control and signal control. Special attention is given to modeling freeway incidents.

Functional structure and information transfer of the INTRAS model are shown in Figure 13.14. The supervisor module is called INTRAS and calls the other modules as needed. The other modules are identified in the figure and are described briefly below.

• PORGIS MODULE	examines all inputs for errors and inconsistencies and generates printed tables of input data.
• LIS MODULE	loads simulation case data into data arrays.
• SIFT MODULE	is the major module that performs all simulation activities and contains over 80 subroutines.
• FUEL MODULE	provides link-specific evaluations of fuel consumption and vehicle emissions.
• POSPRO MODULE	creates a file of simulation statistics for future processing by the SAM module.
• SAM MODULE	performs statistical comparisons between pairs of simulation runs or between a simulation run and field data.
• INCES MODULE	performs all processing of detector data and generated incident detection data, point processing, and MOE reports.
• INPLOT MODULE	prepares vehicle trajectory and MOE contour plots.

The INTRAS program was written in ANSI FORTRAN and operational on an IBM 360/370 computer system under OS/360 and a CDC 6600/7600 computer system equipped with a SCOPE operating system. Refinements and enhancements are being made to the model to make it more user-friendly and to utilize fully the many comprehensive features of the model.

Attention is now turned to the FRECON2 model. The FRECON2 model is an enhanced version of the MACK and FREFLO models. The major enhancements

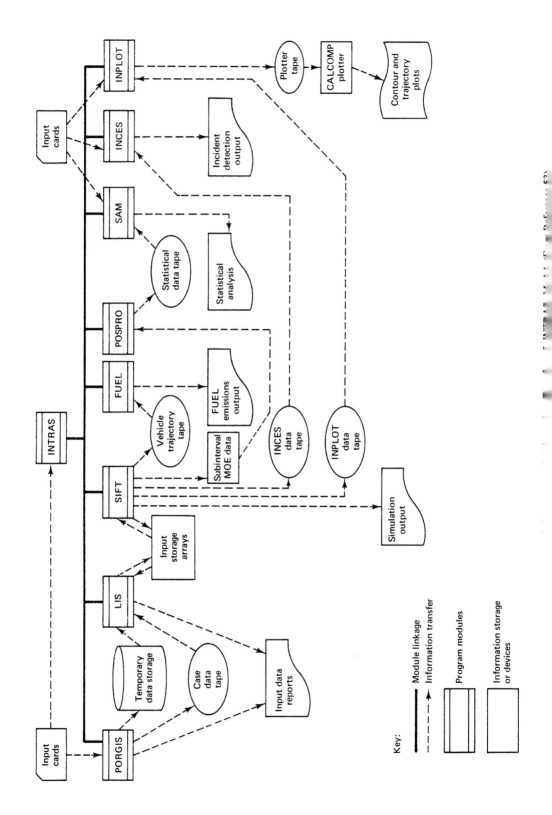

Key:

——— Module linkage

- - - ▶ Information transfer

▭ Program modules

▭ Information storage or devices

include traffic responsive priority entry control, improved means of handling flows and queues at bottleneck locations, and modeling parallel routes with spatial diversion due to entry control. FRECON2 is a dynamic macroscopic freeway simulation model that can simulate freeway performance under normal and incident conditions. The model can generate point detector information for calibration and validation. The model can generate a traffic responsive priority entry control strategy and evaluate its effectiveness. The traffic performance measures include travel times, queue characteristics, delay, fuel consumption, and emissions. The input data required includes subsection geometrics influencing capacity and origin–destination information.

Figure 13.15 shows the structure of the FRECON2 main program. The program starts with the program initialization block, which sets all default values for model inputs and initializes program variables. The program proceeds to read all inputs for a simulation run by executing the read inputs block. Initializations for a simulation run are performed in the next block, followed by a block that prints all model inputs (for confirmation purposes) onto the output file. At this point a check is made to see if the user has requested a simulation. It is often convenient to select the option of no simulation (which will cause the program to bypass the simulation block) to allow checking of a new input deck. If any model input errors exist, the program will bypass the simulation block and print appropriate diagnostic statements.

If simulation has been requested and no input errors exist, the program proceeds to the simulation block. The simulation block contains the state equations. These equations lead to the arterial subsection travel time and the freeway subdivision speed and density values in time. The simulation process includes the accumulation of freeway corridor performance measures and the application of the selected control strategy. When complete, the simulation results are printed onto the same file as the model input confirmation.

The FRECON2 program was written in FORTRAN IV language and is operational on IBM 360 and 370 systems. The *FRECON2—User's Guide* [55] is fairly comprehensive and contains a description of the model structure, identifies and describes each input, identifies and describes each output, lists error messages, and includes a sample problem.

13.4.4 Rural Highways

Both microscopic and macroscopic computer simulation models have been developed and are currently available for rural highway applications. The microscopic models include the TWOPAS [57], TRARR [58], and VTI [59] models. The TRARR model is described as an example of this type of model. The only known macroscopic model for rural highways is the RURAL model [60], and it will be described later in this section.

The microscopic computer simulation model TRARR was developed at the Australian Road Research Board beginning in the late 1970s, and enhanced versions have been developed during the 1980s. The TRARR model is a microscopic, stochastic, time-scanning model designed to predict traffic performance along a two-lane two-way rural highway as a function of geometric design and traffic demand. The predicted measures of performance include:

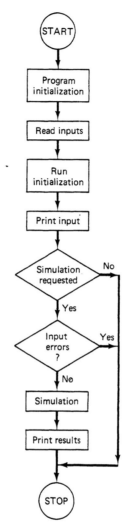

Figure 13.15　FRECON2 Main Program Structure (From Reference 55)

- Distribution of speeds, bunch sizes, and overtakings
- Detailed files of vehicle information passing each observing point
- A plot file for later plotting of vehicle position–time trajectories
- A graphic display file for later presentation of vehicle behavior

　　The study section is divided into subsections and details are given for each subsection, including road grades, curves, barrier lines, sight distance, and auxiliary lanes. The traffic flow and directional split are specified and the flow is classified up to 1: vehicle types.

　　The components of the TRARR simulation model are shown in Figure 13.16 The inputs consist of road details and traffic conditions. Traffic is generated usin;

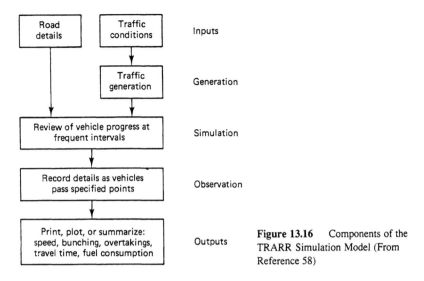

Figure 13.16   Components of the TRARR Simulation Model (From Reference 58)

Monte Carlo techniques. The main component of the model is the simulation component that contains five routines. Every time interval, usually 1 second, the set of five routines are called and they perform the following tasks:

- MANVR routine   determines the state of each vehicle and considers possible changes of state, depending on surrounding vehicles.
- POSIT routine   updates vehicle speeds and positions.
- ENDS routine   generates new vehicles arriving and deletes vehicles leaving the study section.
- RECORD routine   maintains lists of the relative position of vehicles in each lane.
- LVOPP routine   maintains lists of the next opposing vehicles to be met.

The observing component records details of each vehicle's passage at subsection boundaries. The last component is the outputs component, which records the performance measures identified earlier.

The TRARR model is operational on both mainframe and personal computers. The program is well documented with a user guide and program manual. The program is equipped with error messages and all program variables are identified and defined in the program manual. It has been used extensively in Australia and in several other countries.

Attention is now turned to the macroscopic RURAL model. This model is a deterministic model that can simulate two-way traffic flow along a linear rural highway. The predictive performance includes profiles of average speeds and densities of traffic as well as statistical performance for each subsection of the study area, including travel times, travel distances, percent time delay, and level of service. The input requirements include geometric design features and traffic flows. The geometric features of each subsection are entered and five types of rural highways can be handled: freeway, divided and undivided multilane, two-lane, and passing lane sections. Traffic flow is also a required input and includes directional split and vehicle classification: passenger cars, recreation vehicles, and trucks.

A flowchart of the RURAL2 model is shown in Figure 13.17. The supply and demand data are entered into the model, and the data are checked for possible errors. The first subsection is classified by facility type and its performance calculated. The freeway, divided and undivided multilane, and the two-lane submodels follow performance calculation procedures covered in the *Highway Capacity Manual*. The passing

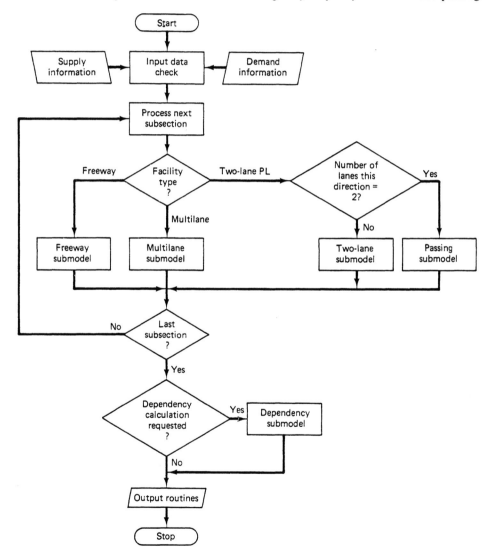

Figure 13.17   Flowchart of RURAL2 Model (From Reference 60)

lane submodel was formulated by the developers based on empirical field studies. The process continues to the next subsection until all subsections have been analyzed. If the interactions between subsections are to be considered, the dependency submode is engaged. The output is then obtained in the form of tables and of directional speed and density profile graphs.

The RURAL model is written in FORTRAN IV and is operational on mainframe and personal computers. The model has been calibrated and validated through field studies on several California rural highways. Recent research efforts have been directed to the improvement of the subsection dependency logic. There are reservations about its use under the combined conditions of steep grades and a high percentage of low-performance vehicles.

13.5 SUMMARY

People look at simulation differently. Strengths and weaknesses are identified in the early portions of this chapter. The history of computer simulation modeling of transportation systems is less than a half-century old. Yet many models are available today.

The procedural elements required in the development of a computer simulation model are described in a step-by-step fashion. This was followed by an example of the development of the FREQ model. Although these procedural elements and examples can serve as a guide, simulation is still an art as well as a science, and requires creativity and hard work on the part of the developer.

Microscopic and macroscopic computer simulation models have been developed and are currently available for use at signalized intersections, arterial networks, freeway corridors, and rural highways. Most of these models have been identified and referenced, and a microscopic and macroscopic model described in each operating environment.

Computer simulation modeling is one analytical technique available for the analysis and assessment of the highway system and its components. This approach can often lead to large investments of time and resources, depending on the availability of an existing model and complexity of the problem. On the other hand, computer simulation modeling may offer the only possible solution to a problem.

13.6 SELECTED PROBLEMS

1. Conduct a literature search of examples where computer simulation models have been developed and applied for one of the following mode transfer locations.
   (a) Parking facilities
   (b) Transit stations
   (c) Airports (land side)
2. Conduct a literature search of examples where computer simulation models have been developed and applied for one of the following modes of transportation.
   (a) Bus systems

  (b) Transit rail systems

  (c) Railroad systems

  (d) Air transport systems

  (e) Waterway systems

  (f) Ferry systems

  (g) Pedestrian-ways

  (h) Elevation systems

3. Consult computer simulation textbooks to identify applications of models in fields outside of transportation.

4. Review the early portion of this chapter dealing with definitions of simulation, and consult computer simulation textbooks for their definitions. What additional insights do you have about the scope of simulation?

5. Computer simulation modeling can be a controversial subject. Study the strengths and reservations of simulation given in Tables 13.1 and 13.2. Attempt to establish guidelines as to when you might use simulation.

6. The 1981 conference on traffic simulation models provided a view of what had happened in the past and a prognosis of things to come. What transitions were occurring at that time, and how well did the prognosis predict what really has happened since the conference?

7. Reread Section 13.2 with particular attention to the flowchart presented in Figure 13.1. Suggest alternative flowcharts of the simulation process.

8. Develop a first-level flowchart for some highway traffic system.

9. Develop an expanded second-level flowchart for one of the simulation models described in Section 13.4.

10. Develop a third-level flowchart for one major subroutine of one of the simulation models described in Section 13.4.

11. It is assumed that a single-file of vehicles enter a traffic system randomly. Draw a flowchart and develop the code for this vehicle generation process using a random number generator.

12. In Problem 11, as each vehicle enters the system, the vehicle must be identified as a truck recreation vehicle, or passenger car. If the average percent of the vehicles of each type are user specified, and if the arrival by type of vehicle occurs randomly, draw a flowchart and develop the code for this vehicle identification processing using a random number generator.

13. It is assumed that a single file of vehicles of one type enter a traffic system. Further, it i assumed that the desired speeds of the vehicles are normally distributed. Draw a flowchar and develop the computer code for this vehicle speed identification process using a random number generator.

14. Conduct a numerical manual check of one of the three subroutines developed in Problems 11, 12, and 13.

15. Identify 5 to 10 currently used simulation languages, including FORTRAN. Discuss the advantages and disadvantages of each.

16. Distinguish among the processes of verification, calibration, and validation.

17. Develop a design of experiment for one of the simulation models described in Section 13.4.

18. Prepare a brief description of one of the signalized intersection simulation model identified but not described in Section 13.4.1.

19. Prepare a brief description of one of the arterial network simulation models identified but not described in Section 13.4.2.

20. Prepare a brief description of one of the freeway corridor simulation models identified but not described in Section 13.4.3.

21. Prepare a brief description of one of the rural highway simulation models identified but not described in Section 13.4.4.

13.7 SELECTED REFERENCES

1. Donald R. Drew, *Traffic Flow Theory and Control,* McGraw-Hill Book Company, New York, 1968, pages 255–297.

2. Claude McMillan and Richard F. Gonzales, *Systems Analysis—A Computer Approach to Decision Models,* Richard D. Irwin, Inc., Homewood, Ill., 1968, pages 1–32.

3. Joe H. Mize and J. Grady Cox, *Essentials of Simulation,* Prentice-Hall, Inc., Englewood Cliffs, N.J., 1968, pages 1–10.

4. Martin Wohl and Brian V. Martin, *Traffic System Analysis for Engineers and Planners,* McGraw-Hill Book Company, New York, 1967, pages 496–536.

5. Thomas H. Naylor, Joseph L. Balintfy, Donald S. Burdick, and Kong Chu, *Computer Simulation Techniques,* John Wiley & Sons, Inc., New York, 1968, pages 1–22.

6. Robert E. Shannon, *Systems Simulation—the Art and Science,* Prentice-Hall, Inc., Englewood Cliffs, N.J., 1975, pages 1–14.

7. George W. Evans, Graham F. Wallace, and Georgia L. Sutherland, *Simulation Using Digital Computers,* Prentice-Hall, Inc., Englewood Cliffs, N.J., 1967, pages 1–15.

8. Francis F. Martin, *Computer Modeling and Simulation,* John Wiley & Sons, Inc., New York, 1968, pages 3–12.

9. VanCourt Hare, Jr., *Systems Analysis: A Diagnostic Approach,* Harcourt, Brace, & World, Inc., New York, 1967, pages 358–369.

10. J. W. Schmidt and R. E. Taylor, *Simulation and Analysis of Industrial Systems,* Richard D. Irwin, Inc., Homewood, Ill., 1970, pages 3–9.

11. D. L. Gerlough and D. G. Capelle, *An Introduction to Traffic Flow Theory,* Highway Research Board, Special Report 79, HRB, Washington, D.C., 1964, Chapter 4, pages 97–118.

12. Daniel L. Gerlough and Matthew J. Huber, *Traffic Flow Theory,* Transportation Research Board, Special Report 163, TRB, Washington, D.C., 1975, Chapter 9, pages 175–196.

13. Phyllis Fox and Frederick Lehman, Digital Computer Simulation of Automobile Traffic, *Traffic Quarterly,* Vol. 21, No. 1, January 1967, pages 53–66.

14. Menahem Eldor and Adolf D. May, *Selected References on Application of Computer Simulation to Transportation Systems,* ITTE Library Reference 43, University of California, Berkeley, Calif., July 1975, 31 pages.

15. Wai-Ki Yip and Adolf D. May, *Corrections and Addenda to Library Reference 43,* Library Reference 43A, Institute of Transportation Studies, Berkeley, Calif., July 1977, 8 pages.

16. M. R. Wigan, *Applications of Simulation to Traffic Problems—An Annotated Bibliography,* Road Research Bibliography 102/MRW, Transport and Road Research Laboratory, Crowthorne, England, 1969, 27 pages.

17. Organization for Economic Cooperation and Development, *Road Research, Area Traffic Control Systems,* OECD, Paris, 1972, 69 pages.

18. Transportation Research Board, *The Application of Traffic Simulation Models,* Transportation Research Board, Special Report 194, TRB, Washington, D.C., May 1981, 114 pages.

19. William A. Stock, Richard C. Blankenhorn, and Adolf D. May, *The FREQ3 Freeway Model,* ITTE Report 73-1, University of California, Berkeley, Calif., 1973, 247 pages.

20. Menahem Eldor and Adolf D. May, *Cost-Effectiveness Evaluation of Freeway Design Alternatives,* ITTE Report 73-2, University of California, Berkeley, Calif., June 1973.

21. Jin J. Wang, *Analysis of Freeway On-Ramp Control Strategies,* Dissertation, University of California, Berkeley, Calif., 1973.

22. Khosrow Ovaici, *Simulation of Freeway Priority Strategies,* Dissertation, University of California, Berkeley, Calif., 1975, 471 pages.

23. Abraham J. Kruger and Adolf D. May, *The Analysis and Evaluation of Selected Impacts of Traffic Management Strategies on Freeways,* UCB-ITS-SR-76-4, University of California, Berkeley, Calif., October 1976, 116 pages.

24. Abraham J. Kruger, *Further Analysis and Evaluation of Selected Impacts of Traffic Management Strategies on Freeways,* Dissertation, University of California, October 1977, 679 pages.

25. Matthys P. Cilliers, *FREQ6PL—A Freeway Priority Lane Simulation Model,* Dissertation, University of California, September 1978, 333 pages.

26. Paul P. Jovanis, Wai-ki Yip, and Adolf D. May, *FREQ6PE—A Freeway Priority Entry Control Simulation Model,* UCB-ITS-RR-78-9, University of California, Berkeley, Calif., November 1978, 516 pages.

27. David B. Roden, Walter Okitsu, and Adolf D. May, *FREQ7PE—A Freeway Corridor Simulation Model,* UCB-ITS-RR-80-4, University of California, Berkeley, Calif., June 1980, 340 pages.

28. Tsutomu Imada and Adolf D. May, *FREQ8PL—A Freeway Priority Lane Simulation Model,* UCB-ITS-TD-85-1, University of California, Berkeley, Calif., March 1985, 68 pages.

29. Tsutomu Imada and Adolf D. May, *FREQ8PE—A Freeway Corridor Simulation and Ramp Metering Optimization Model,* UCB-ITS-RR-85-10, University of California, Berkeley, Calif., June 1985, 263 pages.

30. Adolf D. May and Vincent Wong, *An Education Program for Freeway Simulation and Optimization,* Institute of Transportation Studies and Texas State Department of Highways and Public Transportation, December 1987 and March 1988.

31. *Etude de Faisabilite d'un Système de Gestion de Circulation pour le Corridor Autoroutier* SNC/DeLuc/Co.—Entreprise, Montreal, Canada, March 1987.

32. Alexander Skabardonis, *Computer Programs for Traffic Operations,* UCB-ITS-TD-84-3, University of California, Berkeley, Calif., August 1984, 85 pages.

33. A. E. Radwan et al., *Comparative Assessment of Computer Program for Traffic Signal Planning, Design and Operations,* Volumes I to III, Arizona State University, Tempe, Ariz. 1986.

34. Adolf D. May, Freeway Simulation Models—Revisited, *Arizona Department of Transportation Symposium on Freeway Surveillance and Control Proceedings,* October 1986.

35. Federal Highway Administration, *Traffic Network Analysis with NETSIM—A User Guide* FHWA-IP-80-3, FHWA, Washington, D.C., January 1980.

36. A. Hansson, SIGSIM: A Simulation Model for Signalized Intersections, *Proceedings of International Symposium of Traffic Control Systems*, Berkeley, August 1979.

37. Clyde E. Lee, Robert F. Inman, and Wylie M. Sanders, *User-Friendly TEXAS Model—Guide to Data Entry*, Research Report 361-IF, University of Texas, Austin, Tex., November 1986.

38. Ignacio Sanchez, Michael J. Cassidy, and Adolf D. May, *CALSIG Software User's Manual*, UCB-ITS-TD-88-1, University of California, Berkeley, Calif., February 1988, 38 pages.

39. A Hansson, *CAPCAL Computer Program—Manual*, VBB (Consulting Engineers, Architects, and Economists), Göteborg, Sweden, October 1980.

40. *Comprehensive Analysis Program for Single Signalized Intersections*, Mohle, Grover and Associates, La Habra, Calif., 1986.

41. Hobih Chen and Joe Lee, POSIT, An Interactive Timing Optimization Program, in *Proceedings of National Conference on Microcomputers*, American Society of Civil Engineers, New York, 1985.

42. Rahmi Akcelik, *SIDRA Version 2.2 program: User Guide*, Australian Road Research Board, Report AIR 1142-1, ARRB, Victoria, Australia, 1986, 52 pages.

43. *SIGNAL85 Tutorial/Reference Manual*, Barton-Aschman Associates, Evanston, Ill., 1986.

44. *SOAP/M - Program Documentation*, University of Florida, Gainesville, Fla., September 1981.

45. Scott W. Sibley, NETSIM for Microcomputers, in *Public Roads*, Vol. 49, No. 2, Washington, D.C., September 1985.

46. M. D. Kelson et al., *Optimal Signal Timing for Arterial Signal Systems*, Volumes 1 to 3, Federal Highway Administration, Washington, D.C., December 1980.

47. Blair G. Marsden, Edmund Chin-Ping Chang, and B. Ray Derr, *The Design of the Passer II-84 Microcomputer Environment System*, Texas Transportation Institute, 1986.

48. D. J. Messer, D. B. Fambro, and J. M. Turner, *Analysis of Diamond Interchange Operation Program—PASSER-III*, Texas Transportation Institute, Texas A & M University, College Station, Texas, August 1976.

49. E. B. Lieberman, J. Lai, and R. E. Ellington, *SIGOP-III User's Manual*, Federal Highway Administration, Washington, D.C., 1983.

50. *SPAN—Program Documentation*, University of Florida, Gainesville, Fla., September 1980.

51. Erhart C. Schroeder et al., *SSTOP: Off-Line Signal System Optimization Package*, McMaster University, Hamilton, Ontario, Canada, 1979.

52. *The TRANSYT7F User's Manual*, University of Florida, Gainesville, Fla., July 1987.

53. D. A. Wicks and E.B. Lieberman, *Development and Testing of INTRAS*, Volumes 1 to 4, FHWA-RD-80/106, Federal Highway Administration, Washington, D.C., October 1980.

54. Sam Yagar and Colin Leech, *Enhancements in the CORQ2 Freeway—Corridor Simulation Model*, University of Waterloo, Waterloo, Ontario, Canada, May 1986.

55. James F. Campbell, Philip S. Babcock, and Adolf D. May, *FRECON2—User's Guide*, UCB-ITS-TD-84-2, Univeristy of California, Berkeley, Calif., August 1984, 132 pages.

56. R. Plum et al., *KRONOS4: An Interactive Freeway Simulation Program for Personal Computers User's Guide*, University of Minnesota, Minneapolis, Minn., 1985.

57. A. D. St. John and D. W. Harwood, *TWOPAS Guide: A Microscopic Computer Simulation Model of Traffic on Two-Lane, Two-Way Highways*, FHWA DTFH51-82-C-00070, Federal Highway Administration, Washington, D.C., May 1986.

58. C. J. Hoban, G. J. Faucett, and G. K. Robinson, *A Model for Simulating Traffic on Two-Lane Rural Roads: User Guide and Manual for TRARR Version 3.0,* Australian Road Research Board, Technical Manual ATM 10A, ARRB, Victoria, Australia, May 1985, 96 pages.

59. G. Gynnersted et al., *A Model for the Monte Carlo Simulation of Traffic Flow Along Two-Lane Single Carriageway Rural Roads,* Statens Vagoch Trafik Institute (VTI), Borlange, Sweden, 1977.

60. Juan C. Sananez, Laura Wingerd, and Adolf D. May, *A Macroscopic Model for the Analysis of Traffic Operations on Rural Highways: Final Report,* UCB-ITS-RR-85-3, University of California, Berkeley, Calif., February 1985, 372 pages.

APPENDIX A

Headway Tabulations and Parameter Values of Four Measured Time Headway Distributions*

---

*Measured headway distributions taken from reference: Adolf D. May, *Gap Availability Studies*, Highway Research Board Record 72, 1965, pages 105–136.

TABLE A.1 Time Headway Distributions: 10 to 14 Vehicles per Minute Flow Rate (1320 Measured Headways)

Time Headway Group (seconds)	Probabilities of Headways in Time Headway Groups				
	Measured Distribution	Negative Exponential Distribution	Normal Distribution	Pearson Type III Distribution	Composite Distribution
0.0–0.5	0.061	0.095	0.010	0.000	0.007
0.5–1.0	0.066	0.086	0.015	0.044	0.041
1.0–1.5	0.068	0.078	0.022	0.088	0.102
1.5–2.0	0.077	0.071	0.032	0.084	0.102
2.0–2.5	0.066	0.063	0.042	0.078	0.114
2.5–3.0	0.077	0.058	0.053	0.071	0.072
3.0–3.5	0.063	0.052	0.065	0.064	0.059
3.5–4.0	0.061	0.048	0.079	0.058	0.052
4.0–4.5	0.049	0.042	0.083	0.052	0.051
4.5–5.0	0.027	0.039	0.087	0.047	0.042
5.0–5.5	0.030	0.035	0.087	0.042	0.037
5.5–6.0	0.031	0.032	0.083	0.038	0.033
6.0–6.5	0.025	0.028	0.079	0.034	0.030
6.5–7.0	0.024	0.026	0.065	0.030	0.027
7.0–7.5	0.020	0.024	0.053	0.026	0.024
7.5–8.0	0.015	0.021	0.042	0.024	0.022
8.0–8.5	0.017	0.019	0.032	0.021	0.019
8.5–9.0	0.018	0.018	0.022	0.018	0.019
9.0–9.5	0.013	0.015	0.015	0.016	0.015
> 9.5	0.192	0.150	0.023	0.165	0.132
Mean Time Headway	5.0	5.0	5.0	5.0	5.0
Standard Deviation	3.9	5.0	2.3	3.9	3.5

TABLE A.2 Time Headway Distributions: 15 to 19 Vehicles per Minute Flow Rate (3432 Measured Headways)

Time Headway Group (seconds)	Probabilities of Headways in Time Headway Groups				
	Measured Distribution	Negative Exponential Distribution	Normal Distribution	Pearson Type III Distribution	Composite Distribution
0.0–0.5	0.012	0.133	0.013	0.000	0.012
0.5–1.0	0.064	0.116	0.025	0.067	0.058
1.0–1.5	0.114	0.100	0.044	0.127	0.147
1.5–2.0	0.159	0.087	0.067	0.114	0.147
2.0–2.5	0.157	0.075	0.093	0.099	0.148
2.5–3.0	0.130	0.065	0.119	0.085	0..080
3.0–3.5	0.088	0.056	0.129	0.072	0.065
3.5–4.0	0.065	0.049	0.129	0.061	0.054
4.0–4.5	0.043	0.043	0.119	0.052	0.043
4.5–5.0	0.033	0.037	0.093	0.043	0.039
5.0–5.5	0.022	0.032	0.067	0.036	0.033
5.5–6.0	0.019	0.027	0.044	0.030	0.026
6.0–6.5	0.014	0.024	0.025	0.025	0.024
6.5–7.0	0.010	0.021	0.013	0.021	0.020
7.0–7.5	0.012	0.018	0.006	0.018	0.016
7.5–8.0	0.008	0.016	0.002	0.015	0.014
8.0–8.5	0.005	0.013	0.001	0.012	0.012
8.5–9.0	0.007	0.012	0.000	0.010	0.010
9.0–9.5	0.005	0.010	0.000	0.008	0.009
> 9.5	0.033	0.066	0.000	0.105	0.047
Mean Time Headway	3.5	3.5	3.5	3.5	3.5
Standard Deviation	2.6	3.5	1.5	2.6	2.7

TABLE A.3   Time Headway Distributions: 20 to 24 Vehicles per Minute Flow Rate (6327 Measured Headways)

Time Headway Group (seconds)	Probabilities of Headways in Time Headway Groups				
	Measured Distribution	Negative Exponential Distribution	Normal Distribution	Pearson Type III Distribution	Composite Distribution
0.0–0.5	0.015	0.169	0.016	0.000	0.015
0.5–1.0	0.077	0.140	0.038	0.082	0.076
1.0–1.5	0.172	0.117	0.077	0.161	0.191
1.5–2.0	0.194	0.097	0.123	0.146	0.191
2.0–2.5	0.164	0.080	0.168	0.122	0.163
2.5–3.0	0.112	0.067	0.178	0.099	0.085
3.0–3.5	0.081	0.056	0.161	0.078	0.056
3.5–4.0	0.055	0.046	0.114	0.061	0.046
4.0–4.5	0.037	0.039	0.068	0.047	0.035
4.5–5.0	0.025	0.032	0.032	0.036	0.030
5.0–5.5	0.014	0.026	0.013	0.028	0.022
5.5–6.0	0.012	0.022	0.004	0.021	0.018
6.0–6.5	0.009	0.019	0.001	0.016	0.015
6.5–7.0	0.007	0.015	0.000	0.012	0.012
7.0–7.5	0.006	0.013	0.000	0.009	0.010
7.5–8.0	0.004	0.010	0.000	0.007	0.007
8.0–8.5	0.003	0.009	0.000	0.005	0.006
8.5–9.0	0.003	0.007	0.000	0.004	0.005
9.0–9.5	0.002	0.006	0.000	0.003	0.004
> 9.5	0.008	0.030	0.000	0.063	0.015
Mean Time Headway	2.7	2.7	2.7	2.7	2.7
Standard Deviation	1.6	2.7	1.1	1.6	2.0

TABLE A.4 Time Headway Distributions: 25 to 29 Vehicles per Minute Flow Rate (3491 Measured Headways)

Time Headway Group (seconds)	Probabilities of Headways in Time Headway Groups				
	Measured Distribution	Negative Exponential Distribution	Normal Distribution	Pearson Type III Distribution	Composite Distribution
0.0–0.5	0.017	0.204	0.018	0.000	0.018
0.5–1.0	0.106	0.162	0.057	0.107	0.092
1.0–1.5	0.236	0.128	0.127	0.202	0.232
1.5–2.0	0.223	0.103	0.199	0.169	0.232
2.0–2.5	0.158	0.083	0.232	0.129	0.166
2.5–3.0	0.095	0.065	0.190	0.095	0.084
3.0–3.5	0.059	0.051	0.111	0.068	0.047
3.5–4.0	0.035	0.042	0.046	0.048	0.033
4.0–4.5	0.023	0.033	0.014	0.034	0.026
4.5–5.0	0.016	0.026	0.003	0.024	0.019
5.0–5.5	0.010	0.021	0.001	0.016	0.014
5.5–6.0	0.005	0.017	0.000	0.011	0.010
6.0–6.5	0.005	0.013	0.000	0.008	0.008
6.5–7.0	0.005	0.011	0.000	0.005	0.006
7.0–7.5	0.002	0.008	0.000	0.004	0.004
7.5–8.0	0.002	0.007	0.000	0.002	0.003
8.0–8.5	0.000	0.005	0.000	0.002	0.002
8.5–9.0	0.001	0.004	0.000	0.001	0.002
9.0–9.5	0.001	0.004	0.000	0.001	0.001
> 9.5	0.001	0.013	0.000	0.074	0.004
Mean Time Headway	2.2	2.2	2.2	2.2	2.2
Standard Deviation	1.2	2.2	0.85	1.2	1.5

APPENDIX
B

Derivation of Poisson Count Distribution*

*This appendix is reprinted with the permission of the Transportation Research Board and is repro-
duced from the reference: Daniel G. Gerlough and Matthew J. Huber, *Traffic Flow Theory—A Monograph*
Transportation Research Board Special Report 165, 1975, Appendix B, pages 202–203.

Consider a line that can represent in a general case either distance or time; for the present purposes consider it to represent time (Figure B.1). Specifically, consider the occurrence of random arrivals where the average rate of arrival (i.e., probability density) is λ. Let $P_i(t)$ = the probability of i arrivals up to the time t, and $P_n(\Delta t) = \lambda \Delta t$ = the probability of one arrival in the incremental period Δt. Because it is assumed that Δt is of such short duration, the probability of more than one arrival in Δt is negligible; therefore, $(1 - \lambda \Delta t)$ = the probability of no arrival in Δt. Then,

$P_i(t + \Delta t)$ = the probability that i arrivals have taken place to the time $(t + \Delta t)$

$\qquad = [\text{Prob}(i - 1 \text{ arrivals in } t) \cdot \text{Prob}(1 \text{ arrival in } \Delta t)] + [\text{Prob}(i \text{ arrivals in } t)$

$\qquad \cdot \text{Prob}(0 \text{ arrivals in } \Delta t)]$

$P_i(t + \Delta t) = P_{i-1}(t) \cdot P_1(\Delta t) + P_i(t) \cdot P_0(\Delta t)$

$\qquad = P_{i-1}(t)\lambda \Delta t + P_i(t)(1 - \lambda \Delta t)$

$\qquad = [P_{i-1}(t) - P_i(t)](\lambda \Delta t) + P_i(t)$

and

$$\frac{P_i(t + \Delta t) - P_i(t)}{\Delta t} = \lambda[P_{i-1}(t) - P_i(t)]$$

Letting $\Delta t \to 0$,

$$\frac{dP_i(t)}{dt} = \lambda[P_{i-1}(t) - P_i(t)] \qquad (B.1)$$

Now,

$P_{-1}(t) = 0 \qquad$ (i.e., impossible to have < 0)

$P_0(0) = 1 \qquad$ (i.e., no arrivals up to time $t = 0$)

$P_i(0) = 0 \qquad$ for $i \geq 1$ (zero probability of i arrivals at time $t = 0$)

Setting $i = 0$ in equation (B.1),

$$\frac{dP_0(t)}{dt} = \lambda[0 - P_0(t)]$$

$$\frac{dP_0(t)}{P_0(t)} = -\lambda \, dt$$

$$\ln P_0(t) = -\lambda t + c$$

$$P_0(t) = e^{-\lambda t + c}$$

Since $P_0(0) = 1$ and $1 = e^0 = e^c$, $c = 0$, and

$$P_0(t) = e^{-\lambda t}$$

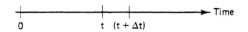

$$\xrightarrow{\qquad\qquad\qquad\qquad} \text{Time}$$

$\quad 0 \qquad\qquad\qquad t \quad (t + \Delta t)$

Figure B.1   Schematic Representation of Uniform Probability Density

Setting $i = 1$ in equation (B.1) and inserting the above value for $P_0(t)$,

$$\frac{dP_1(t)}{dt} = \lambda[e^{-\lambda t} - P_1(t)]$$

$$\frac{dP_1(t)}{dt} + \lambda P_1(t) = \lambda e^{-\lambda t}$$

Using method of operators for solving this differential equation*

$$(D + \lambda)P_1(t) = \lambda e^{-\lambda t}$$

$$P_1(t) = \frac{1}{D + \lambda} \lambda e^{-\lambda t}$$

$$= (\lambda t)e^{-\lambda t} + C_2 e^{-\lambda t}$$

But

$$P_1(0) = 0 \qquad \therefore C_2 = 0$$

$$\therefore P_1(t) = (\lambda t)e^{-\lambda t}$$

For $i = 2$,

$$\frac{dP_2(t)}{dt} = \lambda[P_1(t) - P_2(t)]$$

$$\frac{dP_2(t)}{dt} + \lambda P_2(t) = \lambda P_1(t) = \lambda(\lambda t)e^{-\lambda t}$$

$$P_2(t) = \frac{1}{D + \lambda} \lambda(\lambda t)e^{-\lambda t}$$

$$= \frac{\lambda^2 t^2}{2} e^{-\lambda t} + C_3 e^{-\lambda t}$$

But

$$P_2(0) = 0 \qquad \therefore C_3 = 0$$

$$P_2(t) = \frac{(\lambda t)^2 e^{-\lambda t}}{2!}$$

Similarly,

$$P_3(t) = \frac{(\lambda t)^3 e^{-\lambda t}}{3!}$$

$$P_4(t) = \frac{(\lambda t)^4 e^{-\lambda t}}{4!}$$

*Any standard method may be used for solution of this differential equation. The method of operator is particularly simple; see any standard text. The form $y = [1/(D + A)] u(x)$ results in a solution

$$y = e^{-Ax} \int e^{Ax} u(x) \, dx + c e^{-Ax}$$

$$P_x(t) = \frac{(\lambda t)^x e^{-\lambda t}}{x!}$$

If $\lambda t = m$, the result is the most familiar form of the Poisson distribution:

$$P(x) = \frac{m^x e^{-m}}{x!}$$

This relationship states the probability that exactly x arrivals will occur during an interval (of length t) when the mean number of arrivals is m (per interval of t).

B.1 Population Mean of Poisson Distribution

In the foregoing m is defined as the mean *arrival rate*. To determine the mean value of the distribution, begin with the definition of the population mean μ for a discrete distribution:

$$\mu = \sum_{x=0}^{\infty} xP(x), \qquad \text{for } P(x) = \frac{f(x)}{\sum_{x=0}^{\infty} f(x)} \tag{B.2}$$

where $f(x)$ is the frequency of occurrence of x. For the Poisson distribution, substitute

$$P(x) = \frac{m^x e^{-m}}{m!}$$

Thus

$$\mu = \sum_{x=0}^{\infty} \frac{xm^x e^{-m}}{m!} \tag{B.3}$$

$$= 0 + me^{-m} + \frac{2m^2 e^{-m}}{2!} + \frac{3m^3 e^{-m}}{3!} \cdots$$

$$= me^{-m}\left[1 + m + \frac{m^2}{2!} + \frac{m^3}{3!} \cdots\right]$$

$$= me^{-m}e^m$$

$$= m \tag{B.4}$$

B.2 Population Variance of Poisson Distribution

By definition, the *population* variance, σ^2, may be expressed:

$$\sigma^2 = \frac{\Sigma f(x)(x-\mu)^2}{\Sigma f(x)} \tag{B.5}$$

$$= \sum_{x=0}^{\infty} (x-\mu)^2 P(x) \tag{B.6}$$

Because the population mean is m, this variance may be stated:

$$\sigma^2 = \Sigma\,(x-m)^2 P\,(x)$$

$$= \Sigma\,(x^2 - 2xm + m^2)P\,(x)$$

$$= \Sigma\,x^2 P\,(x) - 2m\,\Sigma\,xP\,(x) + m^2\,\Sigma P\,(x)$$

The last term reduces to m^2 because $\Sigma P\,(x) = 1$. The middle term reduces to $-2m^2$ because $\Sigma xP\,(x)$ has been shown equal to m in the derivation of the population mean. The first term may be reduced by the following steps:

$$\Sigma\,x^2 P\,(x) = \Sigma\,[x\,(x-1) + x\,]P\,(x)$$

$$= \Sigma\,x\,(x-1)P\,(x) + \Sigma\,xP\,(x)$$

$$= A + B$$

$$B = \Sigma\,xP\,(x) = m$$

$$A = \sum_{x=0}^{\infty} x\,(x-1)\frac{m^x e^{-m}}{x!}$$

$$= \left[0 + 0 + \frac{2m^2 e^{-m}}{2!} + \frac{6m^3 e^{-m}}{3!} + \frac{12m^4 e^{-m}}{4!} + \cdots \right]$$

$$= m^2 e^{-m}\left[1 + m + \frac{m^2}{2!} + \cdots \right]$$

$$= m^2 e^{-m} e^m = m^2$$

$$\Sigma\,x^2 P\,(x) = m^2 + m$$

$$\sigma^2 = [m^2 + m] - [2m^2] + [m^2]$$

$$\sigma^2 = m \qquad\qquad\qquad\qquad\qquad\qquad\qquad\qquad\qquad \text{(B.7}$$

Thus, for the Poisson distribution the population variance equals the population mean.

APPENDIX C

Exponential Function Tables

Exponential Functions

x	e^x	$Log_{10}(e^x)$	e^{-x}	x	e^x	$Log_{10}(e^x)$	e^{-x}
0.00	1.0000	0.00000	1.000000	0.50	1.6487	0.21715	0.606531
0.01	1.0101	0.00434	0.990050	0.51	1.6653	0.22149	0.600496
0.02	1.0202	0.00869	0.980199	0.52	1.6820	0.22583	0.594521
0.03	1.0305	0.01303	0.970446	0.53	1.6989	0.23018	0.588605
0.04	1.0408	0.01737	0.960789	0.54	1.7160	0.23452	0.582748
0.05	1.0513	0.02171	0.951229	0.55	1.7333	0.23886	0.576950
0.06	1.0618	0.02606	0.941765	0.56	1.7507	0.24320	0.571209
0.07	1.0725	0.03040	0.932394	0.57	1.7683	0.24755	0.565525
0.08	1.0833	0.03474	0.923116	0.58	1.7860	0.25189	0.559898
0.09	1.0942	0.03909⁻	0.913931	0.59	1.8040	0.25623	0.554327
0.10	1.1052	0.04343	0.904837	0.60	1.8221	0.26058	0.548812
0.11	1.1163	0.04777	0.895834	0.61	1.8404	0.26492	0.543351
0.12	1.1275	0.05212	0.886920	0.62	1.8589	0.26926	0.537944
0.13	1.1388	0.05646	0.878095	0.63	1.8776	0.27361	0.532592
0.14	1.1503	0.06080	0.869358	0.64	1.8965	0.27795	0.527292
0.15	1.1618	0.06514	0.860708	0.65	1.9155	0.28229	0.522046
0.16	1.1735	0.06949	0.852144	0.66	1.9348	0.28663	0.516851
0.17	1.1853	0.07383	0.843665	0.67	1.9542	0.29098	0.511709
0.18	1.1972	0.07817	0.835270	0.68	1.9739	0.29532	0.506617
0.19	1.2092	0.08252	0.826959	0.69	1.9937	0.29966	0.501576
0.20	1.2214	0.08686	0.818731	0.70	2.0138	0.30401	0.496585
0.21	1.2337	0.09120	0.810584	0.71	2.0340	0.30835	0.491644
0.22	1.2461	0.09554	0.802519	0.72	2.0544	0.31269	0.486752
0.23	1.2586	0.09989	0.794534	0.73	2.0751	0.31703	0.481909
0.24	1.2712	0.10423	0.786628	0.74	2.0959	0.32138	0.477114
0.25	1.2840	0.10857	0.778801	0.75	2.1170	0.32572	0.472367
0.26	1.2969	0.11292	0.771052	0.76	2.1383	0.33006	0.467666
0.27	1.3100	0.11726	0.763379	0.77	2.1598	0.33441	0.463013
0.28	1.3231	0.12160	0.755784	0.78	2.1815	0.33875	0.458406
0.29	1.3364	0.12595	0.748264	0.79	2.2034	0.34309	0.453845
0.30	1.3499	0.13029	0.740818	0.80	2.2255	0.34744	0.449329
0.31	1.3634	0.13463	0.733447	0.81	2.2479	0.35178	0.444858
0.32	1.3771	0.13897	0.726149	0.82	2.2705	0.35612	0.440432
0.33	1.3910	0.14332	0.718924	0.83	2.2933	0.36046	0.436049
0.34	1.4049	0.14766	0.711770	0.84	2.3164	0.36481	0.431711
0.35	1.4191	0.15200	0.704688	0.85	2.3396	0.36915	0.427415
0.36	1.4333	0.15635	0.697676	0.86	2.3632	0.37319	0.423162
0.37	1.4477	0.16069	0.690734	0.87	2.3869	0.37784	0.418952
0.38	1.4623	0.16503	0.683861	0.88	2.4109	0.38218	0.414783
0.39	1.4770	0.16937	0.677057	0.89	2.4351	0.38652	0.410656
0.40	1.4918	0.17372	0.670320	0.90	2.4596	0.39087	0.406570
0.41	1.5068	0.17806	0.663650	0.91	2.4843	0.39521	0.402524
0.42	1.5220	0.18240	0.657047	0.92	2.5093	0.39955	0.398519
0.43	1.5373	0.18675	0.650509	0.93	2.5345	0.40389	0.394554
0.44	1.5527	0.19109	0.644036	0.94	2.5600	0.40824	0.390628
0.45	1.5683	0.19543	0.637628	0.95	2.5857	0.41258	0.386741
0.46	1.5841	0.19978	0.631284	0.96	2.6117	0.41692	0.382893
0.47	1.6000	0.20412	0.625002	0.97	2.6379	0.42127	0.379083
0.48	1.6161	0.20846	0.618783	0.98	2.6645	0.42561	0.375311
0.49	1.6323	0.21280	0.612626	0.99	2.6912	0.42995	0.371577

Exponential Functions *(continued)*

x	e^x	$\text{Log}_{10}(e^x)$	e^{-x}	x	e^x	$\text{Log}_{10}(e^x)$	e^{-x}
1.00	2.7183	0.43429	0.367879	1.50	4.4817	0.65144	0.223130
1.01	2.7456	0.43864	0.364219	1.51	4.5267	0.65578	0.220910
1.02	2.7732	0.44298	0.360595	1.52	4.5722	0.66013	0.218712
1.03	2.8011	0.44732	0.357007	1.53	4.6182	0.66447	0.216536
1.04	2.8292	0.45167	0.353455	1.54	4.6646	0.66881	0.214381
1.05	2.8577	0.45601	0.349938	1.55	4.7115	0.67316	0.212248
1.06	2.8864	0.46035	0.346456	1.56	4.7588	0.67750	0.210136
1.07	2.9154	0.46470	0.343009	1.57	4.8066	0.68184	0.208045
1.08	2.9447	0.46904	0.339596	1.58	4.8550	0.68619	0.205975
1.09	2.9743	0.47338	0.336216	1.59	4.9037	0.69053	0.203926
1.10	3.0042	0.47772	0.332871	1.60	4.9530	0.69487	0.201897
1.11	3.0344	0.48207	0.329559	1.61	5.0028	0.69921	0.199888
1.12	3.0649	0.48641	0.326280	1.62	5.0531	0.70356	0.197899
1.13	3.0957	0.49075	0.323033	1.63	5.1039	0.70790	0.195930
1.14	3.1268	0.49510	0.319819	1.64	5.1552	0.71224	0.193980
1.15	3.1582	0.49944	0.316637	1.65	5.2070	0.71659	0.192050
1.16	3.1899	0.50378	0.313486	1.66	5.2593	0.72093	0.190139
1.17	3.2220	0.50812	0.310367	1.67	5.3122	0.72527	0.188247
1.18	3.2544	0.51247	0.307279	1.68	5.3656	0.72961	0.186374
1.19	3.2871	0.51681	0.304221	1.69	5.4195	0.73396	0.184520
1.20	3.3201	0.52115	0.301194	1.70	5.4739	0.73830	0.182684
1.21	3.3535	0.52550	0.298197	1.71	5.5290	0.74264	0.180866
1.22	3.3872	0.52984	0.295230	1.72	5.5845	0.74699	0.179066
1.23	3.4212	0.53418	0.292293	1.73	5.6407	0.75133	0.177284
1.24	3.4556	0.53853	0.289384	1.74	5.6973	0.75567	0.175520
1.25	3.4903	0.54287	0.286505	1.75	5.7546	0.76002	0.173774
1.26	3.5254	0.54721	0.283654	1.76	5.8124	0.76436	0.172045
1.27	3.5609	0.55155	0.280832	1.77	5.8709	0.76870	0.170333
1.28	3.5966	0.55590	0.278037	1.78	5.9299	0.77304	0.168638
1.29	3.6328	0.56024	0.275271	1.79	5.9895	0.77739	0.166960
1.30	3.6693	0.56458	0.272532	1.80	6.0496	0.78173	0.165299
1.31	3.7062	0.56893	0.269820	1.81	6.1104	0.78607	0.163654
1.32	3.7434	0.57327	0.267135	1.82	6.1719	0.79042	0.162026
1.33	3.7810	0.57761	0.264477	1.83	6.2339	0.79476	0.160414
1.34	3.8190	0.58195	0.261846	1.84	6.2965	0.79910	0.158817
1.35	3.8574	0.58630	0.259240	1.85	6.3598	0.80344	0.157237
1.36	3.8962	0.59064	0.256661	1.86	6.4237	0.80779	0.155673
1.37	3.9354	0.59498	0.254107	1.87	6.4883	0.81213	0.154124
1.38	3.9749	0.59933	0.251579	1.88	6.5535	0.81647	0.152590
1.39	4.0149	0.60367	0.249075	1.89	6.6194	0.82082	0.151072
1.40	4.0552	0.60801	0.246597	1.90	6.6859	0.82516	0.149569
1.41	4.0960	0.61236	0.244143	1.91	6.7531	0.82950	0.148080
1.42	4.1371	0.61670	0.241714	1.92	6.8210	0.83385	0.146607
1.43	4.1787	0.62104	0.239309	1.93	6.8895	0.83819	0.145148
1.44	4.2207	0.62538	0.236928	1.94	6.9588	0.84253	0.143704
1.45	4.2631	0.62973	0.234570	1.95	7.0287	0.84687	0.142274
1.46	4.3060	0.63407	0.232236	1.96	7.0993	0.85122	0.140858
1.47	4.3492	0.63841	0.229925	1.97	7.1707	0.85556	0.139457
1.48	4.3929	0.64276	0.227638	1.98	7.2427	0.85990	0.138069
1.49	4.4371	0.64710	0.225373	1.99	7.3155	0.86425	0.136695

Exponential Functions *(continued)*

x	e^x	$\text{Log}_{10}(e^x)$	e^{-x}	x	e^x	$\text{Log}_{10}(e^x)$	e^{-x}
2.00	7.3891	0.86859	0.135335	2.50	12.182	1.08574	0.082085
2.01	7.4633	0.87293	0.133989	2.51	12.305	1.09008	0.081268
2.02	7.5383	0.87727	0.132655	2.52	12.429	1.09442	0.080460
2.03	7.6141	0.88162	0.131336	2.53	12.554	1.09877	0.079659
2.04	7.6906	0.88596	0.130029	2.54	12.680	1.10311	0.078866
2.05	7.7679	0.89030	0.128735	2.55	12.807	1.10745	0.078082
2.06	7.8460	0.89465	0.127454	2.56	12.936	1.11179	0.077305
2.07	7.9248	0.89899	0.126186	2.57	13.066	1.11614	0.076536
2.08	8.0045	0.90333	0.124930	2.58	13.197	1.12048	0.075774
2.09	8.0849	0.90768	0.123687	2.59	13.330	1.12482	0.075020
2.10	8.1662	0.91202	0.122456	2.60	13.464	1.12917	0.074274
2.11	8.2482	0.91636	0.121238	2.61	13.599	1.13351	0.073535
2.12	8.3311	0.92070	0.120032	2.62	13.736	1.13785	0.072803
2.13	8.4149	0.92505	0.118837	2.63	13.875	1.14219	0.072078
2.14	8.4994	0.92939	0.117655	2.64	14.013	1.14654	0.071361
2.15	8.5849	0.93373	0.116484	2.65	14.154	1.15088	0.070651
2.16	8.6711	0.93808	0.115325	2.66	14.296	1.15522	0.069948
2.17	8.7583	0.94242	0.114178	2.67	14.440	1.15957	0.069252
2.18	8.8463	0.94676	0.113042	2.68	14.585	1.16391	0.068563
2.19	8.9352	0.95110	0.111917	2.69	14.732	1.16825	0.067881
2.20	9.0250	0.95545	0.110803	2.70	14.880	1.17260	0.067206
2.21	9.1157	0.95979	0.109701	2.71	15.029	1.17694	0.066537
2.22	9.2073	0.96413	0.108609	2.72	15.180	1.18128	0.065875
2.23	9.2999	0.96848	0.107528	2.73	15.333	1.18562	0.065219
2.24	9.3933	0.97282	0.106459	2.74	15.487	1.18997	0.064570
2.25	9.4877	0.97716	0.105399	2.75	15.643	1.19431	0.063928
2.26	9.5831	0.98151	0.104350	2.76	15.800	1.19865	0.063292
2.27	9.6794	0.98585	0.103312	2.77	15.959	1.20300	0.062662
2.28	9.7767	0.99019	0.102284	2.78	16.119	1.20734	0.062039
2.29	9.8749	0.99453	0.101266	2.79	16.281	1.21168	0.061421
2.30	9.9742	0.99888	0.100259	2.80	16.445	1.21602	0.060810
2.31	10.074	1.00322	0.099261	2.81	16.610	1.22037	0.060205
2.32	10.176	1.00756	0.098274	2.82	16.777	1.22471	0.059606
2.33	10.278	1.01191	0.097296	2.83	16.945	1.22905	0.059013
2.34	10.381	1.01625	0.096328	2.84	17.116	1.23340	0.058426
2.35	10.486	1.02059	0.095369	2.85	17.288	1.23774	0.057844
2.36	10.591	1.02493	0.094420	2.86	17.462	1.24208	0.057269
2.37	10.697	1.02928	0.093481	2.87	17.637	1.24643	0.056699
2.38	10.805	1.03362	0.092551	2.88	17.814	1.25077	0.056135
2.39	10.913	1.03796	0.091630	2.89	17.993	1.25511	0.055576
2.40	11.023	1.04231	0.090718	2.90	18.174	1.25945	0.055023
2.41	11.134	1.04665	0.089815	2.91	18.357	1.26380	0.054476
2.42	11.246	1.05099	0.088922	2.92	18.541	1.26814	0.053934
2.43	11.359	1.05534	0.088037	2.93	18.728	1.27248	0.053397
2.44	11.473	1.05968	0.087161	2.94	18.916	1.27683	0.052866
2.45	11.588	1.06402	0.086294	2.95	19.106	1.28117	0.052340
2.46	11.705	1.06836	0.085435	2.96	19.298	1.28551	0.051819
2.47	11.822	1.07271	0.084585	2.97	19.492	1.28985	0.051303
2.48	11.941	1.07705	0.083743	2.98	19.688	1.29420	0.050793
2.49	12.061	1.08139	0.082910	2.99	19.886	1.29854	0.050287

Exponential Functions *(continued)*

x	e^x	$\text{Log}_{10}(e^x)$	e^{-x}	x	e^x	$\text{Log}_{10}(e^x)$	e^{-x}
3.00	20.086	1.30288	0.049787	3.50	33.115	1.52003	0.030197
3.01	20.287	1.30723	0.049292	3.51	33.448	1.52437	0.029897
3.02	20.491	1.31157	0.048801	3.52	33.784	1.52872	0.029599
3.03	20.697	1.31591	0.048316	3.53	34.124	1.53306	0.029305
3.04	20.905	1.32026	0.047835	3.54	34.467	1.53740	0.029013
3.05	21.115	1.32460	0.047359	3.55	34.813	1.54175	0.028725
3.06	21.328	1.32894	0.046888	3.56	35.163	1.54609	0.028439
3.07	21.542	1.33328	0.046421	3.57	35.517	1.55034	0.028156
3.08	21.758	1.33763	0.045959	3.58	35.874	1.55477	0.027876
3.09	21.977	1.34197	0.045502	3.59	36.234	1.55912	0.027598
3.10	22.198	1.34631	0.045049	3.60	36.598	1.56346	0.027324
3.11	22.421	1.35066	0.044601	3.61	36.966	1.56780	0.027052
3.12	22.616	1.35500	0.044157	3.62	37.338	1.57215	0.026783
3.13	22.874	1.35934	0.043718	3.63	37.713	1.57649	0.026516
3.14	23.104	1.36368	0.043283	3.64	38.092	1.58083	0.026252
3.15	23.336	1.36803	0.042852	3.65	38.475	1.58517	0.025991
3.16	23.571	1.37237	0.042426	3.66	38.861	1.58952	0.025733
3.17	23.807	1.37671	0.042004	3.67	39.252	1.59386	0.025476
3.18	24.047	1.38106	0.041586	3.68	39.646	1.59820	0.025223
3.19	24.288	1.38540	0.041172	3.69	40.045	1.60255	0.024972
3.20	24.533	1.38974	0.040762	3.70	40.447	1.60689	0.024724
3.21	24.779	1.39409	0.040357	3.71	40.854	1.61123	0.024478
3.22	25.028	1.39843	0.039955	3.72	41.264	1.61558	0.024234
3.23	25.280	1.40277	0.039557	3.73	41.679	1.61992	0.023993
3.24	25.534	1.40711	0.039164	3.74	42.098	1.62426	0.023754
3.25	25.790	1.41146	0.038774	3.75	42.521	1.62860	0.023518
3.26	26.050	1.41580	0.038388	3.76	42.948	1.63295	0.023284
3.27	26.311	1.42014	0.038006	3.77	43.380	1.63729	0.023052
3.28	26.576	1.42449	0.037628	3.78	43.816	1.64163	0.022823
3.29	26.843	1.42883	0.037254	3.79	44.256	1.64598	0.022596
3.30	27.133	1.43317	0.036883	3.80	44.701	1.65032	0.022371
3.31	27.385	1.43751	0.036516	3.81	45.150	1.65466	0.022148
3.32	27.660	1.44186	0.036153	3.82	45.604	1.65900	0.021928
3.33	27.938	1.44620	0.035793	3.83	46.063	1.66335	0.021710
3.34	28.219	1.45054	0.035437	3.84	46.525	1.66769	0.021494
3.35	28.503	1.45489	0.035084	3.85	46.993	1.67203	0.021280
3.36	28.789	1.45923	0.034735	3.86	47.465	1.67638	0.021068
3.37	29.079	1.46357	0.034390	3.87	47.942	1.68072	0.020858
3.38	29.371	1.46792	0.034047	3.88	48.424	1.68506	0.020651
3.39	29.666	1.47226	0.033709	3.89	48.911	1.68941	0.020445
3.40	29.964	1.47660	0.033373	3.90	49.402	1.69375	0.020242
3.41	30.265	1.48094	0.033041	3.91	49.899	1.69809	0.020041
3.42	30.569	1.48529	0.032712	3.92	50.400	1.70243	0.019841
3.43	30.877	1.48963	0.032387	3.93	50.907	1.70678	0.019644
3.44	31.187	1.49397	0.032065	3.94	51.419	1.71112	0.019448
3.45	31.500	1.49832	0.031746	3.95	51.935	1.71546	0.019255
3.46	31.817	1.50266	0.031430	3.96	52.457	1.71981	0.019063
3.47	32.137	1.50700	0.031117	3.97	52.985	1.72415	0.018873
3.48	32.460	1.51134	0.030807	3.98	53.517	1.72849	0.018686
3.49	32.786	1.51569	0.030501	3.99	54.055	1.73283	0.018500

Exponential Functions *(continued)*

x	e^x	$\mathrm{Log}_{10}(e^x)$	e^{-x}	x	e^x	$\mathrm{Log}_{10}(e^x)$	e^{-x}
4.00	54.598	1.73718	0.018316	4.50	90.017	1.95433	0.011109
4.01	55.147	1.74152	0.018133	4.51	90.922	1.95867	0.010098
4.02	55.701	1.74586	0.017953	4.52	91.836	1.96301	0.010839
4.03	56.261	1.75021	0.017774	4.53	92.759	1.96735	0.010781
4.04	56.826	1.75455	0.017597	4.54	93.691	1.97170	0.010673
4.05	57.397	1.75889	0.017422	4.55	94.632	1.97604	0.010567
4.06	57.974	1.76324	0.017249	4.56	95.583	1.98038	0.010462
4.07	58.557	1.76758	0.017077	4.57	96.544	1.98473	0.010358
4.08	59.145	1.77192	0.016907	4.58	97.514	1.98907	0.010255
4.09	59.740	1.77626	0.016739	4.59	98.494	1.99341	0.010153
4.10	60.340	1.78061	0.016573	4.60	99.484	1.99776	0.010052
4.11	60.947	1.78495	0.016408	4.61	100.48	2.00210	0.009952
4.12	61.559	1.78929	0.016245	4.62	101.49	2.00644	0.009853
4.13	62.178	1.79364	0.016083	4.63	102.51	2.01078	0.009755
4.14	62.803	1.79798	0.015923	4.64	103.54	2.01513	0.009658
4.15	63.434	1.80232	0.015764	4.65	104.58	2.01947	0.009562
4.16	64.072	1.80667	0.015608	4.66	105.64	2.02381	0.009466
4.17	64.715	1.81101	0.015452	4.67	106.70	2.02816	0.009372
4.18	65.366	1.81535	0.015299	4.68	107.77	2.03250	0.009279
4.19	66.023	1.81969	0.015146	4.69	108.85	2.03684	0.009187
4.20	66.686	1.82404	0.014996	4.70	109.95	2.04118	0.009095
4.21	67.357	1.82838	0.014846	4.71	111.05	2.04583	0.009005
4.22	68.033	1.83272	0.014699	4.72	112.17	2.04987	0.008915
4.23	68.717	1.83707	0.014552	4.73	113.30	2.05421	0.008826
4.24	69.408	1.84141	0.014408	4.74	114.43	2.05856	0.008739
4.25	70.105	1.84575	0.014264	4.75	115.58	2.06290	0.008652
4.26	70.810	1.85009	0.014122	4.76	116.75	2.06724	0.008566
4.27	71.522	1.85444	0.013982	4.77	117.92	2.07158	0.008180
4.28	72.240	1.85878	0.013843	4.78	119.10	2.07593	0.008396
4.29	72.966	1.86312	0.013705	4.79	120.30	2.08027	0.008312
4.30	73.700	1.86747	0.013569	4.80	121.51	2.08461	0.008230
4.31	74.440	1.87181	0.013434	4.81	122.73	2.08896	0.008148
4.32	75.189	1.87615	0.013300	4.82	123.97	2.09330	0.006067
4.33	75.944	1.88050	0.013168	4.83	125.21	2.09764	0.007987
4.34	76.708	1.88484	0.013037	4.84	126.47	2.10199	0.007907
4.35	77.478	1.88918	0.012907	4.85	127.74	2.10633	0.007828
4.36	78.257	1.89352	0.012778	4.86	129.02	2.11067	0.007750
4.37	79.044	1.89787	0.012651	4.87	130.32	2.11501	0.007673
4.38	79.838	1.90221	0.012525	4.88	131.63	2.11936	0.007597
4.39	80.640	1.90655	0.012401	4.89	132.95	2.12370	0.007521
4.40	81.451	1.91090	0.012277	4.90	134.29	2.12804	0.007447
4.41	82.289	1.91524	0.012155	4.91	135.64	2.13239	0.007372
4.42	83.096	1.91958	0.012034	4.92	137.00	2.13673	0.007299
4.43	83.931	1.92392	0.011914	4.93	138.38	2.14107	0.007227
4.44	84.775	1.92827	0.011796	4.94	139.77	2.14541	0.007155
4.45	85.627	1.93261	0.011679	4.95	141.17	2.14976	0.007083
4.46	86.488	1.93695	0.011562	4.96	142.59	2.15410	0.007013
4.47	87.357	1.94130	0.011447	4.97	144.03	2.15844	0.006943
4.48	88.235	1.94564	0.011333	4.98	145.47	2.16279	0.006874
4.49	89.121	1.94998	0.011221	4.99	146.94	2.16713	0.006806

Exponential Functions *(continued)*

x	e^x	$Log_{10}(e^x)$	e^{-x}	x	e^x	$Log_{10}(e^x)$	e^{-x}
5.00	148.41	2.17147	0.006738	5.50	244.69	2.38862	0.0040868
5.01	149.90	2.17582	0.006671	5.55	257.24	2.41033	0.0038875
5.02	151.41	2.18016	0.006605	5.60	270.43	2.43205	0.0036979
5.03	152.93	2.18450	0.006539	5.65	284.29	2.45376	0.0035175
5.04	154.47	2.18884	0.006474	5.70	298.87	2.47548	0.0033460
5.05	156.02	2.19319	0.006409	5.75	314.19	2.49719	0.0031828
5.06	157.59	2.19753	0.006346	5.80	330.30	2.51891	0.0030276
5.07	159.17	2.20187	0.006282	5.85	347.23	2.54062	0.0028799
5.08	160.77	2.20622	0.006220	5.90	365.04	2.56234	0.0027394
5.09	162.39	2.21056	0.006158	5.95	383.75	2.58405	0.0026058
5.10	164.02	2.21490	0.006097	6.00	403.43	2.60577	0.0024788
5.11	165.67	2.21924	0.006036	6.05	424.11	2.62748	0.0023579
5.12	167.34	2.22359	0.005976	6.10	445.86	2.64920	0.0022429
5.13	169.02	2.22793	0.005917	6.15	468.72	2.67091	0.0021335
5.14	170.72	2.23227	0.005858	6.20	492.75	2.69263	0.0020294
5.15	172.43	2.23662	0.005799	6.25	518.01	2.71434	0.0019305
5.16	174.16	2.24096	0.005742	6.30	544.57	2.73606	0.0018363
5.17	175.91	2.24530	0.005685	6.35	572.49	2.75777	0.0017467
5.18	177.68	2.24965	0.005628	6.40	601.85	2.77948	0.0016616
5.19	179.47	2.25399	0.005572	6.45	632.70	2.80120	0.0015805
5.20	181.27	2.25833	0.005517	6.50	665.14	2.82291	0.0015034
5.21	183.09	2.26267	0.005462	6.55	699.24	2.84463	0.0014301
5.22	184.93	2.26702	0.005407	6.60	735.10	2.86634	0.0013604
5.23	186.79	2.27136	0.005354	6.65	772.78	2.88806	0.0012940
5.24	188.67	2.27570	0.005300	6.70	812.41	2.90977	0.0012309
5.25	190.57	2.28005	0.005248	6.75	854.06	2.93149	0.0011709
5.26	192.48	2.28439	0.005195	6.80	897.85	2.95320	0.0011138
5.27	194.42	2.28873	0.005144	6.85	943.88	2.97492	0.0010595
5.28	196.37	2.29307	0.005092	6.90	992.27	2.99663	0.0010078
5.29	198.34	2.29742	0.005042	6.95	1043.1	3.01835	0.0009586
5.30	200.34	2.30176	0.004992	7.00	1096.6	3.04006	0.0009119
5.31	202.35	2.30610	0.004942	7.05	1152.9	3.06178	0.0008674
5.32	204.38	2.31045	0.004893	7.10	1212.0	3.08349	0.0008251
5.33	206.44	2.31479	0.004844	7.15	1274.1	3.10521	0.0007840
5.34	208.51	2.31913	0.004796	7.20	1339.4	3.12692	0.0007466
5.35	210.61	2.32348	0.004748	7.25	1408.1	3.14863	0.0007102
5.36	212.72	2.32782	0.004701	7.30	1480.3	3.17035	0.0006755
5.37	214.86	2.33216	0.004654	7.35	1556.2	3.19206	0.0006426
5.38	217.02	2.33650	0.004608	7.40	1636.0	3.21378	0.0006113
5.39	219.20	2.34085	0.004562	7.45	1719.9	3.23549	0.0005814
5.40	221.41	2.34519	0.004517	7.50	1808.0	3.25721	0.0005531
5.41	223.63	2.34953	0.004472	7.55	1900.7	3.27892	0.0005261
5.42	225.88	2.35388	0.004427	7.60	1998.2	3.30064	0.0005005
5.43	228.15	2.35822	0.004383	7.65	2100.6	3.32235	0.0004760
5.44	230.44	2.36256	0.004339	7.70	2208.3	3.33407	0.0004528
5.45	232.76	2.36690	0.004296	7.75	2321.6	3.36578	0.0004307
5.46	235.10	2.37125	0.004254	7.80	2440.6	3.38750	0.0004097
5.47	237.46	2.37559	0.004211	7.85	2565.7	3.40921	0.0003898
5.48	239.85	2.37993	0.004169	7.90	2697.3	3.43093	0.0003707
5.49	242.26	2.38428	0.004128	7.95	2835.6	3.45264	0.0003527

Exponential Functions *(continued)*

x	e^x	$\text{Log}_{10}(e^x)$	e^{-x}	x	e^x	$\text{Log}_{10}(e^x)$	e^{-x}
8.00	2981.0	3.47436	0.0003355	9.00	8103.1	3.90865	0.0001234
8.05	3133.8	3.49607	0.0003191	9.05	8518.5	3.93037	0.0001174
8.10	3294.5	3.51779	0.0003035	9.10	8955.3	3.95208	0.0001117
8.15	3463.4	3.53950	0.0002887	9.15	9414.4	3.97379	0.0001062
8.20	3641.0	3.56121	0.0002747	9.20	9897.1	3.99551	0.0001010
8.25	3827.6	3.58293	0.0002613	9.25	10405	4.01722	0.0000961
8.30	4023.9	3.60464	0.0002485	9.30	10938	4.03894	0.0000914
8.35	4230.2	3.62636	0.0002364	9.35	11499	4.06065	0.0000870
8.40	4447.1	3.64807.	0.0002249	9.40	12088	4.08237	0.0000827
8.45	4675.1	3.66979	0.0002139	9.45	12708	4.10408	0.0000787
8.50	4914.8	3.69150	0.0002035	9.50	13360	4.12580	0.0000749
8.55	5166.8	3.71322	0.0001935	9.55	14045	4.14751	0.0000712
8.60	5431.7	3.73493	0.0001841	9.60	14765	4.16923	0.0000677
8.65	5710.1	3.75665	0.0001751	9.65	15522	4.19094	0.0000614
8.70	6002.9	3.77836	0.0001666	9.70	16318	4.21266	0.0000613
8.75	6310.7	3.80008	0.0001585	9.75	17154	4.23437	0.0000583
8.80	6634.2	3.82179	0.0001507	9.80	18034	4.25609	0.0000555
8.85	6974.4	3.84351	0.0001434	9.85	18958	4.27780	0.0000527
8.90	7332.0	3.86522	0.0001364	9.90	19930	4.29952	0.0000502
8.95	7707.9	3.88694	0.0001297	9.95	20952	4.32123	0.0000177
9.00	8103.1	3.90865	0.0001234	10.00	22026	4.34294	0.0000454

APPENDIX D

Calculation Procedures for Normal Distribution

D.1 Example Problem

Given:

$$\bar{t} = 2.2 \text{ seconds}$$

$$s = 0.85 \text{ seconds}$$

$$t = 1.5 \text{ seconds}$$

$$t + \Delta t = 2.0 \text{ seconds}$$

$$\Sigma \dot{F} = 3491 \text{ headways}$$

Required:

(a) Probability of headway between t and $\bar{t}$, and $t + \Delta t$ and $\bar{t}$
(b) Probability of headway between t and $t + \Delta t$
(c) Frequency of headways between t and $t + \Delta t$

Solution:

(a) $P(t \leq h < \bar{t})$

$$z = 2.2 - 1.5 = 0.7$$

$$\frac{z}{s} = \frac{0.7}{0.85} = 0.82$$

$$P(1.5 \leq h < 2.2) = 0.2939$$

$P(t + \Delta t \leq h < \bar{t})$

$$z = 2.2 - 2.0 = 0.2$$

$$\frac{z}{s} = \frac{0.2}{0.85} = 0.24$$

$$P(2.0 \leq h < 2.2) = 0.0948$$

(b) $P(t \leq h < t + \Delta t)$

$$P(1.5 \leq h < 2.0) = 0.2939 - 0.0948$$

$$= 0.1991$$

(c) $F(t \leq h < t + \Delta t)$

$$F(1.5 \leq h < 2.0) = 0.1991 \times 3491$$

$$= 695$$

z/s	Second Decimal Place in z/s									
	0.00	0.01	0.02	0.03	0.04	0.05	0.06	0.07	0.08	0.09
0.0	0.0000	0.0040	0.0080	0.0120	0.0160	0.0199	0.0239	0.0279	0.0319	0.0359
0.1	0.0398	0.0438	0.0478	0.0517	0.0557	0.0596	0.0636	0.0675	0.0714	0.0753
0.2	0.0793	0.0832	0.0871	0.0910	0.0948	0.0987	0.1026	0.1064	0.1103	0.1141
0.3	0.1179	0.1217	0.1255	0.1293	0.1331	0.1368	0.1406	0.1443	0.1480	0.1517
0.4	0.1554	0.1591	0.1628	0.1664	0.1700	0.1736	0.1772	0.1808	0.1844	0.1879
0.5	0.1915	0.1950	0.1985	0.2019	0.2054	0.2088	0.2123	0.2157	0.2190	0.2224
0.6	0.2257	0.2291	0.2324	0.2357	0.2389	0.2422	0.2454	0.2486	0.2517	0.2549
0.7	0.2580	0.2611	0.2642	0.2673	0.2704	0.2734	0.2764	0.2794	0.2823	0.2852
0.8	0.2881	0.2910	0.2939	0.2967	0.2995	0.3023	0.3051	0.3078	0.3106	0.3133
0.9	0.3159	0.3186	0.3212	0.3238	0.3264	0.3289	0.3315	0.3340	0.3365	0.3389
1.0	0.3413	0.3438	0.3461	0.3485	0.3508	0.3531	0.3554	0.3577	0.3599	0.3621
1.1	0.3643	0.3665	0.3686	0.3708	0.3729	0.3749	0.3770	0.3790	0.3810	0.3830
1.2	0.3849	0.3869	0.3888	0.3907	0.3925	0.3944	0.3962	0.3980	0.3997	0.4015
1.3	0.4032	0.4049	0.4066	0.4082	0.4099	0.4115	0.4131	0.4147	0.4162	0.4177
1.4	0.4192	0.4207	0.4222	0.4236	0.4251	0.4265	0.4279	0.4292	0.4306	0.4319
1.5	0.4332	0.4345	0.4357	0.4370	0.4382	0.4394	0.4406	0.4418	0.4429	0.4441
1.6	0.4452	0.4463	0.4474	0.4484	0.4495	0.4505	0.4515	0.4525	0.4535	0.4545
1.7	0.4554	0.4564	0.4573	0.4582	0.4591	0.4599	0.4608	0.4616	0.4625	0.4633
1.8	0.4641	0.4649	0.4656	0.4664	0.4671	0.4678	0.4686	0.4693	0.4699	0.4706
1.9	0.4713	0.4719	0.4726	0.4732	0.4738	0.4744	0.4750	0.4756	0.4761	0.4767
2.0	0.4772	0.4778	0.4783	0.4788	0.4793	0.4798	0.4803	0.4808	0.4812	0.4817
2.1	0.4821	0.4826	0.4830	0.4834	0.4838	0.4842	0.4846	0.4850	0.4854	0.4857
2.2	0.4861	0.4864	0.4868	0.4871	0.4875	0.4878	0.4881	0.4884	0.4887	0.4890
2.3	0.4893	0.4896	0.4898	0.4901	0.4904	0.4906	0.4909	0.4911	0.4913	0.4916
2.4	0.4918	0.4920	0.4922	0.4925	0.4927	0.4929	0.4931	0.4932	0.4934	0.4936
2.5	0.4938	0.4940	0.4941	0.4943	0.4945	0.4946	0.4948	0.4949	0.4951	0.4952
2.6	0.4953	0.4955	0.4956	0.4957	0.4959	0.4960	0.4961	0.4962	0.4963	0.4964
2.7	0.4965	0.4966	0.4967	0.4968	0.4969	0.4970	0.4971	0.4972	0.4973	0.4974
2.8	0.4974	0.4975	0.4976	0.4977	0.4977	0.4978	0.4979	0.4979	0.4980	0.4981
2.9	0.4981	0.4982	0.4982	0.4983	0.4984	0.4984	0.4985	0.4985	0.4986	0.4986
3.0	0.4987	0.4987	0.4987	0.4988	0.4988	0.4989	0.4989	0.4989	0.4990	0.4990
3.1	0.4990	0.4991	0.4991	0.4991	0.4992	0.4992	0.4992	0.4992	0.4993	0.4993
3.2	0.4993	0.4993	0.4994	0.4994	0.4994	0.4994	0.4994	0.4995	0.4995	0.4995
3.3	0.4995	0.4995	0.4995	0.4996	0.4996	0.4996	0.4996	0.4996	0.4996	0.4997
3.4	0.4997	0.4997	0.4997	0.4997	0.4997	0.4997	0.4997	0.4997	0.4997	0.4998
3.5	0.4998									
4.0	0.49997									
4.5	0.499997									
5.0	0.4999997									

Source: Standard Mathematical Tables, 15th ed. Reprinted by permission of CRC Press, Boca Raton, Fla.

APPENDIX E

Gamma Function Table and Example Problem*

*The table and example problem are reprinted with the permission of the publishers and are reprⴰ duced from the reference: *Handbook of Chemistry and Physics*, 63rd Edition, CRC Press Inc., 1982–198 p. A-106.

E.1 Example Problem

$$\Gamma(K) = (K - 1)\Gamma(K - 1) \qquad \text{let } (K) = 4.785$$

$$\Gamma(4.785) = 3.785\Gamma(3.785)$$

$$= 3.785 \times 2.785\Gamma(2.785)$$

$$= 3.785 \times 2.785 \times 1.785\Gamma(1.785)$$

$$= 3.785 \times 2.785 \times 1.785 \times 0.92750$$

$$\doteq 17.45$$

E.2 Gamma Function[†]

$$\text{Values of } \Gamma(n) = \int_0^\infty e^{-x} x^{n-1} \, dx; \; \Gamma(n + 1) = n\Gamma(n)$$

n	$\Gamma(n)$	n	$\Gamma(n)$	n	$\Gamma(n)$	n	$\Gamma(n)$
1.00	1.00000	1.25	0.90640	1.50	0.88623	1.75	0.91906
1.01	0.99433	1.26	0.90440	1.51	0.88659	1.76	0.92137
1.02	0.98884	1.27	0.90250	1.52	0.88704	1.77	0.92376
1.03	0.98355	1.28	0.90072	1.53	0.88757	1.78	0.92623
1.04	0.97844	1.29	0.89904	1.54	0.88818	1.79	0.92877
1.05	0.97350	1.30	0.89747	1.55	0.88887	1.80	0.93138
1.06	0.96874	1.31	0.89600	1.56	0.88964	1.81	0.93408
1.07	0.96415	1.32	0.89464	1.57	0.89049	1.82	0.93685
1.08	0.95973	1.33	0.89338	1.58	0.89442	1.83	0.93969
1.09	0.95546	1.34	0.89222	1.59	0.89243	1.84	0.94261
1.10	0.95135	1.35	0.89115	1.60	0.89352	1.85	0.94561
1.11	0.94739	1.36	0.89018	1.61	0.89468	1.86	0.94869
1.12	0.94359	1.37	0.88931	1.62	0.89592	1.87	0.95184
1.13	0.93993	1.38	0.88854	1.63	0.89724	1.88	0.95507
1.14	0.93642	1.39	0.88785	1.64	0.89864	1.89	0.95838
1.15	0.93304	1.40	0.88726	1.65	0.90012	1.90	0.96177
1.16	0.92980	1.41	0.88676	1.66	0.90167	1.91	0.96523
1.17	0.92670	1.42	0.88636	1.67	0.90330	1.92	0.96878
1.18	0.92373	1.43	0.88604	1.68	0.90500	1.93	0.97240
1.19	0.92088	1.44	0.88580	1.69	0.90678	1.94	0.97610
1.20	0.91817	1.45	0.88565	1.70	0.90864	1.95	0.97988
1.21	0.91558	1.46	0.88560	1.71	0.91057	1.96	0.98374
1.22	0.91311	1.47	0.88563	1.72	0.91258	1.97	0.98768
1.23	0.91075	1.48	0.88575	1.73	0.91466	1.98	0.99171
1.24	0.90852	1.49	0.88595	1.74	0.91683	1.99	0.99581
						2.00	1.00000

$$x^x e^{-x} \sqrt{\frac{2\pi}{x}} \left(1 + \frac{1}{12x} + \frac{1}{288x^2} - \frac{139}{51,840x^3} - \frac{571}{2,488,320x^4} + \cdots \right)$$

[†]For large positive values of x, $\Gamma(x)$ approximates the asymptotic series.

APPENDIX F

Chi-Square Table Values

					α				
df	0.995	0.990	0.975	0.950	0.900	0.100	0.050	0.025	0.010	0.005
1	0.0000393	0.000157	0.000982	0.00393	0.0158	2.71	3.84	5.02	6.64	7.88
2	0.0100	0.0201	0.0506	0.103	0.211	4.61	6.00	7.38	9.21	10.6
3	0.0717	0.115	0.216	0.352	0.584	6.25	7.82	9.35	11.4	12.9
4	0.207	0.297	0.484	0.711	1.0636	7.78	9.50	11.1	13.3	14.9
5	0.412	0.554	0.831	1.15	1.61	9.24	11.1	12.8	15.1	16.8
6	0.676	0.872	1.24	1.64	2.20	10.6	12.6	14.5	16.8	18.6
7	0.990	1.24	1.69	2.17	2.83	12.0	14.1	16.0	18.5	20.3
8	1.34	1.65	2.18	2.73	3.49	13.4	15.5	17.5	20.1	22.0
9	1.73	2.09	2.70	3.33	4.17	14.7	17.0	19.0	21.7	23.6
10	2.16	2.56	3.25	3.94	4.87	16.0	18.3	20.5	23.2	25.2
11	2.60	3.05	3.82	4.58	5.58	17.2	19.7	21.9	24.7	26.8
12	3.07	3.57	4.40	5.23	6.30	18.6	21.0	23.3	26.2	28.3
13	3.57	4.11	5.01	5.90	7.04	19.8	22.4	24.7	27.7	29.8
14	4.07	4.66	5.63	6.57	7.79	21.1	23.7	26.1	29.1	31.3
15	4.60	5.23	6.26	7.26	8.55	22.3	25.0	27.5	30.6	32.8
16	5.14	5.81	6.91	7.96	9.31	23.5	26.3	28.9	32.0	34.3
17	5.70	6.41	7.56	8.67	10.1	24.8	27.6	30.2	33.4	35.7
18	6.26	7.01	8.23	9.39	10.9	26.0	28.9	31.5	34.8	37.2
19	6.84	7.63	8.91	10.1	11.7	27.2	30.1	32.9	36.2	38.6
20	7.43	8.26	9.59	10.9	12.4	28.4	31.4	34.2	37.6	40.0
21	8.03	8.90	10.3	11.6	13.2	29.6	32.7	35.5	39.0	41.4
22	8.64	9.54	11.0	12.3	14.0	30.8	33.9	36.8	40.3	42.8
23	9.26	10.2	11.0	13.1	14.9	32.0	35.2	38.1	41.6	44.2
24	9.89	10.9	12.4	13.9	15.7	33.2	36.4	39.4	43.0	45.6
25	10.5	11.5	13.1	14.6	16.5	34.4	37.7	40.7	44.3	46.9
26	11.2	12.2	13.8	15.4	17.3	35.6	38.9	41.9	45.6	48.3
27	11.8	12.9	14.6	16.2	18.1	36.7	40.1	43.2	47.0	49.7
28	12.5	13.6	15.3	16.9	18.9	37.9	41.3	44.5	48.3	51.0
29	13.1	14.3	16.1	17.7	19.8	39.1	42.6	45.7	49.6	52.3
30	13.8	15.0	16.8	18.5	20.6	40.3	43.8	47.0	50.9	53.7
40	20.7	22.2	24.4	26.5	29.1	51.8	55.8	59.3	63.7	66.8
50	28.0	29.7	32.4	34.8	37.7	63.2	67.5	71.4	76.2	79.5
60	35.5	37.5	40.5	43.2	46.5	74.4	79.1	83.3	88.4	92.0
70	43.3	45.4	48.8	51.8	55.3	85.5	90.5	95.0	100.0	104.0
80	51.2	53.5	57.2	60.4	64.3	96.6	102.0	107.0	112.0	116.0
90	59.2	61.8	65.7	69.1	73.3	108.0	113.0	118.0	124.0	128.0
100	67.3	70.1	74.2	77.9	82.4	114.0	124.0	130.0	136.0	140.0

Source: Adapted from E. S. Pearson and H. O. Hartley, *Biometrika Tables for Statisticians,* Vol. I (1962), pp. 130–131. Reprinted by permission of the Biometrika Trustees.

APPENDIX G

List of Symbols

a	amber time duration (seconds) [Chapter 3]
	acceleration (or deceleration) rate (miles per hour/second) [Chapter 4]
a_{ij}	proportion of traffic that enters at origin i and is destined to pass through subsection j
A	upstream detection zone in a two-closely-spaced presence-type detector system [Chapter 6]
AADT	annual average daily traffic combining both directions
$A, B, C, \ldots$	
	traffic flow states [Chapter 11]
B	downstream detection zone in a two-closely-spaced presence-type detector system
c	capacity (vehicles per hour)
c_j	lane capacity under ideal conditions for design speed j (vehicles per hour per lane) [Chapters 3 and 9]
	capacity of subsection j [Chapter 8]
C	signal cycle length (seconds) [Chapters 3, 5, 9, and 12]
	product of reaction time (Δt) and sensitivity (α) used in traffic stability analysis [Chapter 6]
C.V.	coefficient of variation (ratio of standard deviation to mean)
d	distance (feet) for vehicle to accelerate (or decelerate) at a rate a from beginning speed μ_i to an ending speed μ_e [Chapter 4]
	average total delay per vehicle (seconds per vehicle) [Chapter 5]
	average stopped delay per vehicle (seconds per vehicle) [Chapter 9]
	individual distance headway (feet) [Chapter 6]
	total stopped delay (seconds per vehicle) [Chapter 9]
$\bar{d}$	average distance headway (feet per vehicle) [Chapters 6 and 7]
	average individual delay (seconds or hours, as noted) [Chapter 12]
d_1	uniform stopped delay (seconds per vehicle)
d_2	incremental stopped delay (seconds per vehicle)
d_M	maximum individual delay (seconds or hours, as noted)
d_{MIN}	minimum individual distance headway (feet)
$\bar{d}_Q$	average individual delay while queue is present (seconds or hours, as noted)
D	ratio of design hour volume in major direction to the two-way design hour volume (DDHV/DHV) [Chapter 3]
	distance from the upstream edge of detection zone A to the upstream edge of detection zone B (feet) [Chapter 6]

pedestrian density (persons per square foot) [Chapter 10]

constant value distribution [Chapter 12]

DDHV   design hour volume in major direction

$(DEM)_{t, t+1}$

demand rate in vehicles per hour between time t and $t + 1$

DHV   design hour volume combining both directions

D_{it} input demand rate at origin i in time slice t

DT0, DT1, DT2, DT3

time duration (hours)

e a constant, Napierian base of logarithms ($e = 2.71828...$)

E   Erlang distribution

f coefficient of friction between tires and pavement

f_a area type factor

f_{bb} local bus blocking factor

f_e environment factor

f_g approach grade factor

f_{HV} heavy vehicle factor

f_{LT} left-turn factor

f_o observed number of observations

f_P driver population factor

f_{pb} parking blockage factor

f_{RT} right-turn factor

f_t theoretical number of observations

$f(t)$ probability density function

$f(\mu_i)$ probability density function of individual speeds

f_w lane width and lateral clearance factor

F frequency

FIFO   first in, first out queue discipline

$F(x)$ frequency of exactly x vehicles arriving in a time interval (t)

g effective duration of the green phase (seconds) [Chapters 3, 5, and 9]

	number of speed groups [Chapter 4]
	grade situation expressed as a decimal (i.e., a 3 percent upgrade is expressed as +0.03) [Chapter 4]
	individual distance gap (feet) [Chapter 6]
G	actual duration of the green phase (seconds) [Chapters 3 and 9]
	generalized distribution [Chapter 12]
h	individual time headway (seconds)
$\bar{h}$	average time headway (seconds per vehicle)
$(h)_{1-2}$	time headway between first and second vehicle (seconds)
h_{MIN}	minimum individual time headway (seconds)
h_t	train headway (hours)
i	a particular speed group i [Chapter 4]
	freeway input origin i [Chapter 8]
	level of service i [Chapter 9]
I	number of intervals [Chapter 2]
	size of the class interval [Chapter 4]
j	design speed j (miles per hour) [Chapters 3 and 9]
	freeway subsection j [Chapter 8]
k	density (vehicles per mile per lane; persons per square foot in pedestrian flow) [Chapters 6, 7, 10, and 11]
	variable equal to [Chapter 12]

$$\frac{\lambda_2 - \mu}{\lambda_2 - \lambda_0}$$

k_{it}	density in subsection i at time t (vehicles per lane-mile)
k_j	jam density (vehicles per mile, usually on a per-lane basis)
k_o	optimum density (vehicles per mile, usually on a per-lane basis)
K	parameter of the Pearson type III distribution which affects the shape of the distribution [Chapter 2]
	ratio of the two-way design hour volume to the two-way annual average daily traffic (DHV/AADT) [Chapter 3]
	constant, representing average vehicle length plus length of detection zone (feet) [Chapter 6]
	exponent of the spacing term in the generalized car-following equation

ℓ_e ending lost time (seconds)

ℓ_s starting lost time (seconds)

L physical length of a vehicle (feet)

L_D length of detection zone (feet)

L_i length of subsection i (miles)

L_V length of individual vehicle (feet)

$\overline{L_V}$ average vehicle length (feet)

m average number of vehicles arriving in a time interval t [Chapters 2 and 3]

         exponent of the speed term in the generalized car-following equation [Chapters 6 and 10]

         number of observations during the selected time period [Chapter 7]

M pedestrian space (average square feet per person) [Chapter 10]

         random distribution [Chapter 12]

MAX R_i
         maximum allowable metering rate for freeway input origin i

MIN R_i
         minimum allowable metering rate for freeway input origin i

M_s number of opposing vehicles met when test vehicle is traveling south

n number of degrees of freedom [Chapter 2]

         required sample size [Chapter 4]

         number of observations [Chapter 5]

         subscript denoting the lead vehicle while following vehicles are denoted as $n + 1$, $n + 2$, etc. [Chapter 6]

         number of subsections in the system [Chapter 7]

         model parameter intended to add generality to single-regime models [Chapter 10]

N number of observations [Chapters 2, 4, and 6]

         number of lanes in one direction [Chapters 3 and 9]

         number of vehicles passing over the detector in time period T [Chapter 7]

         number of input origins [Chapter 8]

         number of vehicles (or persons) entering or leaving in flow state [Chapter 11]

         number of vehicles per cycle in vehicles [Chapter 12]

N_i number of lanes in subsection i

N_Q number of vehicles queued in vehicles

O_N	number of vehicles overtaking the test vehicle when test vehicle is traveling north
p	number of parameters
P	probability expressed as a fraction [Chapter 2]
	probability of an observation between some speed, μ_x, and the mean population, $\overline{U}$ [Chapter 4]
% OCC	
	percent occupancy
PF	progression factor
$P(h \geq t)$	
	probability of a time headway h equal to or greater than time interval t
PHF	peak hour factor (ratio of hourly flow rate to peak 15-minute rate of flow)
P_N	number of vehicles passed by test vehicle when test vehicle is traveling north
P_{Np}	proportion of vehicles not in platoons
PN_Q	percent of vehicles queued
$P(0)$	probability of no vehicles arriving in a time interval t
P_p	proportion of vehicles in platoons
Pt_Q	percent time queue is present
$P(x)$	probability of exacting x vehicles arriving in a time interval t
q	flow rate (vehicles per hour) [Chapters 3, 6, and 7]
	flow (vehicles per hour per lane; persons per minute per foot of width for pedestrian flow) [Chapters 10 and 11]
$q_{EXCt, t+1}$	
	excess or unserved flow rate at the bottleneck between time t and $t + 1$ in vehicles per hour
q_m	flow parameter representing maximum flow or capacity (vehicles per hour per lane)
$q_{t, t+1}$	flow rate between time t and $t + 1$ (vehicles per hour)
$\overline{Q}$	average length (vehicles)
Q_M	maximum queue length (vehicles)
$\overline{Q}_Q$	average queue length while queue is present (vehicles)
	effective red time (seconds)
R	actual red time (seconds)
	standard deviation of a sample [Chapters 2, 4, and 5]

	saturation flow rate (vehicles per hour of green) [Chapters 3, 5, and 9]
$\bar{s}$	mean service time (seconds per vehicle)
$\hat{s}$	standard deviation of the difference of the means (miles per hour)
s^2	sample variance
s^2_{sms}	variance of space-mean-speed [(miles per hour)^2]
$s_{\bar{x}}$	standard error of the mean (miles per hour) [Chapter 4]
$s_{\bar{\mu}}$	standard error of the mean (miles per hour) [Chapter 5]
S	minimum stopping distance (feet) [Chapter 4]
	pedestrian speed (feet per minute) [Chapter 10]
SF_i	maximum service flow rate for level of service i
SIRO	served in random order queue descipline
t	selected time interval [Chapters 2 and 3]
	time in seconds for vehicle to accelerate (or decelerate) at rate a from beginning speed (μ_b) to an ending speed (μ_e) [Chapter 4]
	value for the selected confidence level using the t distribution [Chapters 4 and 5]
	instant in time (seconds) [Chapter 6]
$\bar{t}$	mean time headway (seconds per vehicle)
t_i	travel time rate for vehicle i (usually expressed in minutes per mile) [Chapter 5]
	time of occurrence i (seconds) [Chapter 11]
$\bar{t}_o$	average vehicle occupancy time (seconds)
t_{occ}	individual vehicle occupancy time (seconds)
t_{off}	instant in time that the rear edge of the vehicle is no longer detected (seconds)
t_{on}	instant in time that the leading edge of the vehicle is detected (seconds)
t_Q or TQ	
	time duration of queue (seconds or hours, as noted)
t_R	duration of reduced service rate (hours)
T	sampling time interval
TD	total delay (vehicle-seconds or vehicle-hours, as noted)
TI	time interval between time t and $t = 1$ (hours) [Chapter 7]
	time interval during which the arrival rate exceeds the service rate (hours) [Chapter 12]
TM	time interval from the start of the study period to when the arrival rate is equal to the service rate (hours)

T_N	travel time for northbound test vehicle (minutes)
T_O	total occupancy time in time period T (hours)
T_S	travel time for southbound test vehicle (minutes)
TTT	total travel time expended in the system during the selected time period T (vehicle-hours)
$\bar{U}$	population mean speed (miles per hour)
v	pedestrian flow (persons per minute per foot of width)
$\dfrac{v}{c}$	volume–capacity ratio (when v is hourly volume)
v_{it}	average number of vehicles in subsection i during a time interval (TI)
v_t	number of vehicles in the system at time t
V	hourly flow rate (vehicles per hour)
$\bar{V}$	average number of vehicles in the system during the selected time period
$(V_{EXC})_t$	number of excess vehicles in the system at time t
V_N	northbound flow (vehicles per hour)
x	number of vehicles arriving in a time interval being investigated [Chapter 3]
	degree saturation or volume–capacity ratio [Chapter 5]
	distance postion of front edge of vehicle (feet) [Chapter 6]
$\dot{x}$	speed of individual vehicle (feet per second)
$\bar{x}$	time mean speed (miles per hour)
$\ddot{x}$	acceleration (or deceleration) rate of individual vehicle (feet per second^2)
X	speed range or deviation between μ_x and $\bar{U}(X = \mu_x - \bar{U})$ [Chapter 4]
	volume–capacity ratio for the lane group [Chapter 9]
X_{it}	allowable input flow rate at origin i in time slice t
	difference between the mean value of a variable and a selected value of a variable (used in calculating normal distributions)
α	parameter of the Pearson type III distribution which affects the shift of the distribution (can also be interpreted as the minimum expected time headway in seconds) [Chapter 2]
	sensitivity parameter in first GM model (seconds^{-1}) [Chapter 6]
α'	sensitivity parameter in fourth GM model (dimensionless)
α_1, α_2	sensitivity parameters in second GM model (seconds^{-1})
$\alpha_{l,m}$	sensitivity parameter in fifth GM model

α_0 sensitivity parameter in third GM model (feet per second)

ΓK gamma function $[(\Gamma K = (K - 1)!]$

δ change in detector signal from state 0 to state 1

Δt driver reaction time (seconds)

ε user-specified allowable error (miles per hour)

λ parameter of the Pearson type III distribution $\left(\lambda = \dfrac{K}{\bar{t} - \alpha}\right)$ [Chapter 2]

   proportion of the cycle that is effectively green for the phase under consideration (g/C) [Chapter 5]

   mean arrival rate (vehicles per hour) [Chapter 12]

$\lambda_0, \lambda_1, \lambda_2, \lambda_3, \lambda_4$
   means arrival rates in vehicles per hour during time durations DT0, DT1, DT2, DT3, DT4

μ space mean speed (miles per hour) [Chapters 6, 7, and 10]

   speed (miles per hour; feet per minute in pedestrian flow) [Chapter 11]

   mean service rate (vehicles per hour) [Chapter 12]

$\bar{\mu}$ sample mean speed (miles per hour)

μ_b speed at the beginning of the acceleration (or deceleration) cycle (miles per hour)

μ_e speed at the end of the acceleration (or deceleration) cycle (miles per hour)

μ_f speed parameter called free-flow speed which occurs when flow approaches zero under free-flow conditions (miles per hour)

μ_i midpoint speed of group i [Chapter 4]

   speed of vehicle i (miles per hour) [Chapters 4 and 5]

μ_0 optimum speed (miles per hour)

μ_R reduced service rate due to presence of an incident or accident (vehicles per hour)

$\bar{\mu}_{SMS}$ space-mean-speed (miles per hour)

$\bar{\mu}_{TMS}$ time-mean-speed (miles per hour)

μ_x selected speed being investigated (miles per hour)

π a constant ($\pi = 3.1416$)

ρ traffic intensity (ratio of mean arrival rate to mean service rate)

σ population standard deviation

σ^2 population variance

χ^2 chi-square value

ω shock wave speed (miles per hour; feet per minute in pedestrian flow)

APPENDIX H

English - Metric Conversion Factors

TABLE H.1   English to Metric Conversion Factors

Measurement	To Convert	To	Multiply By
DISTANCE	feet miles	meters kilometers	0.3048 1.6093
SPEED	feet per second feet per minute miles per hour	meters per second meters per minute kilometers per hour	0.3048 0.3048 1.6093
ACCELERATION	feet per second^2 miles per hour per second	meter per second^2 Kilometers per hour per second	0.3048 1.6093
DENSITY	vehicles per mile person per square foot square feet per person	vehicles per kilometer persons per square meter square meter per person	0.6214 10.7639 0.0929
VARIANCE OF SPEED	(miles per hour)^2	(kilometer per hour)^2	2.5900

TABLE H.2   Metric to English Conversion Factors

Measurement	To Convert	To	Multiply By
DISTANCE	meters kilometers	feet miles	3.2808 0.6214
SPEED	meters per second meters per minute kilometers per hour	feet per second feet per minute miles per hour	3.2808 3.2808 0.6214
ACCELERATION	meter per second^2 Kilometers per hour per second	feet per second^2 miles per hour per second	3.2808 0.6214
DENSITY	vehicles per kilometer persons per square meter square meter per person	vehicles per mile persons per square foot square feet per person	1.6093 0.0929 10.7639
VARIANCE OF SPEED	(kilometers per hour)^2	(miles per hour)^2	0.38610

Index